Sati McKenzie
May 1982.

Phase Transitions
Cargèse 1980

NATO ADVANCED STUDY INSTITUTES SERIES

A series of edited volumes comprising multifaceted studies of contemporary scientific issues by some of the best scientific minds in the world, assembled in cooperation with NATO Scientific Affairs Division.

Series B: Physics

Recent Volumes in this Series

Volume 71 – Atomic and Molecular Collision Theory
edited by Franco A. Gianturco

Volume 72 – Phase Transitions: *Cargèse 1980*
edited by Maurice Lévy, Jean-Claude Le Guillou, and Jean Zinn-Justin

Volume 73 – Scattering Techniques Applied to Supramolecular and Nonequilibrium Systems
edited by Sow-Hsin Chen, Benjamin Chu, and Ralph Nossal

Volume 74 – Rigorous Atomic and Molecular Physics
edited by G. Velo and A. S. Wightman

Volume 75 – Nonlinear Phenomena in Physics and Biology
edited by Richard H. Enns, Billy L. Jones, Robert M. Miura, and Sadanand S. Rangnekar

Volume 76 – Metal Hydrides
edited by Gust Bambakidis

Volume 77 – Nonlinear Phenomena at Phase Transitions and Instabilities
edited by T. Riste

Volume 78 – Excitations in Disordered Systems
edited by M. F. Thorpe

Volume 79 – Artificial Particle Beams in Space Plasma Studies
edited by Bjørn Grandal

This series is published by an international board of publishers in conjunction with NATO Scientific Affairs Division

A Life Sciences B Physics	Plenum Publishing Corporation London and New York
C Mathematical and Physical Sciences	D. Reidel Publishing Company Dordrecht, The Netherlands and Hingham, Massachusetts, USA
D Behavioral and Social Sciences E Applied Sciences	Martinus Nijhoff Publishers The Hague, The Netherlands

Phase Transitions

Cargèse 1980

Edited by

Maurice Lévy and Jean-Claude Le Guillou

Université Pierre et Marie Curie
Paris, France

and

Jean Zinn-Justin

Theoretical Physics Service
C.E.N.
Saclay, France

PLENUM PRESS • NEW YORK AND LONDON
Published in cooperation with NATO Scientific Affairs Division

Library of Congress Cataloging in Publication Data

Cargèse Summer Institute on Phase Transitions (1980 : Cargèse, Corsica)
Phase transitions.

(NATO advanced study institutes series. Series B, Physics ; v. 72)
"Proceedings of the 1980 Cargèse Summer Institute on Phase Transitions, held July 16-31, 1980, in Cargèse, Corsica" – Verso t.p.
Bibliography: p.
Includes index.
1. Phase transitions (Statistical physics) – Congresses. I. Lévy, Maurice, 1922- . II. Le Guillou, Jean-Claude. III. Zinn-Justin, Jean. IV. Title. V. Series.
QC175.16.P5C37 1980 530.1'3 81-15838
ISBN 0-306-40825-2 AACR2

Proceedings of the 1980 Cargèse Summer Institute on Phase Transitions, held July 16-31, 1980, in Cargèse, Corsica

A Division of Plenum Publishing Corporation
233 Spring Street, New York, N.Y. 10013

Printed in the United States of America

PREFACE

The understanding of phase transitions has long been a fundamental problem of statistical mechanics. It has made spectacular progress during the last few years, largely because of the ideas of K.G. Wilson, in applying to an apparently quite different domain the methods of the renormalization group, which had been developped in the framework of the quantum theory of fields. The ability of these theoretical methods to lead to very precise predictions has, in turn, stimulated in the last few years more refined experiments in different areas.

We now have entered a period where the theoretical results yielded by the renormalization group approach are sufficiently precise and can be compared with those of the traditional method of high temperature series expansion on lattices, and with the experimental data.

Although very similar, the results coming from the renormalization group and high temperature analysis seemed to indicate systematic discrepancies between the continuous field theory and lattice models. It was therefore important to appreciate the reliability of the predictions coming from both theoretical schemes, and to compare them to the latest experimental results.

We think that this Cargèse Summer Institute has been very successfull in this respect. Indeed, leading experts in the field, both experimentalists and theoreticians, have gathered and presented detailed analysis of the present situation. In particular, B.G. Nickel has produced longer high temperature series which seem to indicate that the discrepancies between series and renormalization group results have been previously overestimated. Also, the results are in good agreement with the experimental data.

We think that the resulting set of lecture notes presents a pedagogical and rather complete account of the situation in this area of Phase Transitions.

It is a real pleasure to thank all those who have made this summer institute possible, and mostly, all the lecturers and participants, whose collaboration and enthusiasm made (so we hope) this institute a success.

Thanks are due to Nato for its generous financial support, and also the DRET (France) for its grant. We also thank the C.N.R.S.(France), the D.G.R.S.T. (France) and the N.S.F. (USA) for travel grants. Special thanks are due to the Université de Nice for having put at our disposal the facilities of the Institut d'Etudes Scientifiques de Cargèse, and to Miss Marie-France HANSELER for her very helpful collaboration.

M. Lévy

J.C. Le Guillou

J. Zinn-Justin

CONTENTS

STATIC AND DYNAMIC CRITICAL PHENOMENA NEAR THE SUPERFLUID TRANSITION IN ^{4}HE

Guenter Ahlers

Department of Physics
University of California
Santa Barbara, CA 93106

INTRODUCTION

It is widely appreciated that the superfluid transition in liquid ^{4}He is an exceptionally suitable system for detailed experimental investigations of continuous phase transitions. Its merits and limitations have been discussed in detail elsewhere,[1,2] and the experimental results on the static properties have been reviewed in detail.[2] In the present lectures, I would like to provide a summary of some recent developments in our understanding of the static properties and then discuss the behavior of the singularities of transport properties. Before proceeding to this, however, it is useful to review recent more general developments not limited to the superfluid transition which pertain to the nature of confluent singularities. These singularities have a strong bearing on the interpretation of experimental results near phase transitions.

STATICS

A. Confluent Singularities

In the early 1970's, it was recognized that singularities near critical points in real physical systems cannot always be described adequately by pure power laws. Instead, confluent singularities have to be included in order to fit experimental data of high precision and to extract from such data the parameters of the leading singularity. Evidence for the importance of these confluent singularities came, quite independently, from high-temperature series expansions for the Ising model,[3] from experiments on superfluid helium[4] and liquid-gas critical points,[5] and from the renormalization group theory of critical points.[6-9] Thus, a thermodynamic quantity f_i should be written as:

$$f_i = A_i|t|^{-\lambda_i} [1 + a_i|t|^{\Delta} + O(t^{2\Delta}) + \ldots], \tag{1}$$

where $t = (T-T_c)/T_c$. The exponent Δ has since been estimated by various methods,[4,6-9] and its best values [8,9] are[8] $\Delta = 0.493$, 0.521, and 0.550 for $n = 1,2$, and 3 component systems with isotropic interactions. The confluent term $a_i|t|^{\Delta}$ would be unimportant at small t if Δ were large; but the term represents a singular contribution with $\Delta < 1$ which can be appreciable even for rather small $|t|$ (in this argument we assume that the amplitude a_i is not larger than of order unity).

Experimental data which are fitted with the pure power law

$$f_i \simeq A_{i,eff}|t|^{-\lambda_{i,eff}} \tag{2}$$

will yield values of $\lambda_{i,eff}$ given approximately by[2,10]

$$\lambda_{i,eff} \simeq -\partial \ln f_i / \partial \ln|t|$$

$$\simeq \lambda_i - a_i \Delta |\bar{t}_i|^{\Delta} + O(|\bar{t}_i|^{2\Delta}) \tag{3}$$

where $\bar{t}_i$ is some average reduced temperature of the experiment. Thus, experimentally determined effective exponents may differ significantly from the predicted universal values λ_i of the leading exponents.

There are numerous examples of exponent values which were obtained by fitting experimental data to Eq. 2 and which differ appreciably from the expected values of the exponents of the leading singularity. Thus, a good experimental estimate of the susceptibility exponent γ for the cubic ferromagnet EuO has long been[11] $\gamma_{eff} = 1.29$, whereas the expected value for a Heisenberg system[7-9] (even in the presence of dipolar interactions and cubic anisotropy) is close to 1.38. For the measurement, a typical value of $|\bar{t}|^{\Delta}$ was 0.1, and thus Eq. 3 with the measured γ_{eff} and the expected γ yields $a_{\chi} \simeq 1.8$. Thus, the correction amplitude needed to reconcile experiment and theory is no larger than of order unity, as would be expected.

Although the correction amplitudes a_i are non-universal, it was clear very early from the renormalization group analysis of the equation of state by Wegner[6] that there should be universal relations between the a_i's which pertain to different properties of the same system, and that the confluent singularities in all the properties of a given system should involve only a single non-universal (i.e. system-specific) parameter. This was explored in detail recently by Chang and Houghton[12] and by Aharony and Ahlers,[13] who derived universal values for certain ratios a_i/a_j of the amplitudes for different properties. Thus, for instance, to leading order in $\varepsilon = 4-d$,

$$a_\xi/a_\chi = \frac{1}{2} + O(\varepsilon). \qquad (4)$$

Here a_ξ is the confluent singularity amplitude of the correlation length ξ. It is interesting to note that the effective exponents γ_{eff} and ν_{eff}, to leading order, obey the scaling law $\gamma - 2\nu = \eta\nu$; for with Eq. 3

$$\gamma_{eff} - 2\nu_{eff} = \gamma - 2\nu - a_\chi(1 - 2a_\xi/a_\chi)\ \Delta\ |\bar{t}|^\Delta \qquad (5)$$

which, with Eq. 4 yields $\gamma_{eff} - 2\nu_{eff} = \gamma - 2\nu$. It can be shown[13] that effective exponents obey, to leading order in ε, all of the scaling laws, but not the hyper-scaling relations which involve the dimensionality of the system. Thus, the mere fact that experimentally determined effective exponents obey scaling does not necessarily imply that the exponent values are the correct ones for the leading singularity. Calculations to higher order in ε[12] have shown that scaling of effective exponents holds only to lowest order in ε; but the deviations from scaling of the $\lambda_{i,eff}$'s are small and probably not detectable by experimental results available at this time.

It was shown by Aharony and Ahlers[13] that, to leading order in ε, the amplitude ratios a_i/a_j are given by

$$a_i/a_j = (\lambda_i - \lambda_i^0)/(\lambda_j - \lambda_j^0) + O(\varepsilon), \qquad (6)$$

where the λ_i^0 are the mean field values of the λ_i. This, and Eq. 3 above, imply that

$$(\lambda_{i,eff} - \lambda_i)/(\lambda_{j,eff} - \lambda_j) = a_i/a_j, \qquad (7)$$

i.e. a plot of $\lambda_{i,eff}$ vs. $\lambda_{j,eff}$ should fall on a universal straight line of slope a_i/a_j which, to leading order in ε, connects (λ_i,λ_j) with $(\lambda_i^0,\lambda_j^0)$. This is demonstrated very well by experimental[14] and theoretical[8,9,15] results for liquid-gas critical points which are shown in Figure 1. The experimental data were obtained over various ranges of the reduced temperature and thus have different values of γ_{eff} and β_{eff}. The solid line is drawn through the best renormalization group estimate[8] and extrapolates to the mean field point $(\frac{1}{2},1)$. It also passes through the high temperature series result.[15] The PVT measurements, which were obtained on various systems but relatively far away from T_c near $|\bar{t}| \simeq 0.002$

yield the amplitudes $a_M \simeq 1.4$ and $a_\chi \simeq 2.2$, giving the ratio $a_M/a_\chi \simeq 0.63$ which is in good agreement with theoretical estimates.[12,13]

The individual amplitudes a_M and a_χ can be used to estimate the difference between β and γ on the one hand, and β_{eff} and γ_{eff} on the other, for the CO_2 or Xe results which were obtained much closer to T_c. Using $|\bar{t}| \simeq 10^{-5}$ for those data, one has from Eq. 3

$$\beta_{eff}-\beta = 0.002; \; \gamma_{eff}-\gamma = -0.004.$$

For the measurements on CO_2 and Xe, the difference between $\lambda_{i,eff}$ and λ_i is thus quite small because $|\bar{t}|$ is sufficiently small.

B. The Superfluid Density of ^{4}He

Precise results for the superfluid fraction ρ_s/ρ over a wide pressure range and near T_λ were derived from measurements of the

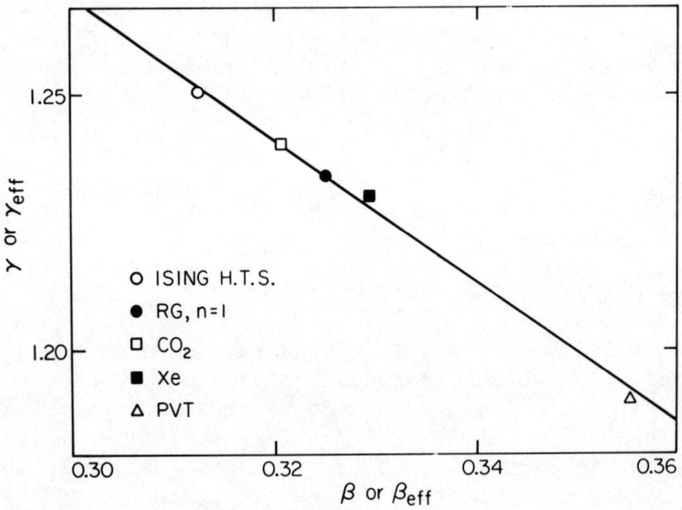

Figure 1. The susceptibility exponent γ as a function of the order parameter exponent β. The line passes through the renormalization group result (solid circle) for Ising systems and the mean field result $\gamma = 1$, $\beta = 1/2$. See Reference 14.

second-sound velocity u_2,[4] using the relation

$$u_2^2 = (S^2T/C_p)(\rho_s/\rho_n)\,[1 + \mathcal{O}(u_2^2/u_1^2)]. \tag{8}$$

Here S is the entropy, C_p the constant pressure specific heat, and $\rho_n = \rho - \rho_s$ the normal-fluid density. The velocity u_1 is the first sound velocity at low frequency. These results for ρ_s provided the first experimental evidence for the necessity of including confluent singular terms, such as those in Eq. 1, in the data analysis. Thus, the data were fit to the equation

$$\rho_s/\rho = k(P)\ |t|^\zeta[1 + a(P)\ |t|^\Delta]. \tag{9}$$

Near vapor pressure, it was found that the amplitude a(P) was very small; and therefore it was assumed that terms of $\mathcal{O}(|t|^{2\Delta})$ would also have a small amplitude and could be neglected in the temperature range $|t| \lesssim 10^{-2}$. Fitting the data to Eq. 9 yielded[2,16] $\zeta = 0.675 \pm 0.001$, and this value was believed to represent accurately the leading singularity.

The exponent ζ is expected to be equal to the exponent ν of the correlation length for spatial fluctuations in the order parameter. Rather accurate renormalization group calculations of the n-vector model exponents were performed by Le Guillou and Zinn-Justin[8], and yielded $\nu = 0.669 \pm 0.002$. More recently, new estimates using different series summation techniques, were obtained by Albert[9] and gave $\nu = 0.672 \pm 0.002$. Although these theoretical values differ only very little from the experimental result, the difference is larger than the error estimates would seem to permit. It therefore became important to re-examine the analysis of the experimental data.

Although Eq. 9 should be adequate to describe the singular part of ρ_s/ρ at least at small P where a(P) is small, it does not allow for any regular temperature dependence of, for example, the leading amplitude k(P) (since a(P) is small, its dependence upon T may surely be neglected). The change of k(P,T) with T over the small range of T ($\approx$ 1%)would of course be expected to be small; but at the level of the small differences between the experimental and theoretical exponent values it is not necessarily negligible. The ρ_s/ρ data were therefore re-analyzed, using the function

$$\rho_s/\rho = k_0(P)[1 + k_1(P)|t|]\ |t|^\zeta[1 + a_0(P)|t|^\Delta]. \tag{10}$$

We would expect the coefficient $k_1(P)$ to be of order unity if it describes a regular temperature dependence of k(P). For this analysis, data with $|t| \leqslant 0.05$ were used, and the value $\Delta = 0.5$ was assumed. It yielded

Table 1. Superfluid density amplitudes near the superfluid transition in ^{4}He (see Eq. 10) assuming $\zeta = 0.6716$ and $\Delta = 0.5$. Only data with $|t| \leqslant .01$ were used.

P [bar]	k_0	k_1	a_0
0.05	2.467	-1.7	0.34
7.23	2.215	-1.9	0.80
12.13	2.042	-3.5	1.37
18.06	1.905	-3.1	1.80
24.10	1.735	-3.4	2.66
24.17	1.755	-5.1	2.74
29.09	1.597	-3.0	3.63

$$\zeta = 0.6716 \pm 0.0004$$

$$k_0 = 2.467 \qquad k_1 = -1.47 \qquad a_0 = 0.32 \tag{11}$$

at vapor pressure. Including a term $k_2\ |t|^2$ in the first bracket of Eq. 10 did not significantly alter these results, and gave $\zeta = 0.6718 \pm 0.0007$. The new result for ζ is in good agreement with either of the renormalization group estimates,[8,9] and agrees particularly well with the result 0.672 ± 0.002 of Albert.[9] The coefficient k_1 indeed turned out to be of order unity, as we expected.

At high pressures, the results for ρ_s/ρ are only for $|t| \leqslant 0.01$ because sufficiently accurate values of S^2T/C_p were not available at larger $|t|$. They were analyzed in terms of Eq. 10, using the vapor pressure value $\zeta = 0.6716$ and the value $\Delta = 0.5$. They gave the amplitudes in Table 1.

At the higher pressures, where a(P) is relatively large, we would expect terms of $\mathcal{O}(|t|^{2\Delta})$ to contribute significantly; but since $2\Delta \approx 1$, these contributions are in effect included in the term $k_1\ |t|$. Indeed although k_1 is not given very accurately by the data with $|t| \leqslant 0.01$, there is a suggestive trend towards higher absolute values of k_1 as P increases.

Having obtained a new, and presumably more reliable, value of ν, it becomes necessary to re-examine the analysis of specific heat and thermal expansion data which yield the exponent α, because α is related to ν by the scaling law

$$3\nu = 2 - \alpha. \tag{12}$$

The new value of ν (or ζ) yields, with Eq. 12, the predicted value

$$2 - 3\nu = -0.015 \pm 0.002 \tag{13}$$

for α. This result does not agree perfectly with the value -0.026 ± 0.004 derived from thermal expansion measurements;[16] but the difference is only 1.8 times the sum of the standard errors of the two estimates and cannot be regarded as significant. In addition, it should be remembered that a regular temperature dependence of the thermal expansion amplitude and background was not taken into consideration in the analysis of those data, although for the relevant temperature range $|t| \leqslant 0.003$ these regular terms will be relatively unimportant. A new analysis of the thermal expansion data has not yet been performed, but is highly desirable.

The result Eq. 13 is in much better agreement with the analysis[16] of direct specific heat measurement[17] which gave $\alpha = -0.016$ at pressures up to 15 bar. At higher pressures the specific heat measurements seem unreliable for reasons discussed elsewhere[1,2].

As implied above, the ratio of the confluent singularity amplitude a_0 (Table 1) to the confluent singularly amplitude a_c^- of the specific heat below T_λ should be universal. On the basis of the older analysis which yielded $\zeta = 0.675$, this ratio varied from -0.21 to -0.13 as the pressure changed from 7 bar to 29 bar. A proper evaluation of this ratio can only be made after a new analysis of the thermal expansion and specific heat data, with a leading exponent which satisfies Eq. 12, has been performed. However, using the values of a_c^- given in Table VIII of Reference 16 with the values of a_0 in Table 1 above gives $a_c^-/a_0 = -0.12$ and independent of pressure. Again, this matter has to be re-examined at a later date.

Finally, it is worth noting that the leading amplitude k_0 in Table 1 changes by a factor 1.545 as the pressure changes from 0.05 to 29.09 bar. With the value 0.675 for ζ, and the temperature independent leading amplitude, this amplitude changed by the factor 1.533. Thus, the relative change in k_0 is nearly the same, and the problem of departures from two-scale-factor universality along the lambda line[1,2] cannot be resolved by including a regular temperature dependence in the amplitude of ρ_s/ρ.

III. DYNAMICS

The transport properties with strong singularities near the superfluid transition are the thermal conductivity λ above T_λ, and the second-sound damping D_2 below T_λ. There now exist very detailed predictions, based upon the renormalization group theory, of the behavior of λ and D_2, and I shall attempt to discuss the relation between these predictions and the experimental data.

A. Thermal Conductivity for $T > T_\lambda$.

Simultaneous with the development in the late 1960's of the dynamic scaling theory of transport properties near the λ-point and the prediction of a divergent thermal conductivity in He-I,[18,19] experimental evidence for a strong singularity in λ was presented by Kerrisk and Keller.[20] Very soon thereafter, rather detailed measurements became available from two separate laboratories.[21-24] These results to a large extent confirmed the predictions of the theory at a semi-quantitative level; but they also raised some important, as yet unresolved, questions. In particular, the measurements by Archibald et al.[22,23] were made in thermal conductivity cells of small,but in comparison to the correlation length macroscopic, spacings d. Surprisingly, they showed a strong size dependence of λ near T_λ. For the small spacings, λ diverged much more rapidly, corresponding to a larger effective exponent, than the theory had predicted. The measurements by Ahlers were made in a cell of very large spacing ($d \simeq 1$ cm), and agreed within their resolution with the theory. To illustrate the size effect, the effective exponents derived from various measurements are shown as a function of d in Figure 2. One sees that for $d \simeq 0.003$ cm the exponent is twice as large as the theory predicts and the measurements for large d reveal. The correlation length in the experimental temperature range is less than 10^{-4} cm, and thus the strong size dependence is not readily understood from a theoretical viewpoint.

Although the effective exponents in Figure 2 are not very precise, the data suggest that true bulk behavior at a highly quantitative level may not be reached until $d \gtrsim 0.3$ cm. This is illustrated by the solid line in the figure, which has no significance other than being a smooth representation of the data. The measurements in the 1 cm cell can surely be regarded as being in the bulk limit; but experiments in such a large cell are difficult because thermal relaxation times for the establishment of steady state gradients become as large as an hour or so. Because of limits on the long-term thermal stability of the apparatus, the resulting scatter in the data of Reference 21 and 24 was thus as large as a few %. Much more precise data were obtained in 1970 in a cell of 0.09 cm spacing[25] (this cell is referred to as Cell D below). These data had a scatter of only about 0.1% except at

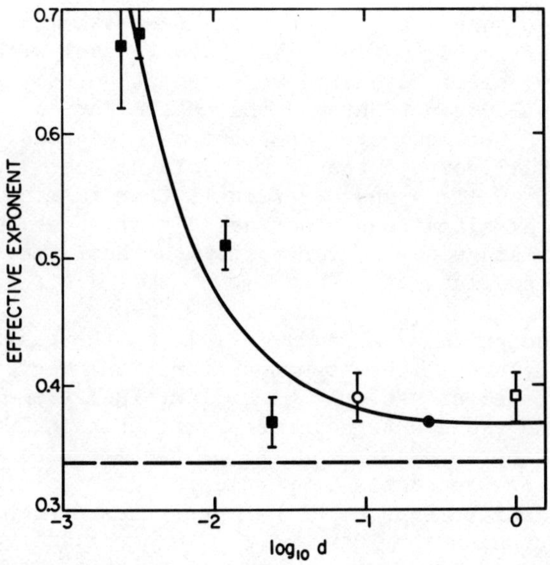

Figure 2. Effective exponent of the thermal conductivity λ for $T > T_\lambda$ near the superfluid transition in ^4He, as a function of the cell height d. Here d is in cm. The solid line is a guide to the eye, and the horizontal dashed line corresponds to $\nu/2$. Solid squares - Reference 23. Solid circle - Reference 26. Open circle - Reference 25. Open Square - Reference 24.

the smallest values of t where temperature resolution of about 10^{-7} K was a limiting factor. But inspection of Figure 2 suggests that these data may not be entirely in the bulk limit and may for that reason differ slightly from the detailed theoretical predictions. A final set of measurements, previously unpublished, was obtained in 1977 by Ahlers and Behringer[26] in a cell of spacing d = 0.26 cm (to be referred to as Cell A). This spacing yielded thermal relaxation times of order 5 minutes, and permitted high precision measurements. On the basis of Figure 2, it is probably close enough to the bulk limit to warrant a very detailed comparison with theoretical predictions. In order to establish this more clearly, as well as because of the intrinsic interest of the unexpectedly large size effect suggested by the previous data, more quantitative measurements in the range $d \leq 0.3$ cm are highly desirable and planned for the future. Accurate measurements at small d are difficult, however, and have to deal with a number of experimental problems involving measurements of d, accurate correction for wall conduction, and the determination of the boundary resistant which becomes more important as d decreases. For the present, we shall compare the results of the theory with the measurements in Cells A and D. We will keep in mind, however, that the experimental situation is not as definitive as one would like for the comparison with the very detailed theoretical predictions which are now available.

It was recognized[25] already in 1970 on the basis of the experimental data that the asymptotic critical region, in which λ can be expressed as a simple power law, is extremely narrow and confined to reduced temperatures smaller than 10^{-5}. Indeed, the data[25] indicate that the entire critical region for λ is much smaller than for the static properties. This is evident particularly from measurements at higher pressures, which show that λ has a minimum near $t = 10^{-3}$ and for larger t increases, presumably because of its regular dependence upon the temperature T (see Figure 6 of Reference 25). Early phenomenological attempts to include a confluent singularity such as that in Eq. 1 in the data analysis[25] for t near 10^{-5} yielded values for the leading exponent of λ which were about 15% larger than the dynamic scaling prediction $\nu/2 = 0.336$.

A number of possible theoretical explanations of the narrow critical region and the large value of the experimental estimate of the leading exponent have been offered. They include an instability of the dynamic scaling fixed point with respect to another, so called weak scaling, fixed point which predicts an exponent for λ greater than $\nu/2$, an anomalously high "background" contribution to λ, and extremely slow transients leading to nearly-vanishing correction exponents which are associated with a very small ratio between the order parameter and the entropy relaxation rate.[27-29] These ideas are reviewed in detail elsewhere[30] and will not be discussed further here.

The possible existence of extremely slow transients renders an analysis in terms of a power law expansion like Eq. 1 impractical because the very small correction exponent implies that very many terms would have to be included. Recently, Hohenberg et al.,[31] and independently Dohm and Folk,[29] suggested that an analysis of the data based directly upon the nonlinear recursion relations which are given by the renormalization group theory should be performed. (The usual power laws are the result of truncation and linearization of these recursion relations.) Such an analysis would automatically include the effect of transients to all orders. Dohm and Folk[29] carried out such an analysis, using the symmetric planar spin model (E) of Halperin et al.[32] This model should be a good approximation to the dynamics of liquid helium; but it neglects the coupling of the specific heat to the dynamics. More recently, Ahlers et al.[30] carried out a similar analysis, using both model E and the asymmetric planar spin model F of Halperin et al.[32] In model F, the coupling of the specific heat to the dynamics is explicitly taken into account, and we would expect model F to yield a better representation of the dynamics near T_λ. This last analysis[30] yielded the following major results:

i. The bare dynamic coupling constant in ^{4}He is a small parameter, leading to a weak coupling regime a high temperatures and a crossover[29] to the scaling regime near T_λ ($t \ll t_c$) at a small reduced temperature $t_c \approx 10^{-3}$. This explains the narrow critical region which is evident from the data in Figure 6 of Reference 25.

ii. The asymmetric model (F) provides a quantitative fit over more than four decades in reduced temperature. The symmetric model (E) is only semiquantitative.

iii. The effect of the weak scaling critical fixed point,[27] which has been invoked[28,29] to explain the data,[25] is of little importance in the experimental temperature range.

The application of the renormalization group theory to the models (E and F) yields the flow equations[27,29,31]

$$\frac{df}{dl} = -\beta_f(w,f); \quad \frac{dw}{dl} = -\beta_w(w,f) \tag{14}$$

for the coupling constant $f \propto 1/\omega_\psi(\ell)\omega_m(\ell)$ and the frequency ratio $w = \omega_\psi(\ell)/\omega_m(\ell)$. Here $\omega_\psi(\ell)$ and $\omega_m(\ell)$ are the characteristic frequencies of an order parameter mode and an entropy mode, and ℓ is a flow parameter related to the reduced temperature by $t = t_0\exp(-\ell/\nu)$. The functions β_w and β_f have been expanded in powers of f, with coefficients which depend on w.[27,28,31] For model F, the singular specific heat is coupled dynamically to f and w through its effective exponent $4v = -d \ln C_p/d\ell$ which can be taken from experiment, and $w(\ell)$ becomes complex. If the real part of $w(\ell)$ obtained by solving Eqs. 14 goes to a positive constant w^* as $\ell \to \infty$ ($t \to 0$), then the system obeys dynamic scaling and $\lambda \sim t^{-\nu/2}$. If $w^* = 0$, then the weak-scaling fixed-point is stable and the leading exponent of λ is greater than $\nu/2$.

A theoretical examination[30] of model E in the high-temperature limit ($t \gg t_0, \ell \to \infty$) revealed the existence of a weak-coupling regime where $f \propto \exp(\ell) \to 0$, $w \to w_\infty$ and $\lambda \to \lambda_\infty > 0$. This behavior is consistent with fits obtained by Dohm and Folk,[29] which yielded small values of f at large t, and with the observation of Ferrell and Bhattacharjee[33] that there exists a high-temperature expansion of λ in the form $\lambda = \lambda_\infty(1 + \lambda_1 t^{-\nu} + \lambda_2 t^{-2\nu} + \ldots)$. The high-temperature limit is physically relevant since one can make rough estimates of f for t near unity using data at high t. One finds $f \ll 1$ near $t \simeq 0.1$, say. Thus, the high-temperature region includes a range of t for which model E is valid ($t < 1$). Similar conclusions pertain to model F, although there the behavior is somewhat more complicated.

For a detailed analysis, it is convenient to define[32]

$$\hat{R}_\lambda(t) = \lambda/g_0(\xi C_p)^{1/2} \tag{15}$$

where g_0 is a bare coupling constant proportional to the entropy. In terms of w and f, to be obtained by solving Eqs. 14, $\hat{R}_\lambda$ is given by[27,29,32]

$$\hat{R}_\lambda = K_d^{1/2}[w(\ell)f(\ell)]^{-1/2}[1 - f(\ell)/4 + O(f^2)] \tag{16}$$

with $K_d^{-1} = 2\pi^2$ in three dimensions. The comparison of the experimental data with the predictions thus consisted of the following procedure.[30] First, values of $\hat{R}_\lambda$ were obtained from the measured λ and C_p, using Eq. 15. The non-linear equations 14 were then integrated numerically, starting at some arbitrary value t_0 (corresponding to $\ell = 0$) of t and with initial conditions f_0 and w_0. The resulting values of $f(\ell)$ and $w(\ell)$ were used in Eq. 16 to obtain $\hat{R}_\lambda$. Finally, the calculated values of $\hat{R}_\lambda$ were compared with the experimental values obtained from Eq. 15, and the integration procedure was iterated with adjusted initial values w_0 and f_0 until a least-squares fit to the data was obtained.

Since the expansion in f of Eqs. 14 is known only to second order, it was desirable to fit the data first in a range of t where $f \ll 1$. For that reason, initially data were fitted over the narrow range $0.003 < t < 0.01$ where f was expected to be small and where the expansion in $f(\ell)$ of Eqs. 14 should be a valid approximation (using data at larger t is not desirable in this procedure because of regular temperature dependences of the parameters which so far have been neglected). The results of this restricted-range fit[30] are shown in Figure 3. Also shown there are w and f. In the case of model F, where w is complex, both the real part w' and the imaginary part w'' are shown. It can be seen that over the range of the fit f is indeed small, and the expansion of Eqs. 14 in f is a controlled approximation. Although model E yields a semi-quantitative fit to the data, the fit with model F is much better and deviations occur only in the range $t \lesssim 2 \times 10^{-4}$ where f is of order unity and the expansion of Eqs. 14 to second order in f breaks down. The crossover from a weak coupling ($f \ll 1$) to a scaling [$f = O(1)$] regime is evident in Figure 3 and occurs in the range $10^{-3} < t < 10^{-2}$. Thus, much

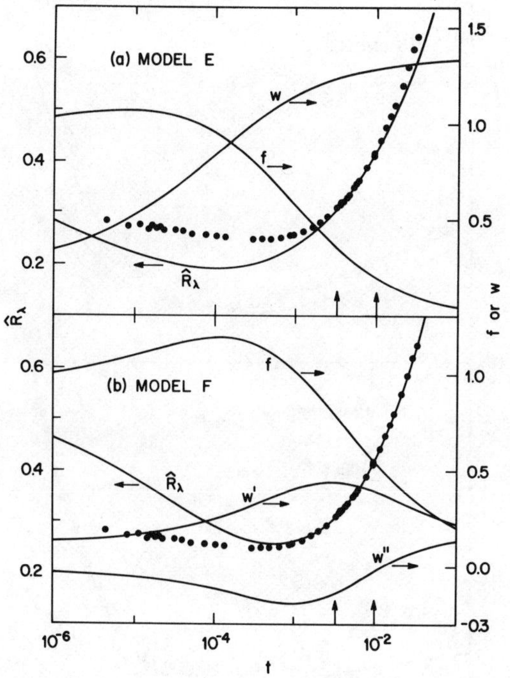

Figure 3. The amplitude $\hat{R}_\lambda$ of the thermal conductivity as a function of reduced temperature obtained from fits in the range $.003 < t < .01$ (between the vertical arrows), using (a) model E and (b) model F. The points are experimental values in cell A. Also shown are the functions f, $w = w'$, and w'' obtained from the fits.

of the experimentally accessible temperature range is strongly influenced by this crossover. On the other hand, the frequency ratio w never becomes small in the experimental range, and the weak-scaling fixed point or the existence of a small w^* therefore are unimportant for the data.

In order to further assess the extent to which model E could represent the data, fits over the broader ranges $10^{-6} < t < 10^{-2}$ and $10^{-6} < t < 10^{-3}$ were carried out[30] and are shown in Figure 4a. Again the deviations from the model are apparent. The fit over

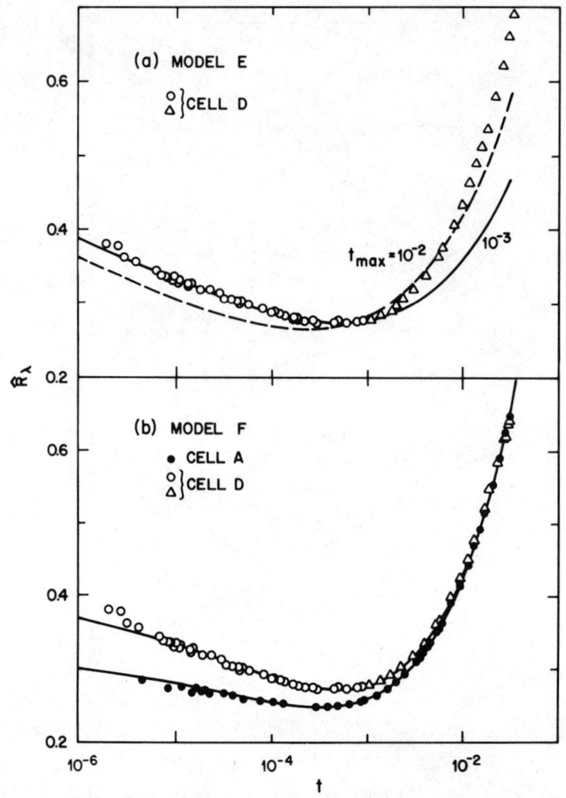

Figure 4. (a) $\hat{R}_\lambda$ fitted to data using model E in the ranges $10^{-6} < t < 10^{-2}$ and $10^{-6} < t < 10^{-3}$. The open circles are from Reference 9 and the open triangles are data generated by a procedure discussed in Reference 30.

(b) Fit of $\hat{R}_\lambda$ data to model F in the range $10^{-6} < t < 10^{-2}$, including the phenomenological term wB_3f^3.

the range $10^{-6} < t < 10^{-3}$ agrees well with a similar fit by Dohm and Folk.[29]

As a means of assessing the influence of the truncation of the expansion of Eqs. 14 on the predictions of the theory, a phenomenological third-order term wB_3f^3 was added to the function

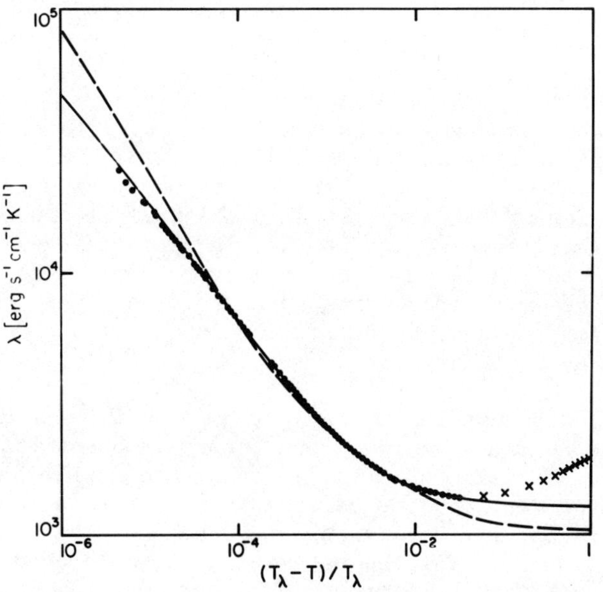

Figure 5. Solid circles - the thermal conductivity of cell A as a function of temperature on logarithmic scales. Crosses - data at high temperatures from Kerrisk and Keller (Reference 20). Solid line - model F fit, including the pheomenological term wB_3f^3. Dashed line - model E fit, including the phenomenological term wB_3f^3.

β_W in Eqs. 14.[30] The parameter B_3 was treated as an additional adjustable parameter. The results for model F are shown in Figure 4b for the data of both cell A and cell D. A small value of B_3 (0.192 for cell A and 0.162 for cell D) yielded excellent fits over four decades in t. In contrast, adding the term wB_3f^3 to model E did not significantly improve the fit.

Finally it is interesting to compare the fits of $\hat{R}_\lambda$ to model E and F over a wide temperature range with λ. This is done in Figure 5, where the theoretical curves (based on fits with B_3) and experimental data for cell A are shown on logarithmic scales.[30] Also shown are measurements by Kerrisk and Keller[20] at high temperatures. Here, too, it is evident that model F provides an excellent fit, with deviations occurring only for $t \gtrsim 0.04$

where the neglect of the regular dependence upon T of the parameters in the models can account for the deviations. Again, the model E fit is seen to be only semiquantitative.

For the data at higher pressures (cell D), fits as good as those in Figure 4b were also obtained. These data do not extend to very large values of t, however, and more experimental work is needed in view of the very detailed theoretical predictions which are now available. Data at the highest pressures are particularly interesting because the crossover temperature t_c moves to smaller values with increasing pressure. Thus, the range of validity of the expansion of Eqs. 14 in f extends to smaller values of t.

Lastly, let me say that one might hope that the remarkable success of the renormalization group theory in quantitatively explaining the thermal conductivity measurements will provide the stimulus for the very tedious calculations of the third order terms in f for Eqs. 14.[34] Although the phenomenological third order term wB_3f^3 could fit the data well, it would be more interesting to see how far third order terms derived from the theory will extend the range in Figure 4b over which the data can be fitted.

B. Second-Sound Damping for $T < T_\lambda$

Although measurements of the second-sound damping D_2 well below the λ-point and at vapor pressure have been available for a long time,[35] results near T_λ and at vapor pressure were obtained by Tyson[36] very soon after the dynamic scaling theory[18,19] had predicted that D_2 should diverge at small wave vector k as $t^{-\nu/2}$. At higher pressure and near T_λ, D_2 has been obtained so far only from Brillouin light scattering measurements.[37,38] Those experiments are done at finite k, and for that reason cannot yield detailed information about the singularity in D_2 near T_λ which exists only for k = 0. Of course, the light scattering does give interesting information about the k-dependence of the dynamics; but I shall not discuss that here.

The measurements by Tyson were in fine agreement with the predictions of the dynamic scaling theory. But in recent years, renormalization group estimates not only of the exponent, but also of the amplitude D_0 of D_2 were made.[39,40] The theoretical values of D_0 tended to be a factor of about five smaller than the measured value. In addition, the light scattering results in the range relatively far from T_λ (where the wave vector was small compared to the inverse correlation length and where a comparison with k = 0 measurements is valid) tended to agree better with the theory than with the small k measurement. This conflict provided the stimulus

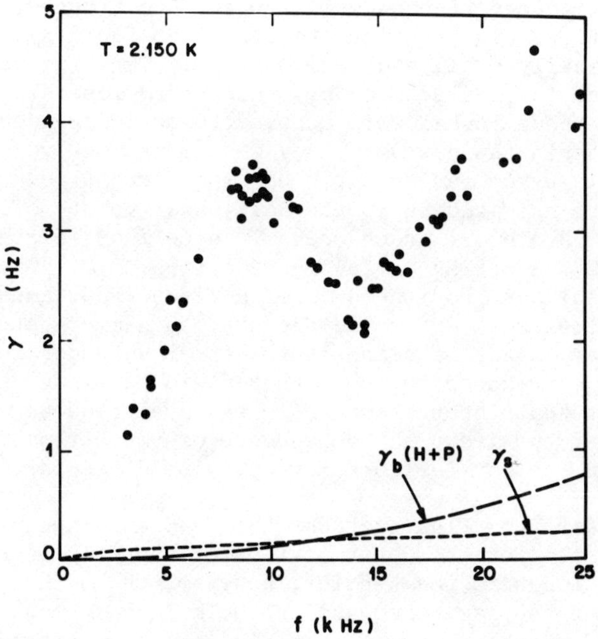

Figure 6. Half widths at half-power points of second-sound resonances in a cell of length 1.40 cm and radius 0.25 cm. The dashed lines are estimates of the bulk and surface loss contributions. At this temperature transducer losses dominate.

for a new set of measurements[41] of D_2 at small k. The new results were indeed smaller by about a factor of five than the early data; and, although they are not of high precision, they agreed well with the then available theoretical estimates.[39,40]

The reason for the difference between the older[36] and newer[41] result is not known but may be due to one of several difficult experimental problems. The older work consisted of measuring resonance line widths of cw cavities in which the second sound was excited by means of a heater and required extrapolation of the widths to zero heater power. In order to eliminate the need for this extrapolation, porous louspeaker and microphone transducers,[42] which are virtually non-dissipative, were used in the later work. In the early work, the widths had to be corrected for diffraction losses because the cavities had no sidewalls. The new experiments employed cavities with lateral walls which eliminated diffraction losses but introduced viscous wall (or surface) losses. The wall-loss contributions can be calculated from two-fluid hydrodynamics[43] and are numerically small and vanish at T_λ. The line widths have to be corrected for reflection (or transducer) losses. In the earlier work, these losses were assumed to be independent of frequency, whereas in the new work they were measured over the entire frequency range by comparing cavities of different lengths. Typical results for the half-widths at the half-power points of resonances generated with Nuclepore transducers in a cell of length 1.40 cm and radius 0.24 cm are shown in Figure 6 for T = 2.150K. The bulk losses γ_b calculated from the data of Reference 25, and the wall (or surface) losses γ_S estimated from hydrodynamics, are shown as dashed lines. At this temperature, the transducer losses dominate, and correspond to the measured widths minus γ_b and γ_S. It is apparent that the transducer losses are frequency dependent; but, of course, the frequency dependence of the carbon-film transducers used in the earlier work may be quite different.

It is clear from Figure 6 that the resonance technique is not very suitable for the determination of the bulk losses at T = 2.15 K. Closer to the λ-point, however, the transducer losses remain of about the same size and the bulk losses grow. A determination of γ_b then becomes possible, although transducer-loss corrections remain appreciable and introduce a sizable uncertainty into the measured γ_b.

Although transducer losses are a source of errors in the measurement of D_2, a more serious problem is lack of adequate temperature stability. The temperature stability in the work of Reference 41 was about $\pm 2 \times 10^{-7}$ K, and was obtained by regulating the temperature with a germanium resistance thermometer. The temperature stability was monitored by observing the amplitude of a second-sound signal at a frequency very close to but slightly different from a second-sound resonance frequency. Over a narrow temperature range, this amplitude is a very sensitive thermometer because u_2 (and thus the resonance frequency) is strongly temperature dependent and because the resonances have a high quality factor. The temperature noise measured thusly is shown in Figure 7, and is seen to be of order $\pm 2 \times 10^{-7}$ K. At a reduced temperature $|t| = 10^{-4}$, for instance, this temperature noise produces random shifts in resonance frequencies of $\delta f/f \simeq 4 \times 10^{-4}$. At a typical frequency of 5 kHz, this corresponds to $\delta f \simeq 2$ Hz and is of the same size as the bulk-loss contribution to the half-width of th resonances. It is clear that better thermometers will have to be developed before highly quantitative second sound damping measurements near T_λ can be made.

The results for D_2 at vapor pressure are collected in Figure 8. The open triangles are from Reference 36 and the open circles

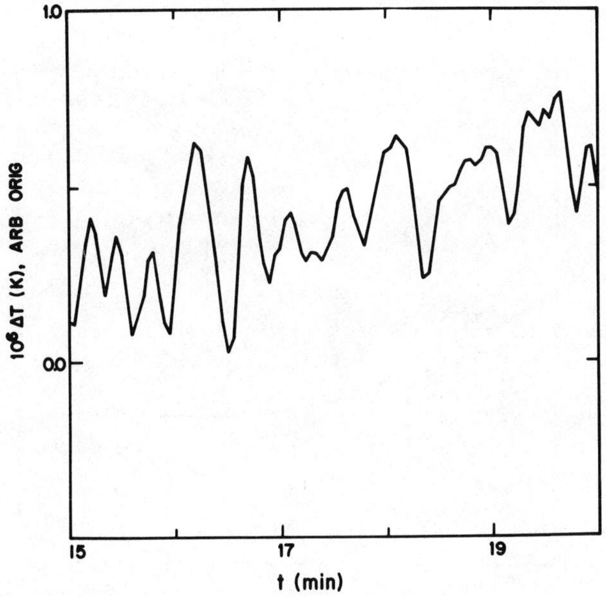

Figure 7. Temperature noise, as determined from a second-sound resonance (see text), for a helium bath regulated with a germanium resistance thermometer.

from Reference 41. Also shown, as solid circles, are the measurements by Hanson and Pellam.[35,44] Considering the possibility of fairly large systematic errors[44] for the smallest values of t of Reference 35, the data by Hanson and Pellam could be consistent with either set of measurements at smaller t. An extrapolation of the Brillouin scattering measurements from higher pressures to vapor pressure[37] is shown for $t = 10^{-3}$ where the light scattering results are still in the hydrodynamic regime. Those results are somewhat higher than the new data,[41] and somewhat lower than the older ones.[36] It is difficult to estimate their uncertainty because they do involve an extrapolation to low pressure.

Originally, the theoretical estimates[39,40] of D_2 from the renormalization group theory did not consider explicitly the slow transients which are known to be associated with the dynamics of the superfluid transition. Nonetheless, they yielded predictions

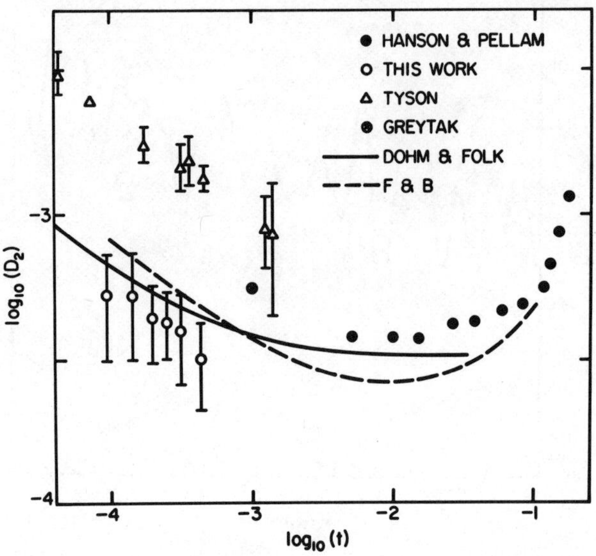

Figure 8. Second-sound damping D_2 (in cm^2sec^{-1}) as a function of the reduced temperature on logarithmic scales. Open triangles - Reference 36. Solid circles - Reference 35. Open circles - Referenc 41. Crossed circle - Reference 37. Solid line - Reference 45. Dashed line - Reference 46.

for D_2 which were of the same size as the recent experimental data.[41] Very recently, the second sound damping has been predicted by Dohm and Folk[45] in terms of the frequency ratio $w(\ell)$ and the coupling constant $f(\ell)$ which were discussed in Section IIIA above. Thus, it is now possible to use the thermal conductivity above T_λ to obtain $w(\ell)$ and $f(\ell)$, and then to calculate D_2 from w and f. The result at vapor pressure, obtained by Dohm and Folk from a fit (using model E) to the thermal conductivity, is shown as a solid line in Figure 8. The agreement with the data[41,35] is seen to be excellent and at this stage the theoretical prediction is probably more reliable than the experimental measurements. The dashed curve in Figure 8 was obtained by Ferrell and Bhattacharjee,[46] and is based on parameters obtained from a fit of the thermal conductivity data to their high-temperature expansion of λ in $t^{-\nu}$. It should be valid only in the weak-coupling regime at relatively large $|t|$, but over the temperature range of the data it also agrees well with the measurements.

Let me conclude by saying that we have witnessed over the last few years a remarkable evolution in our understanding of the dynamics of continuous phase transitions. Theorists are now in a position to predict quantitatively the transport properties both above and below the transition temperature, with only two or three non-universal parameters taken from experiment. These non-universal parameters can be taken from measurements on either side of T_c. Whereas it seemed only a few years ago that the dynamics of the superfluid transition is so complicated that a fully quantitative comparison between theory and experiment might never by possible,[1] these complications are now well under control, and the quantitative explanation of the dynamics of the superfluid transition surely is one of the most spectacular successes of the renormalization group theory. There are, of course, a lot of details to be filled in, both in terms of better experiments and in terms of additional calculations.

This work was supported by NSF Grant #DMR79-23289.

REFERENCES

1. G. Ahlers, Rev. Mod. Phys. 52:482 (1980).
2. G. Ahlers, in "Quantum Liquids," J. Ruvalds and T. Regge, eds., North-Holland, Amsterdam (1978) p. 1.
3. M.Wortis, unpublished (1970); D. M. Saul, M. Wortis and D. Jasnow, Phys. Rev. B11:2571 (1975); W. J. Camp and J. P. Van Dyke, Phys. Rev. B11:2579 (1975); W. J. Camp, D. M. Saul, J. P. Van Dyke, and M. Wortis, Phys. Rev. B14:3990 (1976); W. J. Camp and J. P. Van Dyke, J. Phys.A9:721 (1976); M. Ferer,Phys. Rev. B16:419 (1977).
4. D. S. Greywall and G. Ahlers, Phys. Rev. Lett. 28:1251 (1972); Phys. Rev. A7:2145 (1973).
5. D. Balzarini and K. Ohrn, Phys. Rev. Lett. 29:840 (1972).
6. F. J. Wegner, Phys. Rev. B5:4529 (1972); B6:1891 (1972).
7. E. Brézin, J. C. LeGuillou, and J. Zinn-Justin, Phys. Rev. D8: 2418 (1973); A. D. Bruce and A. Aharony, Phys. Rev. B10:2078 (1974); E. Brézin, Phys. Rev. B8:5330 (1973); J. Swift and M. K. Grover, Phys. Rev. A9:2579 (1974); G. R. Golner and E. K. Riedel, Phys. Lett. 58A:11 (1976); G. A. Baker, B. G. Nickel, M. S. Green and D. I. Meiron, Phys. Rev. Lett. 36:1351 (1976).
8. J. C. LeGuillou and J. Zinn-Justin, Phys. Rev. Lett. 39:95 (1977); Phys. Rev. B21:3976 (1980).
9. D. Z. Albert, Rockefeller University Preprint No. DOE/EY/2232B-204, and to be published.
10. For a discussion of effective exponents, see also G. Ahlers, Lectures at the Banff Summer School on Phase Transitions, August, 1976, unpublished.
11. See, for instance, N. Menyuk, K. Dwight, and T. B. Reed, Phys. Rev. B3:1689 (1971).

12. M. C. Chang and A. Houghton, Phys. Rev. B21:1881 (1980); Phys. Rev. Lett. 44:785 (1980); Phys. Rev. B, in press.
13. A. Aharony and G. Ahlers, Phys. Rev. Lett. 44:782 (1980).
14. R. Hocken and M. R. Moldover, Phys. Rev. Lett. 37:29 (1976); and references therein.
15. A. J. Guttman, J. Phys. A 8:1236, 1249 (1975); D. S. Gaunt and C. Domb, J. Phys. C 3:1442 (1970); C. Domb, in "Phase Transitions and Critical Phenomena," C. Domb and M. S. Green, eds., Academic, New York (1974), Vol. 3, p. 357.
16. K. H. Mueller, G. Ahlers, and F. Pobell, Phys. Rev. B 14:2096 (1976).
17. G. Ahlers, Phys. Rev. A 8:530 (1973).
18. R. A. Ferrell, N. Ményhard, H. Schmidt, F. Schwabl, and P. Szépfalusy, Phys. Rev. Lett. 18:891 (1967); Phys. Lett. 24A:493 (1967); Ann. Phys. (N.Y.) 47:565 (1968).
19. B. I. Halperin and P. C. Hohenberg, Phys. Rev. Lett. 19:700 (1967); Phys. Rev. 177:952 (1969).
20. J. Kerrisk and W. E. Keller, Bull. Am. Phys. Soc. 12:550 (1967); Phys. Rev. 177:341 (1969).
21. G. Ahlers, in "Proceedings Eleventh International Conference on Low Temperature Physics," J. F. Allen, D. M. Finlayson, and D. M. McCall, eds., University of St. Andrews Printing Department (1968), p. 203.
22. M. Archibald, J. M. Mochel, and L. Weaver, in: "Proceedings Eleventh International Conference on Low Temperature Physics," J. F. Allen, D. M. Finlayson, and D. M. McCall, eds., University of St. Andrews Printing Department (1968), p. 211.
23. M. Archibald, J. M. Mochel, and L. Weaver, Phys. Rev. Lett. 21:1156 (1968).
24. G. Ahlers, Phys. Rev. Lett. 21:1159 (1968).
25. G. Ahlers, in: "Proceedings 12th International Conference On Low Temperature Physics," E. Kanda, ed., Academic Press, Japan (1971), p.21.
26. G. Ahlers and R. P. Behringer, unpublished.
27. C. DeDominicis and L. Peliti, Phys. Rev. B 18:353 (1978).
28. R. A. Ferrell and J. K. Bhattacharjee, Phys. Rev. Lett. 42:1638 (1979).
29. V. Dohm and R. Folk, Z. Physik B 40:79 (1980); Phys. Rev. Lett. 46:349 (1981).
30. G. Ahlers, P. C. Hohenberg, and A. Kornblit, Phys. Rev. Lett. 46:493 (1981); and to be published.
31. P. C. Hohenberg, B. I. Halperin and D. R. Nelson, Phys. Rev. B 22:2373 (1980).
32. B. I. Halperin, P. C. Hohenberg, and E. D. Siggia, Phys. Rev. B 13:1299 (1976). See also P. C. Hohenberg and B. I. Halperin, Rev. Mod. Phys. 49:435 (1977).
33. R. A. Ferrell and J. K. Bhattacharjee, in "Proceedings International Conference on Dynamic Critical Phenomena," C. P. Enz, ed., Springer, N.Y. (1979).

34. It should be mentioned that even the second-order terms of model F in their present form are only approximate and based upon model E.

35. W. B. Hanson and J. R. Pellam, Phys. Rev. 95:321 (1954).
36. J. A. Tyson, Phys. Rev. Lett. 21:1235 (1968).
37. T. J. Greytak, in: "Proceedings of the International Conference on Critical Dynamics," C. P. Enz, ed., Springer, Berlin (1979).
38. W. F. Vinen and D. L. Hurd, Advances in Physics 27:533 (1978).
39. E. D. Siggia, Phys. Rev. B 13:3218 (1976); P. C. Hohenberg, E. D. Siggia, and B. I. Halperin, Phys. Rev. B 14:2865 (1976).
40. V. Dohm and R. Folk, Z. Physik B 35:277 (1979).
41. G. Ahlers, Phys. Rev. Lett. 43:1417 (1979).
42. R. Williams, S. E. A. Beaver, J. C. Fraser, R. S. Kagiwada, and I. Rudnick, Phys. Lett. 29A:279 (1969); R. A. Sherlock and D. O. Edwards, Rev. Sci. Instrum. 41:1603 (1970).
43. J. Heiserman and I. Rudnick, J. Low Temp. Phys. 22:481 (1976); and references therin.

44. Hanson and Pellam[35] reported the second-sound attenuation α_2 and the absolute temperature T. In order to convert α_2 to D_2, the second-sound velocity u_2 is needed. The strong dependence upon $T_\lambda - T$ of u_2 introduces a considerable uncertainty into D_2, associated with the uncertainty in the temperature scale of the early work.[35] For the calculation of D_2 from α_2 we assumed that temperatures were reported on the 1948 scale of temperatures.

45. V. Dohm and R. Folk, to be published.
46. R. A. Ferrell and J. K. Bhattacharjee, preprint (1980).

STATUS OF THE EXPERIMENTAL SITUATION IN CRITICAL BINARY FLUIDS

D. Beysens

C.E.N. SACLAY, SPSRM, BP.2, 91190 Gif sur Yvette

To try to summarize the status of the experimental situation in binary mixtures close to their dissolution point is a rather difficult task. A large number of experiments has been performed in the field since more than one century, and especially in the last 15 years. It has therefore been necessary to omit a number of interesting experiments. The justification of our choice has been chiefly the quality of the experimentation, and the consistency of the results in different laboratories. Ten binary mixtures have been selected, which allowed critical exponents, amplitudes and non-analytical corrections to be determined. We have contributed to the experimental study of 6 of them in our laboratory.

In order to have a clearer view of the situation, we will discuss first the main experimental techniques which have been used (section I). Section II will then deal with the experimental results obtained in the ten binary fluids.

EXPERIMENTAL TECHNIQUES

In table I can be found a summary of the main experimental techniques used, together with the temperature range which is accessible to experiment. Some remarks have been added concerning the experimental difficulties.

General

a) Static properties

Mixtures of two chemically different fluids, referred to as a binary fluids, belong to the class of fluid, with space dimensionality d=3 and a scalar order parameter n=1. The order parameter M

Table 1

Main techniques used to determine both exponents $(\gamma, \nu, \beta, \alpha)$ and amplitudes $(C, \xi_o, B, \frac{A}{\alpha})$ of the susceptibility χ, the correlation length ξ, the order parameter M and the specific heat $C_{p,c}$. The accessible $t = \frac{T-T_c}{T_c}$ range, or $q\xi$ range with q the transfert wave-vector, is included, together with some remarks concerning the reliability of the experiments.

Technique (t range / $q\xi$ range)	C	γ	ξ_o, ν	B, β	$\frac{A}{\alpha}, \alpha$	Remarks
Static intensity $(\vec{q}, t)$	—	$10^{-3} - 10^{-1}$ $q\xi \ll 1$	$3 \times 10^{-6} - 3 \times 10^{-2}$ $q\xi \gtrless 1$	—	—	low t : turbidity, multiple scattering corrections high t : weak signal and corrections due to Brillouin, entropy lines and dust.
Turbidity (t)	$3 \times 10^{-6} - 3 \times 10^{-3}$ $q\xi \gtrsim 1$	$3 \times 10^{-4} - 3 \times 10^{-3}$ $q\xi \ll 1$	$3 \times 10^{-6} - 3 \times 10^{-4}$ $q\xi \gtrsim 1$	—	—	low t : possible multiple scattering corrections high t: weak signal
Dynamic intensity Linewidth $(\vec{q}, t)$	—	—	$3 \times 10^{-4} - 3 \times 10^{-2}$ $q\xi \ll 1$	—	—	presently not well assessed formulation
Specific heat (t)	—	—	—	—	$3 \times 10^{-5} - 0.1$	Temperature gradients gravity effects
Density (t)	—	—	—	$3 \times 10^{-5} - 0.1$	$3 \times 10^{-6} - 0.1$	gravity effects ; sensitive method
Visual observation (t)	—	—	—	$3 \times 10^{-5} - 0.1$	—	gravity effects ; possible supercooling in the 2-phases region ?
Refractive index - refractometry - interferometry - Fraunhoffer pattern	—	—	—	$3 \times 10^{-6} - 0.1$	$3 \times 10^{-6} - 0.1$	possible supercooling in the 2-phases region ? very sensitive method.

is the difference between the composition C of one phase and the critical composition C_c (see Fig.1). The main problem here origina-

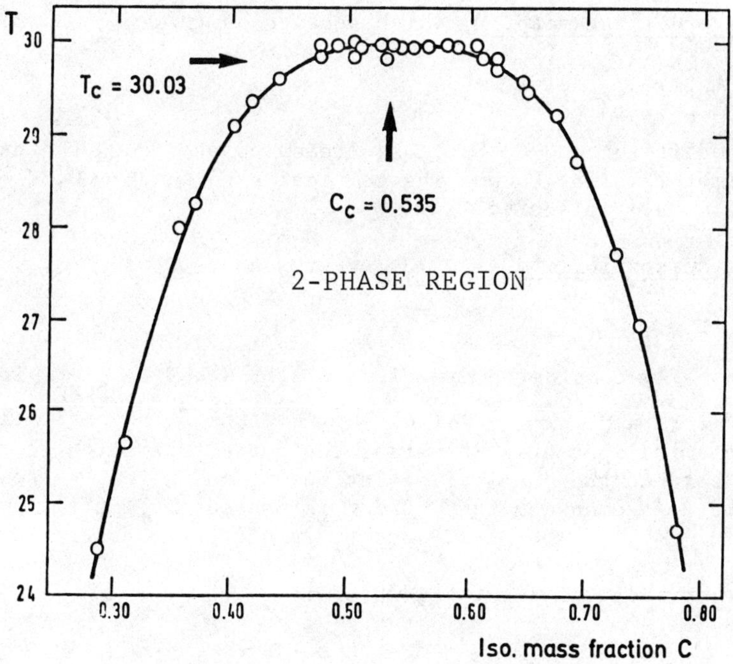

Fig. 1. Coexistence curve determined by the visual method. (Nitroethane-Isooctane, Ref.[1]).

tes from the fact that the exact order parameter is not known, and thus either mass fraction, volume fraction or molar fraction (among numerous possibilities) can be considered. Fortunately, the values of the corresponding exponent and amplitude do not vary very much if one uses different choices for the order parameter. A good choice would be the parameter which makes the coexistence curve the most symmetrical, making the experiment to correspond to the ideal conditions of the 3-d Ising model.

When the temperature T of the fluid approaches the critical temperature T_c, at critical composition C_c, the quantity π which undergoes a divergence can be written as :

$$\pi = P\, t^{-\varpi}\, [1 + a_\pi t^{\omega\nu} + \ldots] + \pi_B,$$

where $t = \left|\frac{T-T_c}{T_c}\right|$ is the reduced temperature. P is the amplitude associated with the exponent ϖ which governs the divergence. Non-analytical corrections with amplitude a_π and exponent $\omega\nu \simeq 0.5$ correct the behavior for large t. π_B is a possible background (regular) contribution.

These are the quantities which can usually be determined by experiments :

The order parameter M, which behaves as :

$$M = Bt^{\beta}[1 + a_M t^{\omega\nu} + \ldots]$$

M is the difference $C-C_c$ but, as already pointed out, the exact order parameter is not known. The critical exponent $\beta \simeq 0.32 - 0.34$, and B is usually close to 1.

The susceptibility χ, which behaves as :

$$\chi = Ct^{-\gamma}[1 + a_\chi t^{\omega\nu} + \ldots] + \chi_B .$$

χ is generally considered dimensionless, $\chi = k_B T_c \cdot \chi_{exp}$. In binary fluids, $\chi_{exp} = \left(\frac{\partial\mu}{\partial c}\right)^{-1}_{p,T}$, i.e. the inverse of the derivative of the chemical potential μ versus the concentration. μ is the difference between the chemical potentials μ_1 and μ_2 of the components, k_B is the Boltzmann constant, and p is pressure. Typically $\gamma \simeq 1.2 - 1.4$, $C \simeq 100\ \mathrm{J.cm^{-3}}$ and χ_B is not detectable.

The correlation length ξ, which is

$$\xi = \xi_0 t^{-\nu}[1 + a_\xi t^{\omega\nu} + \ldots] + \xi_B .$$

The exponent ν is found to be $\simeq 0.62 - 0.64$, and ξ_0 is about 1-4 Å. ξ_B is not visible.

The specific heat $C_{p,c}$, which follows the relation

$$C_{p,c} = \frac{A}{\alpha} t^{-\alpha}[1 + \alpha a_c t^{\omega\nu} + \ldots] + C_B .$$

$C_{p,c}$ is generally considered dimensionless : $C_{p,c} = k_B^{-1} [C_{p,c}]^{exp}$. The specific heat at constant pressure and concentration behaves like the specific heat at constant volume in pure fluids. C_B is a background term, which includes both the regular and the critical contributions. The ratio $\frac{A}{\alpha C_B}$ can range from 3% to 300%[2]. α is weak $\simeq 0.1$.

b) Dynamic properties

We will deal only with dynamic properties related to the statics. The time dependent variations of the order parameter are strongly connected to the size of the fluctuations, and thus to the correlation length. The spectrum corresponding to a given mode of fluctuations (wavenumber q) is generally Lorentzian [3-4]. The half-width Γ behaves as [5-6]

$$\Gamma = R \cdot \frac{k_B T}{6\pi\eta\xi} q^2 H(q\xi) + \Gamma_B .$$

Fig. 2

Here k_B is the Boltzmann constant, η is the ordinary shear viscosity, and $H(q\xi)$ is a complex function of the product $q\xi$ [5]-[7] which is equal to 1 in the region $q\xi << 1$ (hydrodynamic region). R is an universal amplitude factor, $R \simeq 1 - 1.20$. Γ_B is a possible background contribution, which is in most cases undetectable.

<u>(Light) Scattering Techniques : Determination of χ, ξ, Γ</u>

Scattering techniques, which allow the structure factor $S(q,\omega)$ (where ω is the frequency) to be determined, are numerous. The most important are neutron scattering [8], low angle X-ray scattering [9], and light scattering. Light scattering beeing the most simple, convenient, and accurate technique in the case of critical binary mixtures, since the wavelength of visible light is usually of the same order of magnitude as the correlation length, we will consider here only this technique.

Near T_c, the concentration fluctuations give rise to a strong modulation of the density, and therefore of the refractive index. The fluid, when illuminated, looks like milk. Light is scattered many times (multiple scattering phenomenon), the turbidity becomes important, <u>critical opalescence</u> occurs (Fig.2).

Fig. 2. A sample of the Aniline + Cyclohexane mixture enlightened by a laser beam (red). The thin white line is the incident beam (1st scattering) the red halo is its re-scattering (multiple scattering). The blue colour is characteristic of the critical opalescence, and is due to the multiple scattering of the surrounding white light. $T - T_c = 0.03$ K ($t = 10^{-4}$). From Ref.[41].

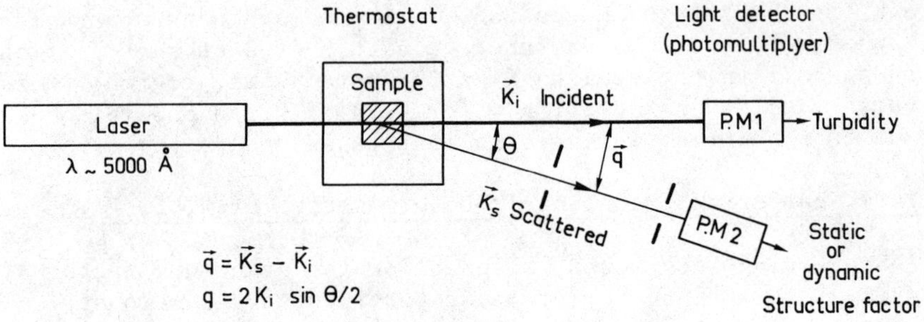

Fig. 3. A typical light scattering experiment.

A schematic of a typical light scattering experiment is shown in Fig.3. The temperature range of the sample which can practically be obtained is from 10^{-3} K to 100 K ($t = 3 \times 10^{-6} - 0.3$), the q-range is from 2×10^5 to 2500 cm^{-1}, so the structure factor can be determined in the range $q\xi \simeq 20 - 5 \times 10^{-4}$. Since the static structure factor $S^s(q,\xi)$ is essentially the Fourier transform of the normalized correlation function G multiplied by χ [10], i.e. $S^s(q,\xi) = \chi \cdot G(q,\xi)$, both the susceptibility χ and the correlation length ξ are obtainable.

To deduce ξ, a good model of G has to be known. The two limits $q\xi >> 1$ and $q\xi << 1$ do not cause problems. In the region $q\xi \simeq 1$, $G(q\xi)$ has been determined numerically [11]. We have successfully fitted [12] these numerical values by the following analytical function, which is valid within 1% accuracy in the range $q\xi = 20 - 5 \times 10^{-4}$:

$$G(X) = \sum_{i=1}^{3} c_i (1 + a_i^2 X^2)^{-b_i} ,$$

with $X = q\xi$, $c_1 = c_2 = -c_3 = 1$; $a_1 = 1.040056$, $a_2 = 1.058947$, $a_3 = 1.053932$; $b_1 = 1 - \eta/2 = 0.98425$, $b_2 = 1.554213$ and $b_3 = 1.627419$.

We will not speak about the determination of the exponent η (see G(X)). Due to its smallness ($\eta \sim 0.03$) in binary fluids, up to now no unambiguous experimental determination exists [78].

Concerning the dynamic structure factor, a spectral analysis of the scattered light allows the linewidth of concentration fluctuations to be determined, and therefore, under certain assumptions, ξ can be measured (see below).

Let us look in greater details at the techniques used :

a) Static scattered intensity versus q and t

The scattered intensity I_S can be put into the following form:

$$I_S(q,\xi) = I_o \cdot (1+t) \cdot \chi_{exp} \cdot G(q\xi)$$

with $\chi_{exp} = C_{exp} t^{-\gamma}[1 + a_\chi t^{\omega\nu} + \ldots]$ and $\xi = \xi_o t^{-\nu}[1 + a_\xi t^{\omega\nu} + \ldots]$. I_o is related to the incident power I_i, to the derivative of the refractive index n with respect to the concentration C, to the local field formulation expressed by the factor Sn(Sn = 1 : Einstein ; $Sn = \frac{9n^2}{(2n^2+1)(n^2+2)}$: Yvon ; $Sn = \frac{3}{n^2+2}$: Rocard) [13], the temperature, the light wavelength in vacuum λ_o, and the observed volume V :

$$I_o = \frac{\pi^2}{\lambda_o^4} \cdot Sn^2 \cdot \left(\frac{\partial n^2}{\partial c}\right)_{p,T} \cdot k_B T_c V \cdot I_i$$

I_s is a function of the two independent parameters q (i.e. the scattering angle θ), and t.

With q fixed and t varying, the region $q\xi \gtrsim 1$ will give ξ.

With t fixed and q varying, ξ will be able to be determined in the region $q\xi \lesssim 1$.

Nevertheless some limitations remain :

i) The system is weakly opalescent (Fig.4) :

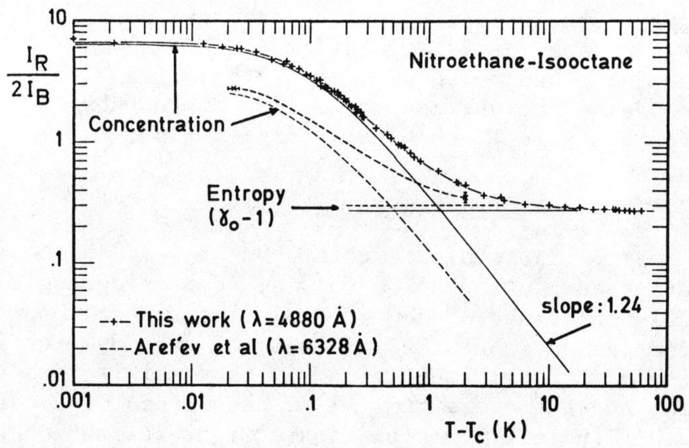

Fig. 4. Spectral intensity versus temperature. I_R represents the sum : Entropy line + $I_s(q,\xi)$. From Ref.[15].

In that case, the region $q\xi \ll 1$, which is a region where t is large, is rather short, due to the parasite contributions of the scattering by dust, or simply the scattering by the entropy and Brillouin lines. Spectral measurements can nevertheless improve the accuracy. On the other hand, in the region $q\xi \gtrsim 1$, no corrections due to turbidity, multiple scattering [14], etc have to be made, which is not the case for strongly opalescent mixtures. ξ can be thus measured with reliability.

ii) The system is strongly opalescent (Fig.5)

The region $q\xi \gtrsim 1$ is not accurately measured ; but the region $q\xi \ll 1$ allows γ to be determined with confidence.

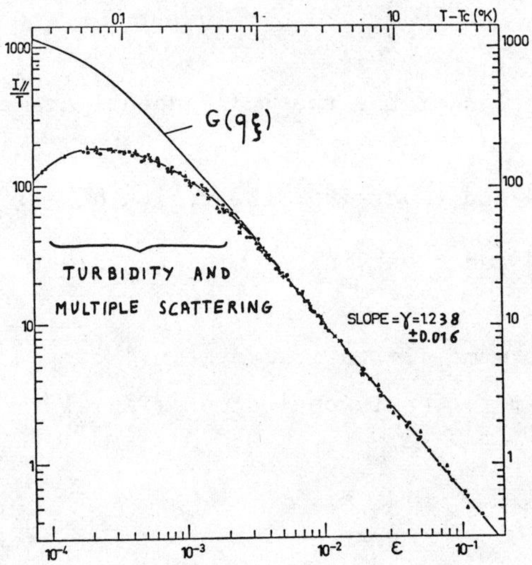

Fig. 5. Temperature dependence of the spectral polarized intensity $I_{//}$ scattered at 90°. In order to show a power law dependence, $I_{//}$ had to be divided by the (T) smooth temperature dependence factor. The solid line represents the corresponding theoretical behaviour. For Ref.[16].

It is nearly impossible to obtain accurately the constant I_o, due to numerous corrections from volume, scattering angle, incident power, etc... and to the unknown formulation of the local field. However, from the ratio of the intensity of the Rayleigh line to the intensity of the Brillouin line, it is possible to have an estimate of I_o, and therefore of χ [15], and without the assumptions concerning the Einstein, Yvon or Rocard light scattering theories.

Finally, another important source of parasitic effects can be found in the sample heating by the laser beam, which is typically of the order of magnitude of 10^{-3} K for an incident light power of 10^{-3} W.

b) Turbidity

The turbidity τ is the rate of decrease of intensity of the incoming beam per unit length. If $\mathcal{T}$ is the transmission rate of the light intensity through a sample of length ℓ, we can write $\mathcal{T} = \exp(-\tau\ell)$. This decrease of the transmitted intensity is simply due to the scattering of light itself. It can be therefore related to the scattered intensity integrated over all solid angles Ω :

$$\tau \propto \int_{4\pi} I_S(q.\xi) \, . \, d\Omega$$

More precisely, can be expressed as[12]:

$$\tau = \tau_o(1+t).C_{exp}.t^{-\gamma}[1 + a_\chi t^{\omega\nu}] \, G'(q\xi)$$

where

$$G'(q\xi) = G'(X_o) =$$

$$\sum_{i=1}^{3} c_i \left\{ [(1+2a_i^2X_o^2)^{\mu_i}-1][1+a_iX_o^2(2-\mu_i)+a_i^4X_o^4(2+\mu_i+\mu_i^2)] - 2\mu_iX_o^2(1+X_o^2) \right\}$$

$$[a_i^6X_o^6\mu_i(1+\mu_i)(2+\mu_i)]^{-1}$$

with $X_o = \sqrt{2}K_o\xi$ and $K_o = \frac{2\pi n}{\lambda}$, the wavevector of incident light. The a_i and and c_i are the same as for $G(X)$, and $\mu_i=1-b_i$. τ_o is functionally similar to I_o :

$$\tau_o = \frac{2\pi^3}{\lambda_o^4} \, . \, Sn \, . \, \left(\frac{\partial n^2}{\partial c}\right)^2 k_BT_c .$$

In contrast to I_o, τ_o can be well determined experimentally. The last item remaining is the uncertainty about Sn. Obviously, it is close to T_c, where scattering is important, that τ is measured accurately, generally in the region $q\xi \lesssim 1$, and the determination of γ is therefore not very reliable in general. From $G'(q\xi)$ we can easily infer ξ .

The limitation of the method is (i) weak scatterer gives a low turbidity and then a poor accuracy ; the useful range in t is small (see Fig.6).

ii) For a strong scatterer (see Fig.7), multiple scattering can be a non negligible contribution if special care is not taken, such as using samples with convenient lengths.

b) Linewidth

The typical frequency range is $\Delta\omega \simeq 0.05 - 10^4$ rd.sec^{-1}. The principle of the method (light beating spectroscopy[10]) is to make interference between the incident light (frequency $\omega_o \sim 10^4$ rd.sec^{-1}) and the scattered light ($\omega_o+\Delta\omega$), and to analyze

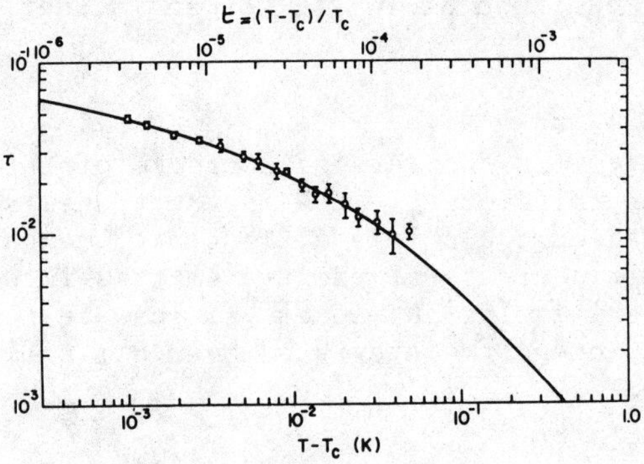

Fig. 6. Turbidity τ as a function of temperature in the Nitroethane-3- methyl pentane system. Full line : theoretical variation. From Ref.[11-b].

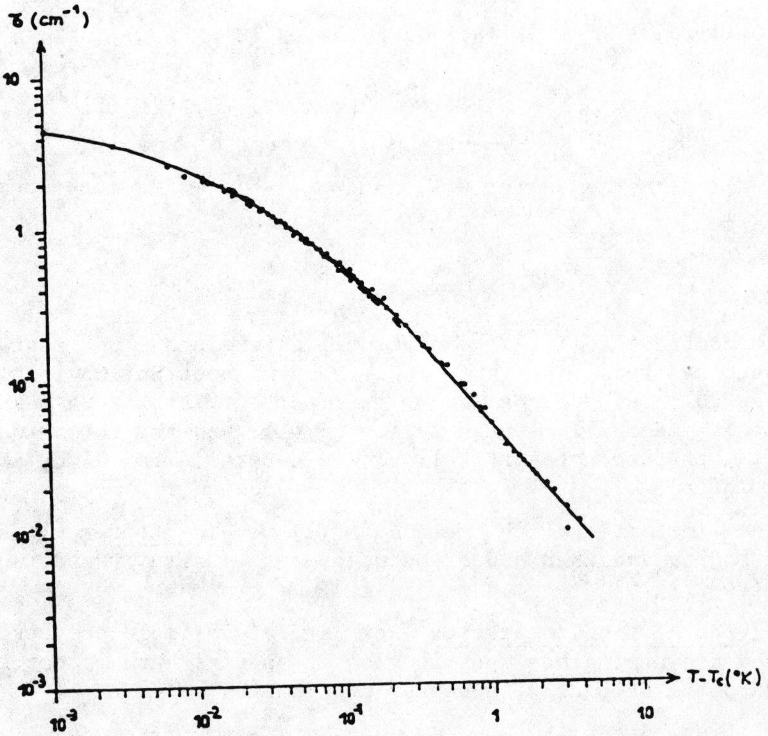

Fig. 7. Turbidity τ of the cyclohexane-Aniline system. From Ref.[18] reanalyzed in Ref.[19] and [12]. Full line : theoretical variation. Data obtained with 5, 10, 20 and 50 mm sample length.

the beating at $\Delta\omega$ on a photomultiplyer cathode. This is called the heterodyne technique. An alternative method is to consider the beating of the scattered light with itself (homodyne technique).

The corresponding spectrum is very close to a Lorentzian [4]. As already pointed out, the linewidth is related to the correlation length by :

$$\Gamma = R \cdot \frac{k_B T}{6\pi\eta\xi} \cdot q^2 \cdot H(q\xi) + \Gamma_B$$

with $\xi = \xi_o t^{-\nu}[1 + a_\xi t^{\omega\nu} + \ldots]$. ξ can be measured only in the region $q\xi << 1$ (practically $\simeq 0.1 - 3 \times 10^{-5}$), region where $H(q,\xi)$ is known $\simeq 1$ and where η is merely the viscosity as measured by a shear flow viscosimeter. The determination of ξ_o requires the knowledge of R. It seems (see Ref.[12] and below) that $R \simeq 1.16$, so the ξ_o amplitude can be determined in principle by this method. Concerning a precise value of the exponent ν, this linewidth technique cannot be regarded as very reliable since up to now a precise formulation of Γ is unknown. Another problem is caused by an eventual additive background. We show nevertheless in Figs.8a) and 8b) determinations of ξ by this method.

We think this linewidth method could be very useful and accurate if a precise formulation of the linewidth would be available.

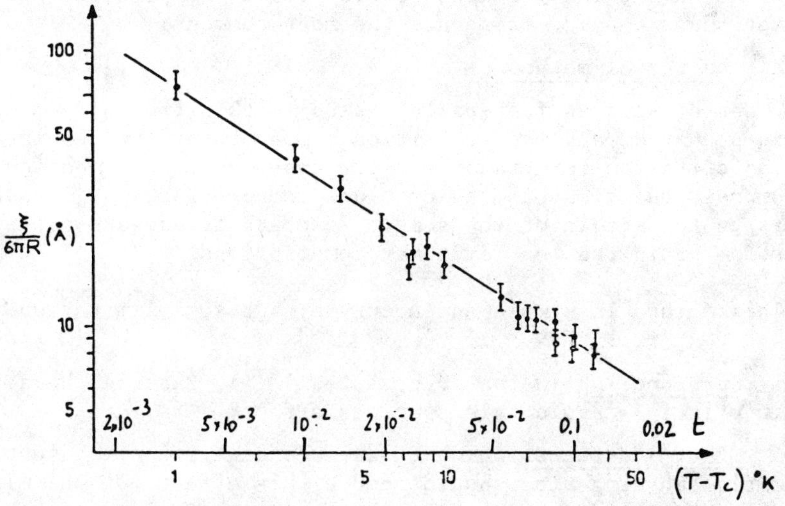

Fig. 8a) First reported determination of a dynamical determination of ξ (Cyclohexane-Aniline mixture). From Ref.[17]. Data reanalyzed in Ref.[12] and [19].

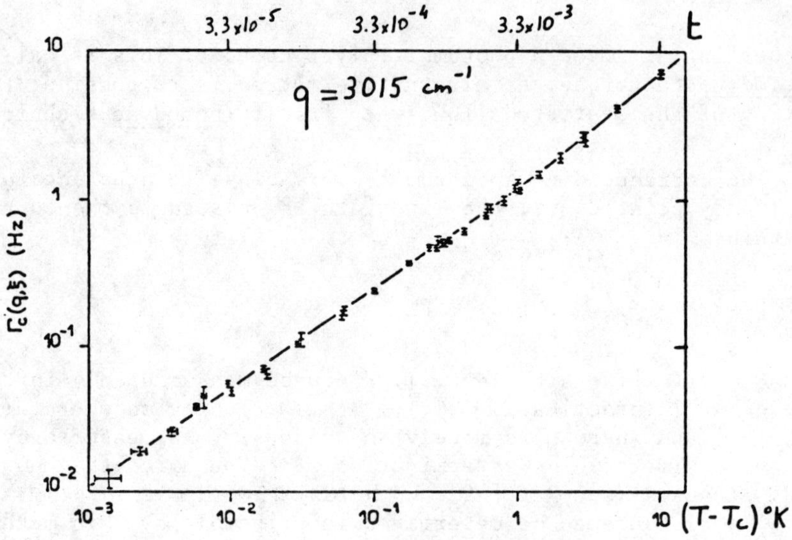

Fig. 8b) Linewidth obtained in the Nitroethane-3-methylpentane system at small angle scattering. From Ref.[19] and [12].

Determination of the Coexistence curve

The principle of the method is to measure the composition of each phase in the inhomogeneous region versus temperature. Numerous methods are used. These are the most common :

a) The visual method

A set of samples (generally sealed) with different composition is prepared, and the transition temperature is determined using as criterion for transition the appearance of a meniscus. For compositions close to the critical composition, the meniscus appears in the middle of the sample, whereas it appears either in the bottom or in the top for other compositions.

This method is simple and accurate, but some limitations exist :

i) Supercooling (metastable states [20]). Shaking the sample does in principle avoid this spurious effect.

ii) Gravity effects, which produce concentration gradients in the sample. Shaking also should reduce this effect. Or one might use a mixture the densities of which are matched. However, the gravity effects are not very important in practice, because their amplitude is high only near T_c, and in this region they take a long time to settle [21].

iii) Very long equilibrium times (Fig.9). Within $T-T_c = 0.01$ K or $t = 3 \times 10^{-5}$, the phase separation process lasts a very long time, several days or more are needed. This is not compatible with shaking to avoid gravity effects and also supercooling. So generally the region $t < 3 \times 10^{-5}$ is not reliable.

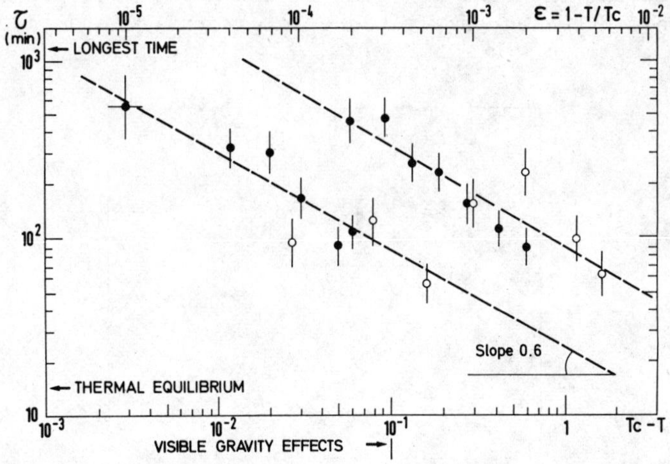

Fig. 9. Temperature variation of the equilibrium time, measured with the time dependence of the refractive index between two temperature levels. Full circles : upper phase. Open circles : lower phase. Nitroethane-Isooctane system Ref.[1]. This mixture has refractive indices closely matched ($\Delta n \simeq 10^{-3}$).

iv) Observation of the meniscus is sometimes very difficult because of the strong opalescence of the mixture. On the other hand, the meniscus can be almost visible when the refractive indices are nearly the same [1]. So special care has to be taken in both cases.

v) Variations of transition temperature. The temperature at which the mixture separates is a function of small impurities in the sample, independently of supercooling effects. Variations lower than 0.01 K ($t = 3 \times 10^{-5}$) can hardly be avoided.

One of the main advantages of this visual method is that the order parameter is really measured (mass fraction by weighing, volume fraction...), and is not inferred using assumptions such as volume additivity, which are not always justified.

b) Density measurements

Measurements are performed on a near critical sample, which includes preliminary measurement to get an estimate of the critical composition. Density of each phase can be measured separately and is directly related to the mass fraction. Although density measurements can be very accurate (see Fig.10), the same limitations as for the visual method apply, except of course the observation of the meniscus and the transition temperature drifts (nevertheless, T_c drifts during the duration time of the experiment are commonly observed in non sealed samples, due to external contaminations).

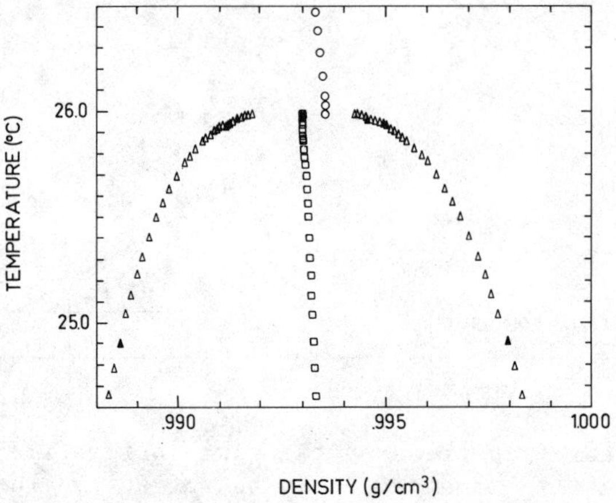

Fig. 10. Coexistence curve of isobutyric acid+water : coexisting densities (▲,△), diameter of the coexistence curve (□), and the density in the one phase region (o). The sample was not at the critical density. From Ref.[22].

Especially gravity effects can lead to distortions of the measurement since the probes have usually sufficient size so that concentration gradients may effect the outcome of the measurement.

b) Refractive index measurements

A near critical sample is here also needed. The refractive index in each phase is proportional to the volume fraction of the components. Absolute [23] or interferometric [1] measurements, and the beautiful Franhofer diffraction pattern technique [24] can be very accurate. The limitations are the same as with the density measurements, except for the concentration gradients induced by gravity, because the probe, which is usually a laser beam, can have a negligible width with respect to the extent of the gravity-induced composition gradients.

The α-Exponent

The measurements of $C_{p,c}$, directly by calorimetry [29], or by light scattering techniques on the thermal diffusivity $\Lambda/C_{p,c}$ [30] (where Λ is the thermal conductivity) has not yet provided an unambiguous determination of the parameters α, A, C_B... because of experimental difficulties associated with gravity effects, temperature instabilities, temperature gradients, etc. An alternative method involves the measurement of the variation of the density ρ or the variation of the volume V near T_c, since both are correlated to $C_{p,c}$ through the thermodynamic relationship

$$C_{p,c} - C_{v,c} = T_c V \alpha_{p,c} \left(\frac{\partial p}{\partial T_c}\right)_{v,c} ,$$

where $\alpha_{p,c} = \frac{1}{V}\left(\frac{\partial V}{\partial T}\right)_{p,c} = -\frac{1}{\rho}\left(\frac{\partial \rho}{\partial T}\right)_{p,c}$. We recall that $C_{v,c}$ has no critical anomaly in binary fluids.

The exponent $\alpha \simeq 0.1$ is small, so the change in the behavior of ρ or V is weak. The best determination up to now gave α within ± 70 % [25]. Moreover gravity effects were generally present since the measurement was integrated over all the sample height (V measurements) or over the probe size (ρ measurements).

All these problems can be ignored if careful interferometric refractive index measurements are made at a given height approximatively in the middle of the sample (Fig.11). A relative accuracy $\frac{\delta n}{n} \sim 10^{-6}$ can then be achieved, which is the ultimate precision if the frequency drifts of the laser are not compensated.

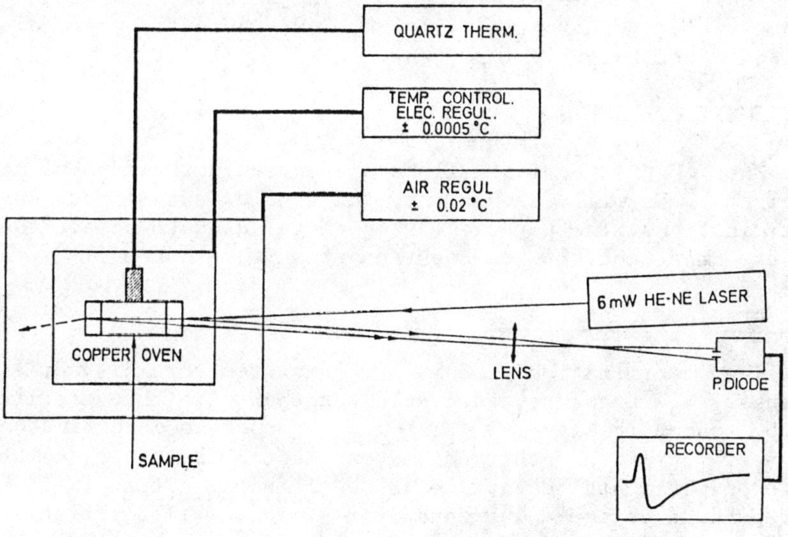

Fig. 11. Experimental setup. (Ref.[26]).

The refractive index variation can be written in the following form :

$$\Delta n = \left(\frac{\partial n}{\partial T}\right)^{reg.}_{p,c} \cdot (T-T_c) + \Delta\rho_c + \Delta F_c .$$

The first term represents the regular thermal variation. The second term is connected to the critical anomaly and the third to the effects of the local field. This last term has been checked to be negligibly small in a system (Nitroethane-Isooctane) where $\Delta\rho_c$ was expected to be negligible. Indeed, it is not only that the contribution ΔF_c could not be noticed but it has been possible to determine $\Delta\rho_c$ and to measure the exponent α [27].

Therefore the variation of the refractive index with t provides a measure of the variation of the specific heat :

$$\Delta n = T_c\left(\frac{\partial n}{\partial T}\right)^{reg.}_{p,c} \cdot t + Rt^{1-\alpha}(1 + a_c t^{\omega\nu} + \ldots)$$

where

$$R = -(n^2-1)\left(\frac{n^2+2}{6n}\right)\frac{}{1-\alpha}\left(\frac{\partial T_c}{\partial p}\right)_{v,c} \cdot \frac{k_B A}{\alpha}$$

R has been obtained assuming the Lorentz-Lorenz formula, which is known to be valid in liquids within ± 1 % [28]. It is therefore straightforward to infer A from R, the coefficient $\left(\frac{\partial T_c}{\partial p}\right)_{v,c}$ being known for many mixtures.

Thus α has been determined with an accuracy less than ± 5 % with a system (Water-Triethylamine) [26] where the critical amplitude was particularily important.

EXPERIMENTAL RESULTS

Most of the results will be presented in the form of tables or figures. We will discuss only the conclusions of each set of determinations. The t-range (or $q\xi$-range) used to obtain the data is very important, it has been reported (when known !).

The Correlation Length ξ (Table II).

The determination of ξ is obtained from turbidity or scattered intensity measurements, therefore together with the determination of the susceptibility χ (see below). In order to determine only the exponent ν, all the other parameters must be set free in the fit of the experimental data. The value of the amplitude ξ_0 is obviously dependent on the value found for ν. The small difference between

TABLE II - Determination of the correlation length ξ

$$\xi = \xi_o t^{-\nu}[1+a_\xi t^{\omega\nu}+\ldots]$$

Systems	Measurement	t range	q ξ range	ξ_o (Å)	ν all free	w_ν	a_ξ
A-C	Turbidity [a)]	$2\ 10^{-5}$-1 10^{-2}	0.08-5	2.45±0.05 [i)]	0.626±0.03	-	-
I-W	Turbidity [a)]	$4.6\ 10^{-6}$-5.10^{-3}	0.16-15	3.620±0.065 [i)]	0.630±0.045	-	-
T-W	Turbidity [a)]	$6.8\ 10^{-7}$-4.10^{-3}	0.08-18	1.25±0.03 [i)]	0.627±0.027	-	-
N-I	Spectral intensity [b)]	$3.3\ 10^{-6}$-0.2	0.017-17	2.42±0.04 [i)]			
	Spectral intensity [c)]	10^{-4} - 10^{-2}	0.08-1.5	2.9±0.4			
N-M	Total intensity [d)]	10^{-6}-3.10^{-2}	0.18-25	2.29±0.10	0.625±0.006 [d)]		
N-H	Turbidity [a)]	10^{-4}-$2.2\,10^{-3}$	0.3-2	3.14±0.06 [i)]	-		
	Total intensity [f)]	10^{-6}-10^{-2}	0.013-0.7	3.64±0.22	-		
L-W	Total intensity [g)]	$3.3\ 10^{-4}$-$3.3\,10^{-3}$	0.02-2.6	2.00±0.2	0.61±0.08		
C-P	Total intensity [e)]	$2.8\ 10^{-4}$-5 10^{-2}	10^{-2}-1	2.28±0.21 [h)]	0.626±0.013		
	Turbidity [e) a)]	$7.7\,10^{-5}$-$1.3\,10^{-2}$	$4.5\ 10^{-2}$-1.16	2.21±0.15 [h) i)]	-		
C-M							
C-N							

a) Ref. [12] ; b) Ref. [15] ; c) Ref. [42] ; d) Ref. [11-b] ; e) Ref. [43] reanalyzed in Ref. [12];
f) Ref. [44] ; g) Ref. [45] ; h) ξ_o questionable, see Table VIII concerning linewidth data : R is very low.
i) ν = 0.6300 ± 0.0008 and γ = 1.2402 ± 0.0009 imposed. (R-G values).

the Renormalization Group (R-G) values, and the High Temperature Series (H-T-S) expansions does not give rise to very different values of ξ_o, due to the experimental uncertainties.

Non-analytical corrections, or background contributions were never found. This may be due to the fact that ξ is generally determined close to T_c.

In Fig.12 the values of ν as given in Table II are plotted. The values obtained by intensity measurements seem to be about 3 times more accurate. The whole data seem closer to the R-G value[31] (0.630) than to the H-T-S value (0.638) [31]. The mean experimental value is 0.625, with a standard deviation of 0.005.

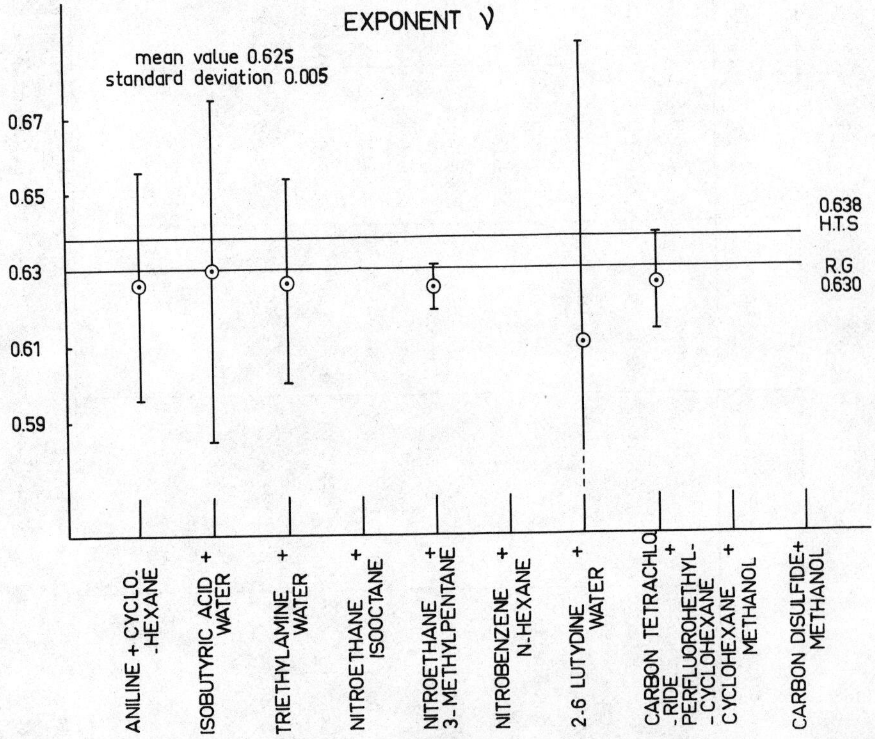

Fig. 12.

The Susceptibility χ (Table III)

The same techniques (scattered intensity, turbidity) as for the determination of ξ have been used to evaluate the amplitude C and the exponent γ. In order to obtain a meaningful value for the exponent γ, all parameters were set free. The value of the amplitude C was obtained -as for the value of ξ_o- with the exponents

Table III
Determination of χ

$$\chi = Ct^{-\gamma}[1 + a_\gamma t^{\omega\nu} + \ldots], \text{ with } C = \left[\left(\frac{\partial\mu}{\partial c}\right)^{E,Y,R}_{p,T}\right]^{-1}.$$

E,Y or R means that the Einsten, Yvon or Rocard formula has been used to calculate the scattered intensity. The value is deduced as follows :

$$\tau(\text{turbidity}) = \frac{2\pi^3}{\lambda_o^4} \cdot R \cdot k_B T_c (1+t) \cdot t^{-\gamma} (1+a_\chi t^{\omega_\nu})\ G'(q\xi)$$

and

$$\left(\frac{\partial\mu}{\partial c}\right)^{E,Y,R}_{p,T} = \frac{Sn^2}{R}\left(\frac{\partial n^2}{\partial c}\right)^2_{p,T} \quad \text{with } Sn = 1\ (E), \frac{9n^2}{(2n^2+1)(n^2+2)}\ (Y) \text{ and } \frac{3}{n^2+2}\ (R).$$

From the additivity of volume and the Lorentz-Lorentz relationship, $\left(\frac{\partial n^2}{\partial c}\right)_{p,T}$ can be evaluated :

$$\left(\frac{\partial n^2}{\partial c}\right)_{p,T} = 3\left(\frac{n^2+2}{3}\right)^2\left[\frac{n_1^2-1}{n_2^2+2}\cdot\frac{\rho}{\rho_1} - \frac{n_2^2-1}{n_2^2+2}\cdot\frac{\rho}{\rho_2}\right] + \frac{\rho_1-\rho_2}{3\rho_1\rho_2}(n^2-1)(n^2+2) .$$

Systems	Measurements	$\left\|\frac{\partial n^2}{\partial c}\right\|_{p,T}$	t range	q ξ range	$(\frac{\partial\mu}{\partial c})^E_{p,T}$ (J.cm^{-3}) a)	$(\frac{\partial\mu}{\partial c})^Y_{p,T}$ (J.cm^{-3}) b)	$(\frac{\partial\mu}{\partial c})^R_{p,T}$ (J.cm^{-3}) c)	γ all free	a_γ	ων
A-C	Turbidity d)	0,48 e)	$2\ 10^{-5}$–$1.3\ 10^{-2}$	0.08-5	116 ± 2	88.9 ± 1.6	59.3 ± 1.1	1.25 ± 0.08	–	–
I-W o)	Turbidity d) o)	0.163 e) 0.21±0.01 f)	$4.6\ 10^{-6}$–$5\ 10^{-3}$	0.16–15	39 ± 3 66 ± 5	28 ± 2 47.6 ± 3.3	15 ± 1 24.6 ± 1.8	1.24 ± 0.1		
	Total intensity g)		$2.6\ 10^{-4}$–$2.6\ 10^{-2}$	10^{-2}–1.7	–	–	–	1.24 ± 0.05	–	–
T-W	Turbidity d)	0.206±0.005 e)	$6.8\ 10^{-7}$–$4\ 10^{-3}$	0.08 – 18	255 ± 19	183 ± 14	94 ± 7	1.20 ± 0.07	–	–
N-I	Spectral intensity h)		$3.3\ 10^{-6}$–0.2	0.017-17		–90 ± 20 r)			–	–
N-M	Total intensity i)		10^{-6}–3.10^{-2}	0.18–25				1.240 ± 0.017	–	–
	Turbidity d)	0.040±0.002 e)	3.10^{-5}–$2\ 10^{-4}$	1.2-4	104 ± 14	86 ± 12	61 ± 9		–	–
N-H	Spectral intensity d)	–	10^{-3}–0.2	0.017–0.5	–	–	–	1.235 ± 0.016	–	–
	Total intensity h)	–	10^{-6}–10^{-2}	0.013–0.7	–	–		1.228 ± 0.018 p)		
	Turbidity d)	0.458±0.012 e)	10^{-4}–$2.2\ 10^{-3}$	0.3-2	127 ± 7	101 ± 6	69 ± 4		–	–
L-W	Total intensity j)	–	$3.3\ 10^{-4}$–$3.3\ 10^{-2}$	0.02–2.6	–	–	–	1.26 ± 0.02	–	–
C-P	Total intensity ℓ)	–	$2.8\ 10^{-4}$–$5\ 10^{-2}$	10^{-2}–1	–	–	–	1.220 ± 0.018	–	–
	Turbidity d)	0.471±0.014 e)	$7.7\ 10^{-5}$–$1.3\ 10^{-2}$	$4.5\ 10^{-2}$–1.16	139 ± 6 m) 125 ± 5 n) q)	117 ± 5 m) 105 ± 5 n) q)	83 ± 4 m) 75 ± 3 n) q)	– –	– –	– –
C-M										
C-N										

(continued)

TABLE III (Continued)

a) Assuming the Einstein formula and $\gamma=1.2402 \pm 0.009$, $\nu=0.6300 \pm 0.0008$ imposed.

b) Assuming the Yvon formula and γ,ν imposed as in a).

c) Assuming the Rocard formula and γ,ν imposed as in a).

d) Ref.[12]

e) From the volume additivity and the Lorentz-Lorenz relationship.

f) Ref.[46]

g) Ref.[47]

h) Ref.[15]

i) Ref.[11-b]

j) Ref.[48]

k) Ref.[44]

l) Ref.[43]

m) from d) with ξ_o imposed between 3.20-2.70 Å as deduced from the linewidth data.

n) from d) with ξ_o imposed to the value (2.28 ± 0.21) Å found in j).

o) $(\partial n^2/\partial c)$ from experiment is different from the evaluation by Lorentz-Lorenz.

p) Accuracy surestimated : see l), we have multiplied it by 3.

q) questionable ? due to the value of ξ_o.

r) independent of E, Y or R.

fixed to their R-G value. The difference between R-G and H-T-S values was negligible with respect to the experimental uncertainties. As stressed above, only the turbidity data are able to give a precise value for C (i.e. $k_B T_c \times 1/\left(\frac{\partial\mu}{\partial c}\right)^0_{p,T}$), and even this last parameter is dependent on the Rayleigh factor formulation (Einstein, Yvon, Rocard). A lack of accuracy is commonly found for the coupling parameter $\left(\frac{\partial n^2}{\partial c}\right)_{p,T}$, and the deviations between its evaluation by the Lorentz-Lorenz formula, assuming the additivity of volume, and the experimental determination, can give deviations of about 50 % in C.

As for ξ, and despite the fact that most of the experiments were carried out far from T_c, non-analytical corrections* have never been evident, neither have the background contributions.

In Fig.13 are plotted the values of γ from Table III. Here again the uncertainty of γ is somewhat larger when the turbidity technique is employed, as explained above.

A mean value of 1.236 (with a standard deviation of 0.008) is found for γ, taking into account the uncertainties of all data. Both R-G and H-T-S values [31] are very close together, and it seems meaningless to make a distinction here.

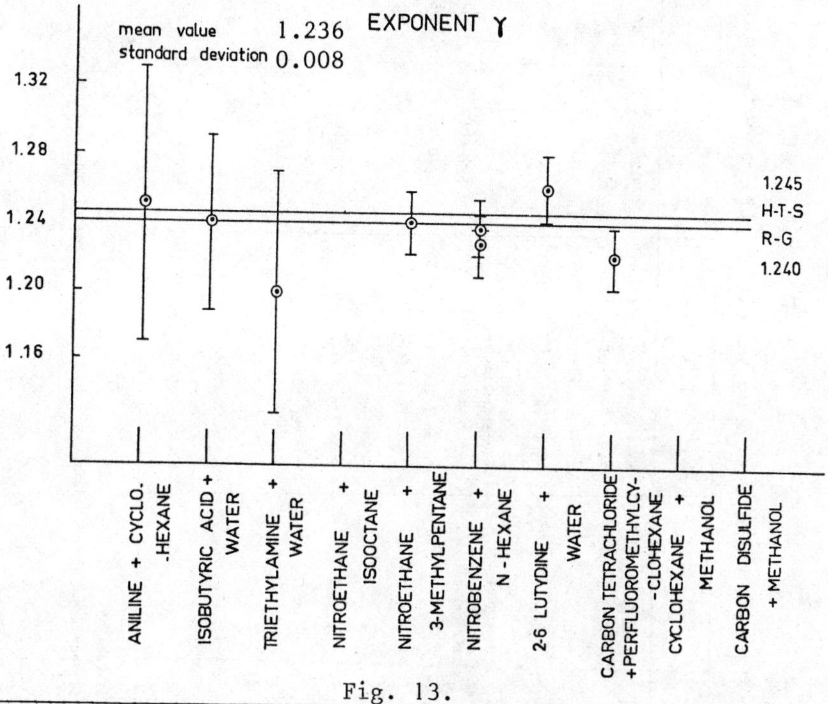

Fig. 13.

* excepted very recently for the Triethylamine-Water system (Ref.[77]).

The Order Parameter M (Table IV)

TABLE IV - Determination of M :

$$|c^{+}-c^{c}| = M = Bt^{\beta}(1 + a_{\beta}^{(1)} t^{\omega\nu} + a_{\beta}^{(2)} t^{2\omega\nu})$$

β and B depend weakly on the choosen parameter (mass, volume or molar fraction)

Systems	Measurement	Order parameter	T_c(°K)	t range from ≃ 2.10^{-5} to :	B(β=0.325 imposed)	β	$a_\beta^{(1)}$	$a_\beta^{(2)}$	ω_ν
A-C	density a)	mass fraction b)	304.05	$9.86\ 10^{-3}$	1.07 ± 0.01 b)	0.340 ± 0.008 a)			
	refractive index d)	volume fraction		$3.28\ 10^{-2}$		0.328 ± 0.007 d)			
I-W	density p) f)	volume fraction	299°.5	$4.33\ 10^{-3}$	1.071 ± 0.023 f)	0.328 ± 0.004 f)			
T-W	visual q) e)	mass fraction	291°.5	$3.43\ 10^{-2}$	1.513 ± 0.003 b)				
N-I	visual g)	{ mass fraction / volume fraction	303.3	1.9×10^{-2}	0.915 ± 0.008 g)	0.318 ± 0.006 g)			
	refractive index g)	volume fraction		6.10^{-3}	0.885 ± 0.015 g)	0.31 ± 0.02 g)			
N-M	density h)	volume fraction	299.5	8.10^{-3}	0.9385 ± 0.0075 b)	0.335 ± 0.005 i)			
		vol ,mol,+mass fract.				0.330 ± 0.006 n)			
N-H	visual j)	mass fraction	293	7×10^{-2}	0.76 ± 0.03 b)				
L-W	visual r) o)	mass fraction	307	$3.25\ 10^{-1}$	0.77 ± 0.03 b)				
C-P	visual k)	volume fraction	301.8	$1.98\ 10^{-4}$	0.905 ± ? k)	0.336 ± 0.02 k)			
	visual l)	volume fraction		$1.32\ 10^{-3}$	0.825 ± 0.0035 l)	0.317 ± 0.005 l)			
					0.8853 ± 0.0033 b)				
C-M	refractive index s) m)	volume fraction	318	$6.28\ 10^{-2}$	0.754 ± 0.04 m)	0.326 ± 0.003 m)			
C-N	visual m)	volume fraction	335.13	$1.97\ 10^{-1}$	0.815 ± 0.045 m)	0.316 ± 0.008 m)	0.385 ± 0.15	-1.21 ± 0.20	0.50 imposed

a) Ref. [49]
b) reanalyzed with β=0.325 imposed, in Ref. [12]
c) Ref. [12]
d) Ref. [24]
e) Ref. [51]
f) Ref. [22]
g) Ref. [1]
h) Ref. [50]
i) mean value of several sets.
j) Ref. [46]
k) Ref. [75]
l) Ref. [76]
m) Ref. [53]
n) Data from Ref. [50] reanalyzed in Ref. [54]
o) Ref. [55]
p) Free of gravity effects
q) Lower critical point
r) For the lower critical point (system with lower and upper critical point).
s) Free of gravity effects.

The amplitude B of the coexistence curve was determined with β imposed to its R-G value 0.325 [31], which is somewhat different from the H-T-S result 0.312 [31]. As stressed above, the choice of the order parameter influences weakly the value of B, as well as the value of the exponent β.

The values of the exponent β are shown in Fig.14 (from Table IV). They are compatible with R-G calculation, and the mean value, calculated with the corresponding uncertainties for each data, is $\beta = 0.326$ (with a standard deviation 0.002).

Non-analytical corrections have been found only for one system (Carbon disulfide + Nitromethane), using the volume fraction as the order parameter and an imposed value of $\omega\nu$ of 0.50.

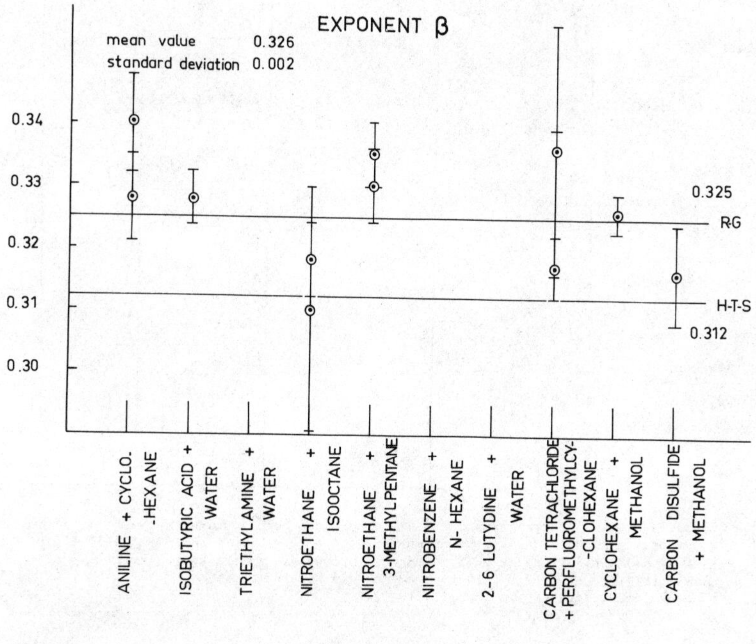

Fig. 14.

The Specific Heat (Table IV)

The amplitude $\frac{A}{\alpha}$ of the diverging part of the specific heat is determined here with an α imposed of 0.110, which is the same for the R-G or the H-T-S calculation [31]. Some experiments have been already interpretated with $\alpha = 0.125$, the resulting difference in

TABLE V

Determination of $C_{p,c} = \frac{At^{-\alpha}}{\alpha}[1 + a_c t^{\omega\nu} + \ldots] + C_B$; C_B is never negligible. $A = \frac{\alpha\rho_c [C^o_{p,c}]^{exp}}{k_B}$; with ρ_c the critical density and $[C^o_{p,c}]^{exp}$ the diverging part of the experimental specific heat.

Systems	Measurement	t range	$A = \alpha\rho_c [C^o_{p,c}]^{exp} k_B^{-1}$ ($\times 10^{21}$ cm^3)	α	$a_\alpha^{(1)}$	ω_ν
A-C [n]	refractive index [a]	10^{-3} - 0.03	1.8 ± ?			
I-W [o]	refractive index [b]	3×10^{-6} - 0.03	0.46 ± 0.03	0.10 ± 0.05		
	volume [c]	5×10^{-6} - 1.2×10^{-3}	0.68 ± ?	0.12 ± 0.36		
	specific heat [d]	10^{-5} - 0.01	0.29 ± ?	0.1 ± 0.1		
T-W [n]	refractive index [b]	3×10^{-6} - 0.06	8.4 ± 0.3	0.113 ± 0.005	−4.67 ± 0.4 [k]	0.50 ± 0.03 [l]
	specific heat [e]	1.5×10^{-5} - 0.07	17 ± 4	0.240 ± 0.085	+1.1 ± 0.5	
N-I [n]	refractive index [b]	1.5×10^{-6} - 0.015	1.8 ± 0.5	0.10 ± 0.02		
N-M	Thermal diffusivity [f]	5×10^{-5} - 0.08	2.4 ± ?	0.059 ± 0.003		
	density [g]	2×10^{-4} - 0.016	3.2 ± 1			
N-H						
L-W						
C-P	specific heat [h]	1.3×10^{-5} - 0.036	1.10 ± 0.07	0.055 ± ?		
	density [m]	3.2×10^{-4} - 0.01	1.10 ± 0.15			
C-M [o]	volume [i]	6×10^{-6} - 0.03	1.8 ± ?	0.11 ± 0.07		
	specific heat [j]	3×10^{-5} - 6×10^{-2}	0.53 ± ?			
C-N						

a) Ref.[19]
b) Ref.[26]
c) Ref.[56]
d) Ref.[57]
e) Ref.[29]
f) Ref.[30]
g) Ref.[58]
h) Ref.[59]
i) Ref.[25]
j) Ref.[60]
k) Non-analytical corrections up to 4th order have been found with the regular term, α and $\omega\nu$ beeing imposed within the known limits.
l) Only α and the linear regular term was imposed within the known limits.
m) Ref.[61] reanalyzed by the same authors in Ref.[62]
n) measurement free of gravity effects
o) free of gravity effects

the amplitude is generally negligible compared with the correspon-ding experimental uncertainty.

Only some experiments give significant results for the exponent α, when fitting is done with all parameters free, including of course the important regular part. The system Triethylamine-Water, which shows a particularily important diverging part with respect to the regular contributions gives the best determination. Some values, although expected to be measured with good accuracy by direct specific measurements or through the thermal diffusivity, are significantly outside the expected values. They have been omitted in Fig.15 where only 4 data could be shown. Three of them used the refractive index technique, the fourth comes from a volume variation technique. The mean value for α has been found $\alpha = 0.112$, with a standard deviation of 0.005. Weighing with the given incertainties has been done.

Non-analytical corrections are important only for the system Triethylamine-Water. They could be detected up to 4th order in $\omega\nu$, whose value was found to be close to the R-G value [31] (0.493) :

$$\omega\nu = 0.50 \pm 0.03 \ .$$

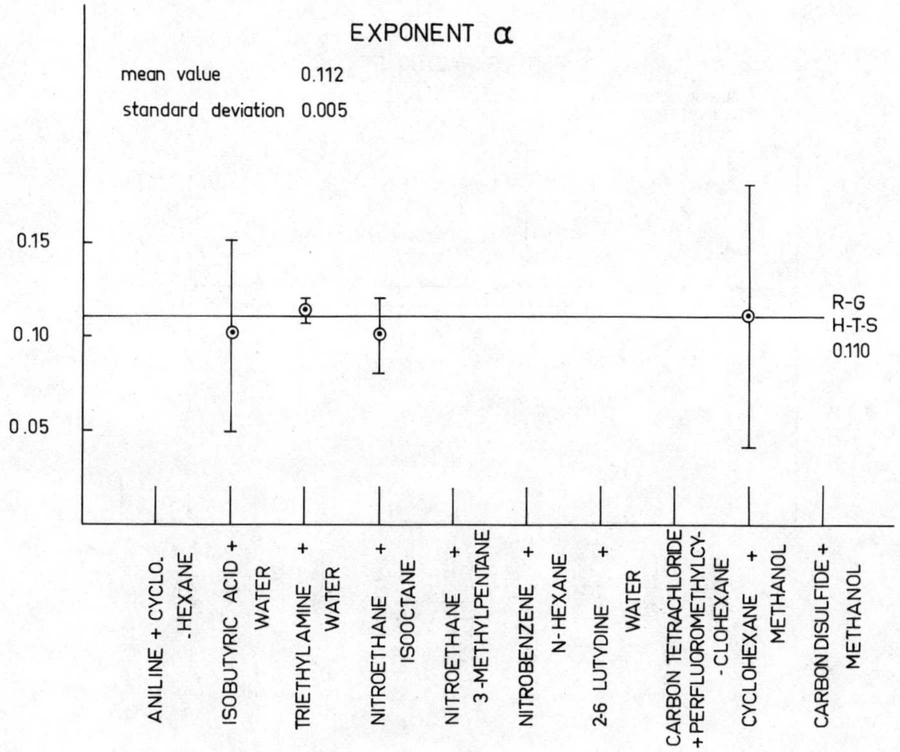

Fig. 15.

Universal Relations Between the Amplitudes

a) Statics

Two universal amplitude ratios can be determined with the experiments reported above. The first, R_ξ^+ [32][33] is concerned with the amplitude $C_{p,c}$ and ξ in the homogeneous phase, it is

$$R_\xi^+ = \xi_o^+ (A^+)^{1/3}, \text{ where } A = \alpha\rho_c [C^o_{p,c}]^{exp} k_B^{-1}$$

(+) means that the above quantities are concerned with the homogeneous region. ρ_c is the critical density, and $[C^o_{p,c}]^{exp}$ is the diverging part of the experimental specific heat. From Tables II and V where the values of ξ_o^+ and A^+ are reported, R_ξ^+ can be estimated straightforward. This has been done in Table VI. The expected value from H-T-S is 0.253 ± 0.001 [34], and from R-G calculation it is 0.2696 ± 0.0008 [33]. The experimental values are in somewhat better agreement with the high values (R-G), as can be seen in Fig.16. The mean value is 0.265, with a standard deviation 0.007.

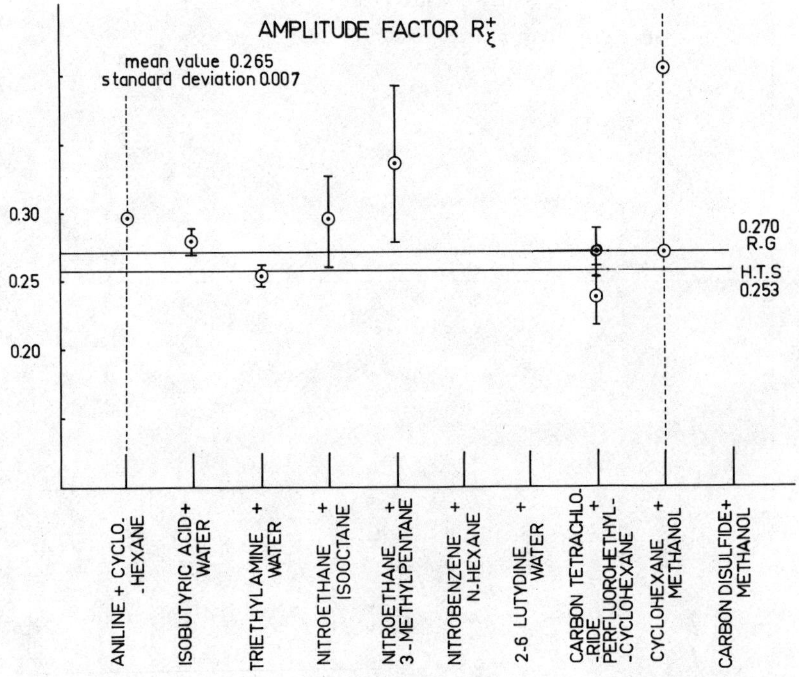

Fig. 16.

Table IV

Experimental data concerning the determination of $R_\xi^+ = \xi_o^+ A^{1/3}$. The R-G approach gives 0.2696 ± 0.0008[33], the H.T.S. value is 0.253 ± 0.001[34].

Systems	experiment	non analytical corrections c)	divergent / regular	$k_B^{-1}\alpha\rho_c[C^o_{p,c}]^{exp}=A$ ($\times 10^{21}$ cm^3) c)	ξ_o^+(Å) d)	R_ξ^+
A-C	refractive index a)	no	0.06	$1.8 \pm ?$	2.45 ± 0.05	$0.298 \pm ?$
I-W	refractive index b)	no	0.281	0.46 ± 0.03	3.620 ± 0.065	0.279 ± 0.011
T-W	refractive index b)	yes	2.70	8.4 ± 0.3	1.25 ± 0.03	0.254 ± 0.009
N-I	refractive index b)	no	0.0377	1.8 ± 0.5	2.42 ± 0.04	0.294 ± 0.032
N-M	density e)	no	0.045	3.2 ± 1	2.29 ± 0.10	0.337 ± 0.05
N-H						
L-W						
C-P	specific heat f)	no	0.0905	1.10 ± 0.07	2.28 ± 0.21 k) or	0.235 ± 0.03 or
	density g)	no		1.10 ± 0.15	2.59 ± 0.08	0.267 ± 0.02
C-M	volume h)	no	0.239	$1.8 \pm ?$	$3.31 \pm ?$ j)	$0\ 403 \pm ?$
	specific heat i)	no	0.032	$0.53 \pm ?$		$0.267 \pm ?$
C-N						

a) Ref.[19]; b) Ref.[26]; c) From Table V ; d) From Table II ; e) Ref.[58] ; f) Ref.[59] ;
g) Ref.[61] reanalyzed by the same authors in Ref.[62]; h) Ref.[25] ; i) Ref.[60] ;
j) From Table VIII and linewidth measurements of ξ_o/R, multiplied by 1.16 ;
k) The value ξ_o = 2.59 Å, obtained as in j), is more probable.

The second universal amplitude factor which can be experimentally determined is the product

$$R_\xi^+ R_c^{-1/3} = \xi_o^+ \left(\frac{B^2}{C^+}\right)^{1/3}, \text{ where } C^+ = k_B T_c\left[1/\left(\frac{\partial\mu}{\partial c}\right)^o_{p,T}\right]$$

We prefer to deal with this quantity rather than with R_c, since R_c involves the parameter A which is not determined for most of the liquids, whereas ξ_o^+ is more widely known.

In Table VII are reported the values of $\xi_o^+\left[\frac{B^2(\partial\mu/\partial c)^o_{p,T}}{kT_c}\right]^{1/3}$. They should be compared with the H-T-S calculation [35] $0.253\times(0.059)^{-1/3}=0.650$ and the R-G value[35] $0.6296\times(0.006)^{-1/3}=0.667$. Data from Tables II, III, IV have been used.

Fig.17 shows the values of $R_\xi^+ R_c^{-1/3}$ according to the Einstein formulation, Fig.18 according to the Yvon formula, and Fig.19 according to Rocard. Despite of important differences caused by the great number of parameters which had to be evaluated, it is clear

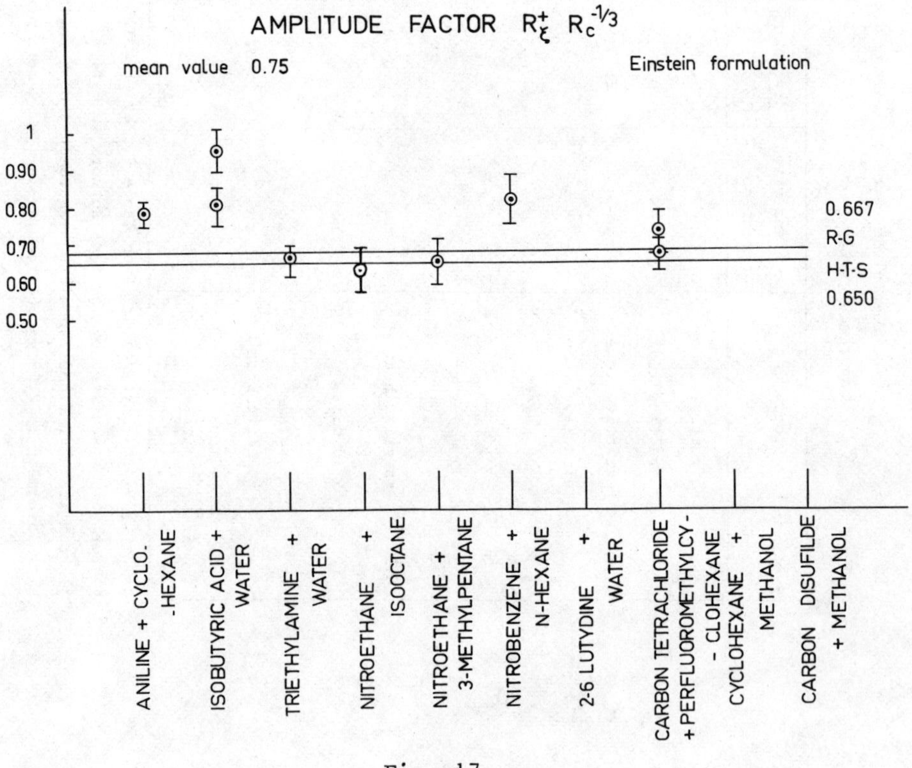

Fig. 17.

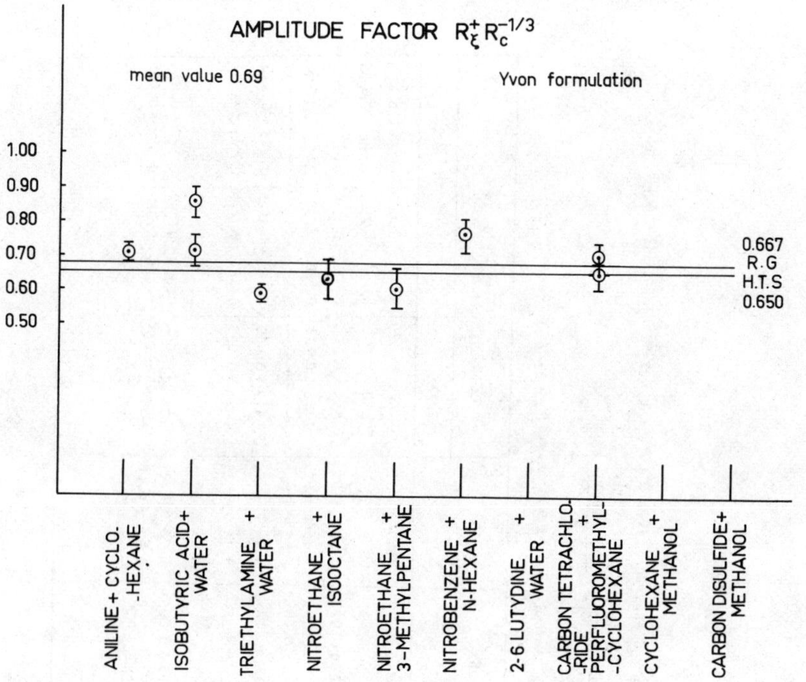
AMPLITUDE FACTOR $R_\xi^+ R_c^{-1/3}$
mean value 0.69
Yvon formulation
1.00
0.90
0.80
0.70
0.60
0.50
0.667
R.G
H.T.S
0.650
ANILINE + CYCLO. .HEXANE
ISOBUTYRIC ACID+ WATER
TRIETHYLAMINE + WATER
NITROETHANE + ISOOCTANE
NITROETHANE + 3-METHYLPENTANE
NITROBENZENE + N-HEXANE
2-6 LUTYDINE + WATER
CARBON TETRACHLO. -RIDE + PERFLUOROMETHYL- -CYCLOHEXANE
CYCLOHEXANE + METHANOL
CARBON DISULFIDE+ METHANOL

Fig. 18

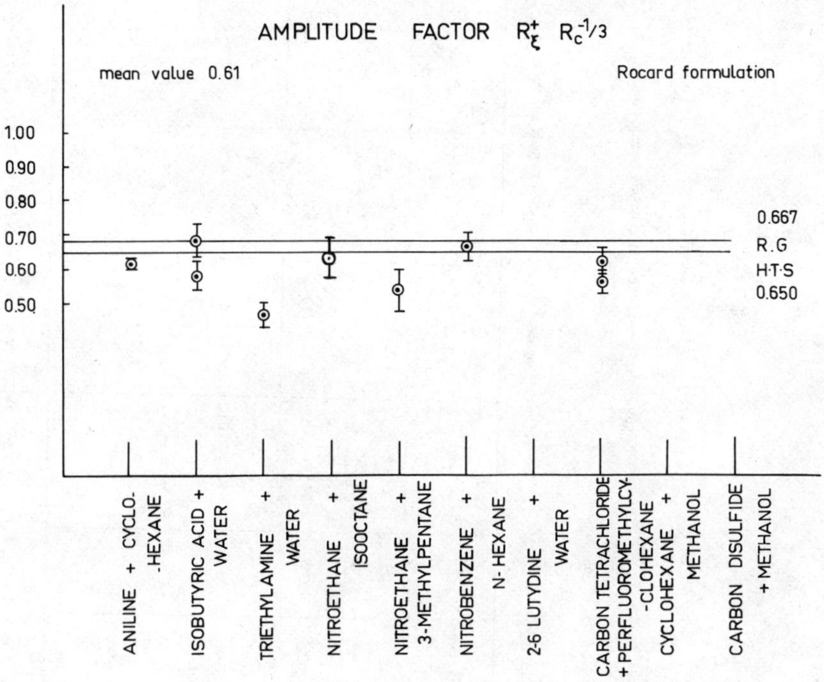
AMPLITUDE FACTOR $R_\xi^+ R_c^{-1/3}$
mean value 0.61
Rocard formulation
1.00
0.90
0.80
0.70
0.60
0.50
0.667
R.G
H.T.S
0.650
ANILINE + CYCLO. .HEXANE
ISOBUTYRIC ACID + WATER
TRIETHYLAMINE + WATER
NITROETHANE + ISOOCTANE
NITROETHANE + 3-METHYLPENTANE
NITROBENZENE + N-HEXANE
2-6 LUTYDINE + WATER
CARBON TETRACHLORIDE + PERFLUOROMETHYLCY- -CLOHEXANE
CYCLOHEXANE + METHANOL
CARBON DISULFIDE + METHANOL

Fig. 19

Table VII

Determination of the universal amplitude $R_\xi^+ R_c^{-1/3}$. The R-G approach gives 0.667, the H-T-S value is 0.650 .

$$R_\xi^+ R_c^{-1/3} = \xi_o^+ \left[B^2 \frac{(\partial\mu/\partial c)^{E,Y,R}_{P,T}}{k_B T_c} \right]^{1/3}$$

E.Y or R means that in the determination of $\left(\frac{\partial\mu}{\partial c}\right)^o_{P,T}$ the formulation of Einstein, Yvon or Rocard has been used.

Systems	T_c (°K)	$\left\|\frac{\partial\varepsilon}{\partial c}\right\|_{T,p}$ a)	$\left(\frac{\partial\mu}{\partial c}\right)^E_{P.T}$ a) ($J.cm^{-3}$)	$\left(\frac{\partial\mu}{\partial c}\right)^Y_{P.T}$ a) ($J.cm^{-3}$)	$\left(\frac{\partial\mu}{\partial c}\right)^R_{P.T.}$ a) ($J.cm^{-3}$)	B b) B = 0.325	ξ_o^+(Å) c) ν = 0.630	$R_\xi^+ R_c^{-1/3}$ E	$R_\xi^+ R_c^{-1/3}$ Y	$R_\xi^+ R_c^{-1/3}$ R
A-C	303	0.48	116 ± 2	88.9 ± 1.6	59.3 ± 1.1	1.07 ± 0.01	2.45 ± 0.05	0.78 ± 0.024	0.71 ± 0.022	0.620 ± 0.010
I-W	299.5	0.163	39 ± 3	28 ± 2	15 ± 1			0.800 ± 0.005	0.716 ± 0.04	0.582 ± 0.035
		0.211 ± 0.01	66 ± 5	47.6 ± 3.3	24.6 ± 1.8	1.071 ± 0.023	3.62 ± 0.065	0.954 ± 0.054	0.855 ± 0.050	0.687 ± 0.040
T-W	291.5	0.206 ± 0.005	255 ± 19	183 ± 14	94.2 ± 7	1.513 ± 0.003	1.25 ± 0.03	0.657 ± 0.033	0.588 ± 0.030	0.471 ± 0.026
N-I	303.31	(3.7±0.1) 10^{-3}	----	90 ± 20 e)	----	0.915 ± 0.008	2.42 ± 0.04	----	0.63 ± 0.06 e)	----
N-M	299.5	0.0403 ± 0.002	104 ± 14	86 ± 12	61 ± 9	0.9385±0.0075	2.32 ± 0.09	0.6519 ± 0.060	0.6119 ± 0.055	0.545 ± 0.055
N-H	293.2	0.458 ± 0.012	127 ± 7	101 ± 6	69 ± 4	0.76 ± 0.03	3.14 ± 0.06	0.826 ± 0.053	0.764 ± 0.05	0.674 ± 0.043
L-W										
C-P	301.8	0.471 ± 0.014	139 ± 6	117 ± 5	83 ± 4	0.885 ± 0.003	2.28 ± 0.21	0.677 ± 0.064	0.639 ± 0.060	0.570 ± 0.054
			125 ± 5	105 ± 5	75 ± 3		2.60 ± 0.08 d)	0.744 ± 0.06	0.698 ± 0.06	0.628 ± 0.060
C-M										
C-N										

a) From Table III ; b) From Table IV ; c) From Table II ; d) ξ_o from $\frac{\xi_o}{R}$ and R = 1.16 (Table VIII) ; e) Independent of the E,Y or R formulation that : i) For all formulations (E,Y,R), the correct order of magnitude is found. ii) The Einstein formulation (mean value 0.75) is further from the expected results than Yvon's (mean value 0.69) or Rocard's (0.61). This is not completely unexpected, since measurements performed with pure fluids far from the critical point[13] [36] or near the critical point[37] lead to the same conclusion.

b) Dynamics

Linewidth measurements in the hydrodynamic region $q\xi \ll 1$, with kinematic viscosity and density measurements in the corresponding temperature range, are necessary to obtain the dynamic amplitude factor R, the correlation length beeing known. From the experimental section above :

$$\Gamma = R \cdot \frac{k_B T}{6\pi\eta\xi} q^2 H(q\xi) \simeq R \cdot \frac{k_B T}{6\pi\eta\xi} q^2 .$$

The small contribution due to $H(q\xi) \simeq 1$ can be accurately determined in the region $q\xi \ll 1$. No background contributions has ever been seen, except may be in the Nitroethane-3-methylpentane system [12][19][38], where they nevertheless remain very small. Note that careful measurements of each of these numerous parameters has to be made if finally an uncertainty of some percent on R is required.

It is useful to separate the amplitude of the correlation length obtained by statics measurements $[\xi_o]$ from the amplitude

$$\left[\frac{\xi_o}{R}\right] = \Gamma^{-1} \cdot \frac{k_B T_c (1+t)}{6\pi\eta(t) . t^{-\nu}} \cdot q^2 \cdot H(q\xi)$$

given by dynamics measurements. The corresponding values can be found in Table VIII where the parameters which enter the determination are also reported. The t- or $q\xi$-range of the measurements have also been reported. The calibration of the viscosity measurements is very important, discrepancies from different authors up to 5-7 % are not unusual.

The expected value from a R-G calculation range from 0.8 to 1.2 [39], the value 1.2 is considered as more probable [40].

The experimental results are drawn in Fig. 20. The values of R lie between 1.0 and 1.20, with a mean value of 1.160, with a standard deviation 0.005, and is therefore in full agreement with the above expectations.

Table VIII

Experimental data concerning the determination of the dynamic amplitude factor R. The expected value from R G is 0.8 - 1.2. The value 1.2 is more probable.

Systems	η/ρ t range	ρ	$\Gamma \to \xi^o(\mathring{A})/R$ $\nu = 0.630$ imposed	t range	$q\xi$ range	$\xi_o(\mathring{A})$ r)	R
A-C	$7.5\ 10^{-5}$ j) $- 0.2$	j)	2.02 ± 0.2 k)	$8.2\ 10^{-3} - 1.3\ 10^{-1}$	$5.2\ 10^{-4} - 7.8\ 10^{-2}$	2.45 ± 0.05	1.20 ± 0.13
I-W	$2.9\ 10^{-4} - 4.5\ 10^{-2}$ d) e)	e) g)	3.08 ± 0.02 f)	$1.33\ 10^{-5} - 6.6\ 10^{-2}$	$6.5\ 10^{-2} - 0.25$ f)	3.620 ± 0.065	1.17 ± 0.03
	$6.68\ 10^{-6} - 2.7\ 10^{-2}$ q)		3.28 ± 0.1 d)	$3.3\ 10^{-3} - 2.8\ 10^{-2}$	$2.10^{-2} - 10$		1.10 ± 0.05
T-W	$3.4\ 10^{-5} - 4.10^{-2}$ d) h)	g)	1.033 ± 0.015	$10^{-3} - 1.7\ 10^{-2}$	$3.1\ 10^{-2} - 0.184$	1.25 ± 0.03	1.21 ± 0.05
N-I	-	-	-	-	-		
N-M			1.94 ± 0.08 d)	$3.3\ 10^{-6} - 3.3\ 10^{-2}$	$0.2 - 6\ 10^{-4}$		1.18 ± 0.05
	$10^{-5} - 2.5\ 10^{-1}$ p)		2.06 ± 0.02 a)	$5\ 10^{-7} - 8.8\ 10^{-4}$	$0.32 - 55$ a)		1.11 ± 0.01
			2.24 ± 0.13 b)	$5.6\ 10^{-5} - 2.7\ 10^{-3}$	$0.1 - 0.7$ b)	2.29 ± 0.10	1.02 ± 0.06 s)
N-H	$7.6\ 10^{-5} - 2.10^{-1}$ i)		2.67 ± 0.04	$7.6\ 10^{-4} - 1.6\ 10^{-2}$	$8.4\ 10^{-2} - 5.8$ d)	3.14 ± 0.06	1.18 ± 0.05 t)
					$3.10^{-4} - 1$ c)	3.64 ± 0.22 c)	1.15 ± 0.05 t)
L-W	$3.10^{-5} - 9\ 10^{-2}$ o)		2.71 ± 0.3 o)	$3.25\ 10^{-6} - 2.3\ 10^{-2}$	$5.10^{-2} - 14$	2.00 ± 0.2	1.35 ± 0.15
C-P	m)		2.25 ± 0.07 n) d)	$2.65\ 10^{-4} - 3.3\ 10^{-2}$	$5.10^{-2} - 12$	2.28 ± 0.21	0.99 ± 0.12 u)
C-M	l)		2.85 ± 0.05 l) r)		$0.08 - 80$		v)
C-N							

a) Ref.[63]
b) Ref.[38]
c) Ref.[64]
d) Ref.[12]
e) Ref.[65]
f) Ref.[66]
g) Ref.[46]
h) Ref.[68]
i) Ref.[69]
j) Ref.[70]
k) Ref.[17]
l) from the same laboratory as a). Viscosity 7% to high. Ref.[71]
m) Ref.[43]
n) Ref.[72]
o) Ref.[45]
p) Ref.[73]
q) Ref.[74]
r) From Table II.

s) From b), the viscosity was too high by 7 %. We have accounted for. The uncertainly seems very low with respect to the other authors.
t) The experimental value was multiplied by 0.95, due to the calibration of the viscosimeter (see d)). Discrepancies exist with Ref. d) for both ξ_o and ξ_o/R, but R agrees well. This may be due to impurities in the sample which is known to increase ξ_o.
u) ξ_o questionable.
v) uncertainly on ξ_o/R estimated from Ref.[71].

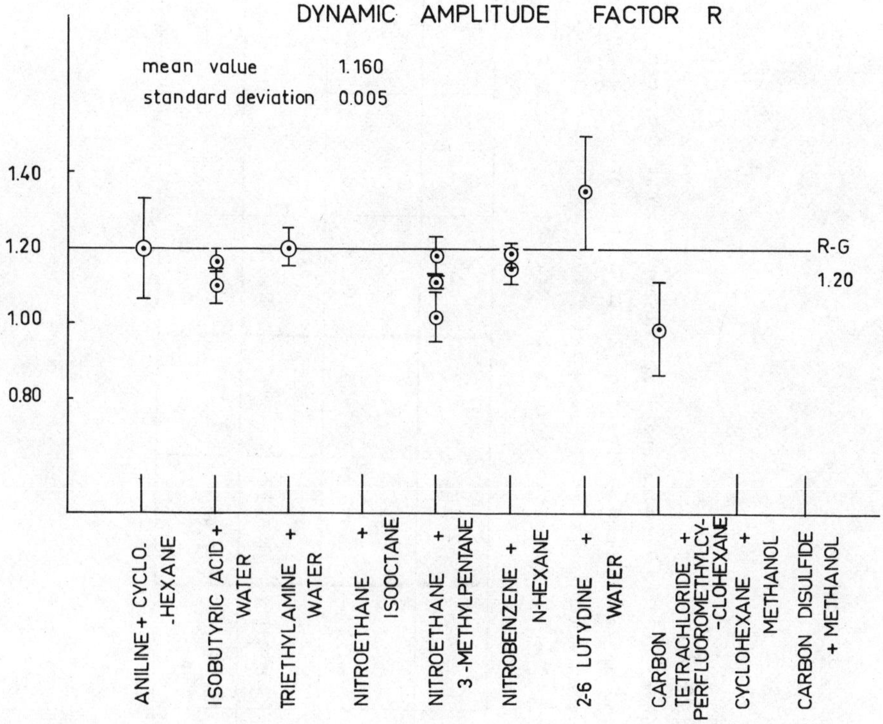

Fig. 20.

CONCLUSION (Table IX)

From this review of experimental techniques and of experimental results concerning both the exponents and the amplitude ratios of ten binary mixtures, we can draw the following conclusions :

i) The critical domain is very large in critical mixtures. Very scarcely are found non-analytical corrections and -or- background contributions.

ii) The experimental difficulties are not so high as in simple fluids near the gaz-liquid critical point ; especially gravity effects are in most cases negligible.

iii) It seems difficult to increase the accuracy of the experimental determination of the exponents by an important factor ; 1 % relative uncertainty seems to be up to now the ultimate precision. On the other hand, the experiments concerning amplitude ratios are not yet numerous, and improvements should be possible.

iv) The agreement with all theoretical calculations is generally very good. A somewhat better agreement is found for the R-G values, especially for the exponents β and ν. The ν-value, combined with the determination of α, allows the hyperscaling to be verified.

Table IX

Comparison between experimental and theoretical values for critical exponents and critical amplitudes. The mean experimental uncertainties have been estimated by us and do not correspond systematically to a standard deviation. It seems that the R-G values are generally in better agreement with experiments, especially for exponents ν and β. *Hyperscaling* (relation $d\nu = 2-\alpha$, where d = dimensionality = 3 here) seems therefore to be verified.

	ν	γ	β	α	$\omega\nu$	$2-\alpha$	3ν	R_ξ^+	$R_\xi^+ R_c^{-1/3}$	R
Experimental	0.615 – 0.635	1.23 – 1.25	0.32 – 0.33	0.105 – 0.115	0.47 – 0.53	1.885 – 1.895	1.845 – 1.905	0.25 – 0.28	0.60 – 0.70	1.1 – 1.2
Renormalization Group	0.6285 – 0.6315	1.239 – 1.243	0.3235 – 0.3265	0.107 – 0.113	0.478 – 0.518	1.887 – 1.893	1.8855 – 1.8945	0.2688 – 0.2704	0.667	1.2
High temperature series	0.636 – 0.640	1.242 – 1.248	0.307 – 0.317	0.110	0.5 – 0.6	1.890	1.908 – 1.920	0.252 – 0.254	0.650	—

Finally, dynamic (linewidth) measurements, which have been scarcely discussed here, could give accurate values for the correlation length if some new theoretical work could be made in this field.

Acknowledgments

I am grateful to A. Bourgou for help in the redaction of this manuscript, and P. Calmettes, P. Bergé, C. Bervillier and P.C. Hohenberg for fruitfull discussions. I thank L. Koschmieder for a critical reading.

REFERENCES

[1] D. Beysens, J. Chem. Phys. 71, 2557 (1979).
[2] D. Beysens, R. Tufeu and Y. Garrabos, J. de Phys. Lett. 40, L-623, (1979).
[3] J. K. Bhattacharjee and R.A. Ferrell, preprint.
[4] H. C. Burstyn, R.F. Chang and J.V. Sengers, Phys. Rev. Lett. 44, 410 (1980).
[5] K. Kawasaki, Ann. Phys. (N.Y) 61, 1 (1970).
[6] E. D. Siggia, B.I. Halperin and P.C. Hohenberg, Phys. Rev. B13, 2110 (1976).
[7] P. Bergé, P. Calmettes, C. Laj, M. Tournarie, and B. Volochine, Phys. Rev. Lett. 24, 1223 (1970).
[8] See for instance G. Parette and R. Kahn, J. Physique 32, Colloque C1, 523 (1971).
[9] See for instance G.W. Brady, D. Mc Intyre, M.E. Myers, and A.M. Wims, J. Chem. Phys. 44, 2197 (1966).
[10] See for instance B.J. Berne and P. Pecora, Dynamic Light Scattering, (Wiley, New-York, 1976).
[11] a) A.J. Bray, Phys. Rev. B14, 1248 (1976). Numerical values in b), R.F. Chang, H. Burstyn, J.V. Sengers and A.J. Bray, Phys. Rev. A19, 866 (1979).
[12] D. Beysens, A. Bourgou and P. Calmettes, to be published.
[13] See for instance D. Beysens, J. Chem. Phys. 64, 2579 (1976).
[14] See for instance D. Beysens, A. Bourgou and H. Charlin, Phys. Lett. A53, 236 (1975) ; L.A. Reith and H.L. Swinney, Phys. Rev. A12, 1094 (1975) ; A.J. Bray and R.F. Chang, Phys. Rev. A12, 2594 (1975) ; for practical evaluations, D. Beysens and G. Zalczer, Opt. Comm. 26, 172 (1978).
[15] Y. Garrabos, G. Zalczer and D. Beysens, to be published.
[16] D. Beysens, A. Bourgou and G. Zalczer, J. de Physique, Colloque, 37, C-1, 225 (1976).
[17] P. Bergé and M. Dubois, Phys. Rev. Lett. 27, 1125 (1971).
[18] P. Calmettes, I. Lagües, and C. Laj, Phys. Rev. Lett. 28, 478 (1972).
[19] P. Calmettes, thesis (Paris VI, 1978) (unpublished).
[20] See, for instance, W.I. Goldburg and J.S. Huang, in "Fluctuations, Instabilities and Phase Transitions", ed. by

T. Riste, NATO Advanced Study Institute Series (Plenum, N.Y, 1975) and ref. therein.
[21] See, for instance, Yu P. Blagoï, V.I. Sokhan and L.A. Pavlichenko, J.E.T.P. Lett. 11, 190 (1970) ; P.C. Hohenberg and M. Bormatz, Phys. Rev. A6, 289 (1972) ; M. Giglio and A. Vendramini, Phys. Rev. Lett. 35, 168 (1975) ; J.C. Greer, T.E. Block, and C.M. Knobler, Phys. Rev. Lett. 34, 250 (1975).
[22] S. C. Greer, Phys. Rev. A14, 1770 (1976).
[23] D. T. Jacobs, D.J. Anthony, R.C. Mockler and W.J. O'Sullivan, Chem. Phys. 20, 219 (1977).
[24] D. A. Balzarini, Can. J. Phys. 52, 499 (1974).
[25] B. A. Scheibner, C.M. Sorensen, D.T. Jacobs, R.C. Mockler and W.J. O'Sullivan, Chem. Phys. 31, 209 (1978).
[26] D. Beysens and A. Bourgou, Phys. Rev. A19, 2407 (1979).
[27] D. Beysens and J. Wesfreid, J. Chem. Phys. 71, 119 (1979).
[28] D. Beysens and P. Calmettes, J. Chem. Phys. 66, 766 (1977).
[29] See for instance J. Thoen, E. Bloemen, and W. Van Daël, J. Chem. Phys. 68, 735 (1978) and ref. therein.
[30] P. Calmettes and C. Laj, Phys. Rev. Lett. 36, 1372 (1976).
[31] J. C. Le Guillou and J. Zinn-Justin, Phys. Rev. B21, 3976 (1980) and ref. therein.
[32] C. Bervillier, Phys. Rev. B14, 4964 (1976).
[33] C. Bervillier, and C. Godreche, Phys. Rev. B21, 5427 (1980).
[34] D. Stauffer, M. Ferer and M. Wortis, Phys. Rev. Lett. 29, 345 (1972).
[35] A. Aharony and P.C. Hohenberg, Phys. Rev. B13, 3081 (1976).
[36] N.B. Rozhdestvenskaya and E.N. Gorbachova, Opt. Comm. 30, 383 (1979).
[37] Y. Garrabos, R. Tufeu and B. Le Neindre, J. Chem. Phys. 68, 495 (1978).
[38] H. C. Burstyn, J.V. Sengers and P. Esfandiari, Phys. Rev. A22, 282 (1980).
[39] P.C. Hohenberg, private communication.
[40] E. D. Siggia, B.I. Halperin and P.C. Hohenberg, Phys. Rev. B13, 2110 (1976).
[41] B. Volochine, thesis (Paris, 1973, unpublished).
[42] I.M. Arefev, I.L. Fabelinskii, M.A. Anisimov, Yu. F. Kiyachenko, and V.P. Voronov, Opt. Commun. 9, 69 (1973).
[43] D. Thiel, B. Chu. A. Stein and G. Allen, J. Chem. Phys. 62, 3689 (1975).
[44] C. C. Lai and S.H. Chen, Phys. Lett. 41A, 259 (1972).
[45] E. Gülari, A.F. Collings, R.L. Schmidt and C.J. Pings, J. Chem. Phys. 56, 6169 (1972).
[46] J. Timmermans, "The Physico-Chemical constant of binary systems in concentrated solutions", Vol.1, Interscience, Pub. Inc. N.Y (1959) and ref. therein.

[47] B. Chu., F.J. Schoenes and W.P. Kao, J. of Am. Chem. Soc. 90, 3042 (1968).
[48] E. Gülari, A.F. Collings, R.L. Schmidt and C.J. Pings, J. Chem. Phys. 56, 6169 (1972).
[49] D. Atack and O.K. Rice, Disc. Trans. Farad. Soc. 23, 2428 (1953), reanalyzed by Wims and al, ref.[50].
[50] A. Wims, D. Mc Intyre and F. Hynne, J. Chem. Phys. 50, 616 (1969).
[51] F. Kohler and O.K. Rice, J. Chem. Phys. 26, 1614 (1957).
[52] See Ref.[59], part I.

[53] E.S.R. Gopal, R. Ramachandra, P. Chandrasekhar, K. Govindarajun, and S.V. Subramanyam, Phys. Rev. Lett. 32, 284 (1974), reanalyzed in ref.[22].
[54] J. Reeder, T.H. Block, and C.M. Knobler, J. Chem. Therm. 8, 133 (1976).
[55] R.J.L. Andon and J.D. Cox, J. Chem. Soc. 4601 (1952).
[56] G. Morrisson and C.M. Knobler, J. Chem. Phys. 65, 5507 (1976).
[57] H. Klein and D. Woermann, Ber. Bunsenges. Phys. Chem. 79, 1180 (1975).
[58] S. C. Greer and R. Hocken, J. Chem. Phys. 63, 5067 (1975).
[59] M. Pelger, H. Klein and D. Woermann, J. Chem. Phys. 67, 5362 (1977).
[60] M. A. Anisimov, A.V. Voronel and T.M. Ovodova, J.E.T.P. Lett. 34, 583 (1972).
[61] H. Klein and D. Woermann, J. Chem. Phys. 62, 2913 (1974).
[62] H. Klein and D. Woermann, J. Chem. Phys. 65, 2913 (1976).
[63] C. M. Sorensen, R.C. Mockler and W.J. O'Sullivan, Phys. Rev. Lett. 40, 777 (1978).
[64] S. H. Chen, C.C. Lai, and J. Rouch, J. Chem. Phys. 68, 1994 (1978) and ref. therein.
[65] D. Woermann and W. Sarholz, Ber. Bunsenges. Physik. Chem. 69, 319 (1965).
[66] G. Zalczer, Private communication. Fit of the data from ref.[67] with $\nu = 0.630$ imposed, giving $\xi_o/R = 3.06 - 3.10$ Å.
[67] B. Chu, S.P. Lee and W. Tscharnuter, Phys. Rev. A7, 353 (1973).
[68] V. K. Semenchenko, E.L. Zorina, Dokl. Akad. Neuk. SSSR 73, 331 (1950), and 84, 1191 (1952), and E.F. Zhuralev, Zh. Obshch. Khim. 31, 363 (1961).
[69] D. Beysens, S.H. Chen, J.P. Chabrat, L. Letamendia, J. Rouch and C. Vaucamps, J. de Physique Lett. 38, L-203 (1977).
[70] G. Arcovito, C. Falori, M. Roberti and L. Mistura, Phys. Rev. Lett. 22, 1040 (1969).
[71] C. M. Sorensen, R.C. Mockler and W.J. O'Sullivan, Phys. Rev. A16, 365 (1977).
[72] B. Chu, D. Thiel, W. Tscharnuter and D. Fenby, J. de Phys. C1, 33, C-111 (1972).
[73] R. Tsaï and D. Mc Intyre, J. Chem. Phys. 60, 937 (1974).

[74] J. C. Allegra, A. Stein, and G.F. Allen, J. Chem. Phys. 55, 1716 (1971).
[75] D. R. Thomson and O.K. Rice, J. Am. Chem. Soc. 86, 3547 (1964) reanalyzed in Ref.[76].
[76] P. Heller, Rep. Prog. Phys. 30, 731 (1967).
[77] A. Bourgou and D. Beysens, to be published.
[78] E. Gürmen, M. Chandrasekhar, P.E. Chumbley, H.D. Bale, D.A. Dolejsi, J.S. Lin and P.W. Schmidt, Phys. Rev. A22, 170 (1980).

THERMODYNAMIC ANOMALIES NEAR THE LIQUID-VAPOR CRITICAL POINT: A REVIEW OF EXPERIMENTS

Michael R. Moldover

Thermophysics Division
National Bureau of Standards
Washington, D.C. 20234

OUTLINE

I INTRODUCTION AND OVERVIEW

From the point of view of testing theoretical developments, liquid-vapor systems are representative of 3-dimensional systems with short-ranged forces and scalar order parameters. Liquid-vapor systems are attractive experimental systems because they can be studied in thermodynamic equilibrium (unlike liquid-liquid systems near consolute points). Liquid-vapor systems are free from frozen-in impurities and long-ranged strain fields that occur in crystals. A field conjugate to the order parameter is accessible in liquid-vapor systems, thus the scaling function for the free energy can be measured. These systems lack the symmetry of certain magnets and Ising models. This presents problems for simple-minded analyses of data. Nature does not tell us in advance which variables to use; however, this problem and its associated asymmetries are of interest in their own right. The primary macroscopic inhomogeneity in liquid-vapor systems is a density gradient over the height of a sample because of the earth's gravitational field. This phenomenon is fully understood and has been exploited for the most delicate measurements of the equation of state.

Liquid-vapor systems, particularly in mixtures, provide a challenge for those interested in applying the understanding gained about critical phenomena in the past 10-15 years to problems of technological interest. There is a large body of data about multiphase multicomponent mixtures in the chemical engineering literature. Is it too optimistic to hope that the powerful concepts such as scaling, universality, and cross-over will assist in organizing and understanding these data and in providing predictions to reduce future data needs?

The primary concern of this review is to describe the thermodynamic anomalies found by experiments near liquid-vapor critical points. Insofar as it is possible, the emphasis will be on those features which can be found most directly from particular experiments. Statistical analyses of data which rely heavily on theoretical input will be minimized, in part, because the detailed conclusions will change as rapidly as the theoretical input does.

A review of the theoretical framework for these experiments will not be attempted here. The concerned reader is referred to other lectures in this volume. (The lecture by Professor Sengers contains a list of recent review papers.)

The remainder of this introduction will be devoted to a summary of the experimental situation. In general, there is rather detailed agreement between the data taken closest to the critical point and the theoretical predictions for asymptotic behavior of 3-dimensional systems with scalar order parameters. Existing experiments do not have the precision to unambiguously resolve the differences between renormalization group (RG) calculations of critical exponents and amplitude ratios and analyses of the high temperature series (HTS) expansions of relevant Ising models. There is no fundamental reason why experiments near liquid-vapor critical points cannot be carried out with sufficient resolution to resolve these differences, should the motivation be present.

At the present time, August 1980, the best experimental information about the vicinity of liquid-vapor critical points is consistent with the following description: The thermodynamic potential should be taken to be $P(\mu,T)$ or possibly $P/T(\mu/T,1/T)$. Within the experimental precision, this potential shows a leading singularity at the critical point which has the same form as the singularity theoretically predicted to occur at the critical point of three-dimensional systems with a scalar order parameter. This singularity is best studied with a variety of optical techniques and constant volume specific heat measurements (C_v). For the fluids studied the singularity in the potential is characterized by the exponents:

$0.10 < \alpha < 0.12$; $0.32 < \beta < 0.33$; and $1.23 < \gamma < 1.27$. Reasonable estimates for the amplitude ratios appearing in the potential are $0.48 < A^+/A^- < 0.50$; $4.5 < \Gamma^+/\Gamma^- < 5.0$; and $1.57 < \Gamma^+ DB^{\delta-1} < 1.75$.

The imprecision in the measured values of the critical exponents and in the scaling function (insofar as the scaling function is known from amplitude ratios) is about as large as the theoretical uncertainty in these quantities. Here we make the naive assumption that the theoretical uncertainty is about as large as the differences between these quantities as recently calculated from either high temperature series[1](HTS) expansions or renormalization group[2] (RG) techniques. Thus the existing experiments cannot be used as a reliable guide to assess which of these two approaches to calculation is more reliable. It is quite possible that further analysis of the existing data may reduce the imprecision of the critical exponents and amplitude ratios. This could occur if stronger assumptions were made about the functions used to fit the data than we have chosen to make here. For example, if the correction-to-scaling exponent, Δ, is assumed to be 0.5, then existing C_v data favor a value of α nearer to 0.11 than 0.12. Thus with a stronger assumption the C_v data favor the RG number over the HTS number in this particular case. It seems certain that better measurements could be made by straightforward extensions of existing techniques. We will discuss this possibility in some detail below.

As one moves out from the critical point, the data are consistent with the simple picture that the thermodynamic potential has the symmetry of decorated lattice gases as proposed by Rehr and Mermin[3]. More general and/or more complicated proposals are not required by the data. The evidence for the subdominant singularity in the potential with $\Delta \sim 0.5$ comes most directly from measurements of the difference between the coexisting densities, $\Delta\rho(t)$ with $\Delta\rho = (\rho_{liquid} - \rho_{vapor})/2\rho_c$; $t = (T - T_c)/T_c$. Supporting evidence comes from equation of state and C_v measurements. The existence of this confluent singularity is consistent with the expansion derived by Wegner from field theoretic considerations[4,5]. The coexistence curve data have the resolution to determine Δ rather accurately; however, the necessary analysis simply has not been done (see below).

At the present time neither optical equation-of-state measurements nor conventional equation-of-state measurements (PVT) provide direct evidence for the presence of confluent singularities. The optical measurements exist in a narrow region $10^{-4} < r < 10^{-5}$ where confluent singularities probably make a very small contribution to the measured quantities. Conventional PVT measurements have sufficient precision to detect a confluent singularity in the range $10^{-1} < r < 10^{-2}$; however, at such large values of r several terms in the Wegner expansion may contribute to the measured quantities. (Here we use the variable r as a measure of the distance from the critical point. Suitable, roughly equivalent, estimates for r can be obtained from the succeptibility or the correlation length: $r=(\chi_T/\chi_T^{\text{ideal gas}})^{-1/\gamma}$ or $r=(\xi/\xi_o)^{-1/\nu}$. The optical and conventional PVT measurements taken together are probably consistent with the presence of confluent singularities. Confirmation of this conjecture would require inter-comparison of the optical and PVT data on an absolute basis. To my knowledge, this tedious and complicated task has not yet been undertaken with the more recent data.

In contrast with the universality expected and observed for critical exponents and amplitude ratios, the nonuniversal amplitude for the specific heat, A^-/R, varies substantially from fluid to fluid. It is near 2 in He^3 and above 10 in classical fluids. The nonuniversal amplitude for the density difference, B, is near 1.0 in He^3 while it is near 1.5 for classical fluids. This amplitude is roughly 1.5 for all the three dimensional nearest-neighbor Ising models. I am not aware of theoretical explanations for these facts.

In the rest of this review I will discuss those experiments which provide the primary evidence for the thermodynamic model I have just outlined. The order of the discussion will be: (1) The selection of thermodynamic variables suggested by C_v and global equation-of-state measurements, (2) the exponents and amplitude ratios that may be determined from C_v measurements, (3) asymptotic critical behavior determined by studies of the density vs. height profile, (4) conventional PVT measurements. In a concluding section, I will emphasize the need and the challenge in applying the understanding we have gained about critical phenomena to problems in the thermodynamics of fluid mixtures.

II. THE CHOICE OF THE POTENTIAL AND ITS VARIABLES

Figure 1 is a drawing of the Gibbs surface of a fluid in the (μ,P,T) space. The fold in the surface which gradually ends at the critical point, is the curve along which coexisting liquid and vapor phases may be found. The projection of this fold on the (P,T) plane is the familiar vapor pressure curve of the fluid. The projections of the fold on the (μ,T) on (μ,P) planes do not have special names. Yang and Yang[6] pointed out that the singularity in the shape of the coexistence curve at the critical point could be accurately studied by determining the curvature of its projections on the (μ,T) and (P,T) planes using C_v data and the thermodynamic identity:

$$\frac{\rho C_v}{T} = -\rho\left(\frac{\partial^2\mu}{\partial T^2}\right)_v + \left(\frac{\partial^2 P}{\partial T^2}\right)_v \tag{1}$$

Here ρC_v is the specific heat per unit volume of fluid measured at constant volume. In two phase states μ and P are independent of the total volume of (say) a mole so that the partial derivatives in

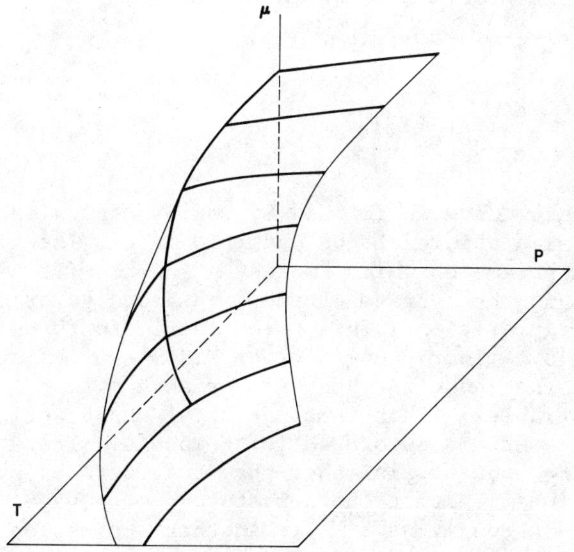

Fig. 1. The Gibbs surface for a liquid. Several isotherms are shown on the surface. The coexistence curve is the locus of points at which the slopes of the isotherms change discontinuously.

equation (1) become total derivatives. The relations used to analyze the data become:

$$\frac{d^2\mu}{dT^2} = \frac{d}{d\rho}\left(\frac{-\rho C_v}{T}\right); \quad \frac{d^2P}{dT^2} = \frac{d}{dV}\left(\frac{C_v}{T}\right) \tag{2}$$

This analysis[7-9] has been carried out for He^3, He^4, and Argon. In each case the conclusion from C_v data alone is that d^2P/dT^2 has the α-like divergence that C_v has along the critical isochore while $d^2\mu/dT^2$ has a weaker singularity or none at all. These results mean that a satisfactory representation of the singularity in C_v data can be obtained by taking the pressure (or P/T) to be the thermodynamic potential in the form:

$$\frac{P}{P_c} = |x|^{\beta(\delta+1)} g_0\left(\frac{y}{|x|^{\beta\delta}}\right) + P_0(x,y) \tag{3}$$

Here the scaling function, g_0, and the function P_0 are both analytic in in their arguments. The scaling variables, x and y, are, to lowest order, linearly related to the laboratory variables μ and T:

$$x = \frac{T-T_c}{T_c} + q\,\frac{\rho_c}{P_c}\,(\mu-\mu_o(T))$$

$$y = \frac{\rho_c}{P_c}\,(\mu-\mu_o(T)) \qquad (4)$$

The coexistence curve is defined by y=o or equivalently by $\mu=\mu_o(t)$ where μ_o is also assumed to be analytic in t. The constant q is a nonuniversal parameter which "mixes" μ and t. (If $q\neq 0$, then $(\partial^2\mu/\partial T^2)_{\rho_c}$ and the average of the liquid and vapor densities, ρ_D, have $t^{1-\alpha}$ singularities.) This potential leads to an equation of state which is a minor "revision" of Widom's original scaling proposal[10]. (He chose $x\equiv t$ and his variable y is defined in terms of a density variable.) This same potential can describe the asymmetries that occur in certain decorated Ising models. There have been many suggestions for potentials other than that defined by Eqs. (3) and (4). Some suggestions[11] permit the curvature of the vapor pressure, d^2P/dT^2, to diverge as strongly as $t^{-\alpha-\beta}$. Another suggestion by Griffiths and Wheeler[12], that all the field variables be treated on an equal basis (rather than singling out the pressure as a potential), has great geometric appeal. These alternatives are not required by the existing data at liquid-vapor critical points.

An example of the analysis of C_v data which has led to these conclusions is shown in Figure 2. The upper solid curve in both halves of the figure is C_v/R for He^4 at the critical density, $\rho=\rho_c$. The curves are split into two branches near T_c to indicate the errors in the original data[8]. In Figure 2, (left), the lower solid curve is $(-T/R)/(d^2\mu/dT^2)$ as determined from differentiating the C_v data with respect to density. This lower curve shows no evidence of any singularity near T_c. The difference between the two solid curves is $(TV/R)(d^2P/dT^2)$. It is clear that the C_v singularity at ρ_c, namely $C_v \sim t^{-\alpha}$, is also present in d^2P/dT^2. Thermodynamic inequalities[13] permit d^2P/dT^2 to diverge as strongly as $t^{-\alpha-\beta}$. There is no clear evidence for such a strong divergence. The fractional error in computing the curvature of the vapor pressure from C_v data is given by:

$$\delta\left(\frac{d^2P}{dT^2}\right) = \frac{\delta_T}{Bt^{\beta+1}} \qquad (5)$$

Here δ_T is the fractional error in temperature measurement, a quantity which can be less than 10^{-7}. The parameters B and β are the same ones which describe the difference of the coexisting densities: $\rho_{liquid}-\rho_{vapor} = 2\rho_c Bt^{\beta}$.

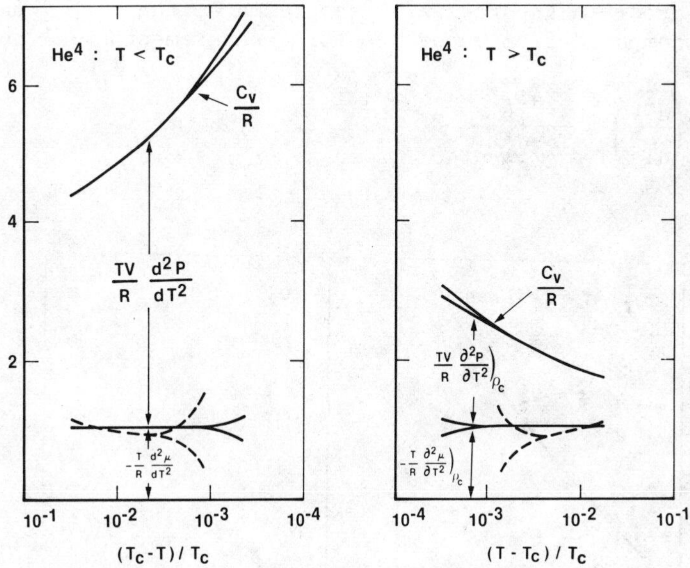

Fig. 2. Analysis of C_v for He^4. The upper solid curve is C_v/R at $\rho=\rho_c$. (It is split into two curves near T_c to indicate the scatter in the data.) The lower solid curves are $-(T/R)(\partial^2\mu/\partial T^2)_{\rho_c}$ as determined from differentiating the C_v data with respect to density. The dashed curves are $-T/R(\partial^2\mu/\partial T^2)_{\rho_c}$ as determined from a combination of C_v and vapor pressure data[8,14].

Another method has been used to search for possible singularities in $d^2\mu/dT^2$. For both He^3 and He^4, C_v data has been combined with vapor pressure data to search for a possible singularity in $d^2\mu/dT^2$. In both cases the published method has been to first integrate the C_v data along the critical isochore and then to subtract the dP/dT data from the integral[14,15]. The result of this subtraction is $d\mu/dT$, or $\partial\mu/\partial T$ at $\rho=\rho_c$ for $t>0$ as a function of T. This quantity must then be differentiated either analytically or graphically to find $d^2\mu/dT^2$, the lowest order temperature derivative of μ which might be divergent. An equivalent procedure, which we have illustrated with the dashed curves for He^4 on Figure 2 and for CO_2 on Figure 3 is to first differentiate the vapor pressure data twice to compute $T(d^2P/dT^2)$ and then subtract $T(d^2P/dT^2)$ from C_v to find $-T(d^2\mu/dT^2)$. The dashed curves in Figure 2 are near the lower solid curves. Thus, this procedure which combines C_v and vapor pressure data tends to confirm the conclusion obtained from C_v data alone, namely $d^2\mu/dT^2$ is less singular than d^2P/dT^2. (The He^4 data are from Refs. 8 and 14; the CO_2 data from Refs. 16 and 17.)

It is evident that differentiating the pressure data twice leads to a rapid loss of information as T_c is approached. If δ_p is the fractional error in the pressure measurements, the fractional error in d^2P/dT^2 computed from vapor pressure measurements will be approximately

$$\delta\left(\frac{d^2P}{dT^2}\right) = \frac{4\delta_P}{\frac{T^2}{P^2}\frac{d^2P}{dT^2}\,t^2} \tag{6}$$

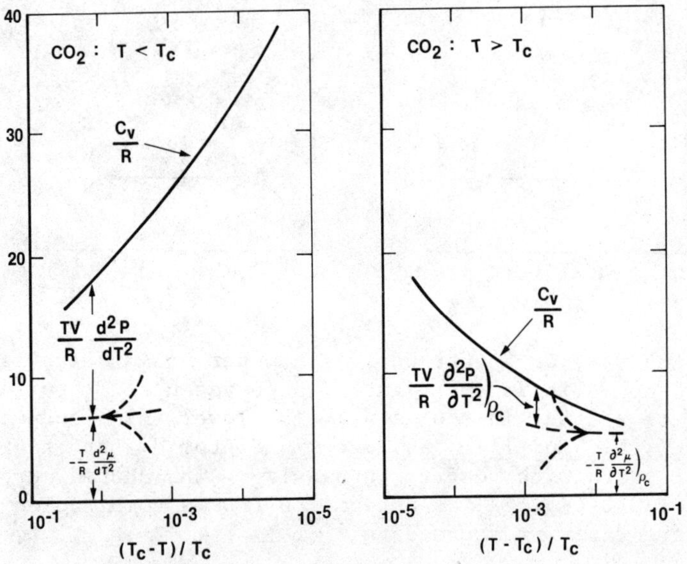

Fig. 3. Analysis of C_v for CO_2. The solid curves are C_v/R at $\rho=\rho_c$. The scatter in the C_v data is about the width of the curve. The dashed curves are obtained by subtracting $(TV/R)(\partial^2P/\partial T^2)_{\rho_c}$ from the upper curve[16,17].

For the He^4 data shown, a differential pressure measurement technique was used leading to the unusually small value $\delta_P = 5 \times 10^{-7}$. For the CO_2 data shown in Figure 3, $\delta_P = 2\text{x}10^{-5}$. The information about d^2P/dT^2 degrades very rapidly for $|t| < 10^{-2}$. Thus it is almost hopeless to look for divergences in d^2P/dT^2 and $d^2\mu/dT^2$ from pressure measurements alone. Nevertheless it is quite informative to look at these quantities equally far above and below T_c. Thus we form $Q(t) = F(t) - F(-t)$. From Figure 3 it is easy to see Q is much more singular when F is chosen to be $(\partial^2P/\partial T^2)_{\rho_c}$ than when F is chosen to be $(\partial^2\mu/\partial T^2)_{\rho_c}$.

The precise C_v data we have considered so far became available in the late 1960's and is concentrated in the region $|t| < 10^{-2}$. In hindsight we can find evidence for the same choice of potential and scaling variables in earlier data taken further from T_c.

In Figure 4, we reproduce a graph of the coexisting densities and volumes of Argon from Levelt Sengers[18]. The density plot is obviously more symmetrical. This symmetry suggests that a good choice for a laboratory variable which will play the role of an order parameter is the difference between the density and its average value (that is $\rho-\rho_D(t)$ where $\rho_D(t)$ is the so-called rectilinear diameter). In fact, we will see this choice can be significantly improved. Figure 5 is a representation of the equation of state of steam[19] over the wide range of states most commonly encountered in engineering practice. On Figure 5, the dashed curve below T_c is $\rho_D(t)$. It departs noticeably from ρ_c as the temperature is reduced. In contrast, the dashed curve above T_c in Figure 5 is the locus of inflection points in the P-ρ isotherms. This curve of high isothermal compressibility remains close to ρ_c even at temperatures quite far above T_c. Thus the dashed curve seems to have a weak singularity at T_c in the ρ-P and ρ-T planes. In a symmetrical magnetic system, this locus of maximum susceptibility would correspond to the locus of zero applied field and would exhibit no singularity in the M-T plane at T_c. For a real fluid the available evidence suggests that the weak singularity in the dashed curve can be removed, at least to lowest order, by choosing as the density variable a particular linear combination of $\rho-\rho_D(t)$ and $\sigma-\sigma_D(t)$ (here σ is the entropy per unit volume). Such combined density variables result when both scaling fields, x and y, are taken to be analytic functions of t and μ as in Eq. (4). We will return to this point below when we discuss equation-of-state data.

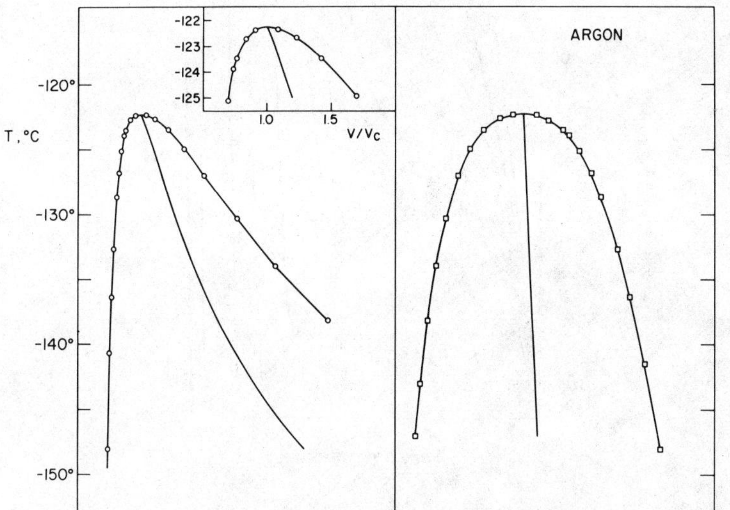

Fig. 4. The coexisting densities and volumes of argon. (Ref. 18)

III. LIQUID-VAPOR CRITICAL BEHAVIOR FROM C_v DATA

In this section I will comment on the measurements of the critical exponent α characterizing the C_v divergence near the liquid-vapor critical point. The very diverse experimental techniques used by various authors will not be discussed. This section concludes with a few remarks concerning the prospects for future measurements.

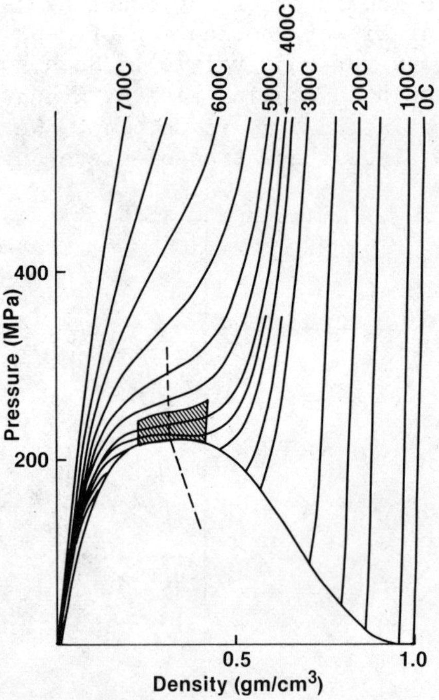

Fig. 5. The equation of state of steam. Data in the shaded region near the critical point has been fit to thermodynamic models based on RG theory. The dashed curve is the line of symmetry of the model.

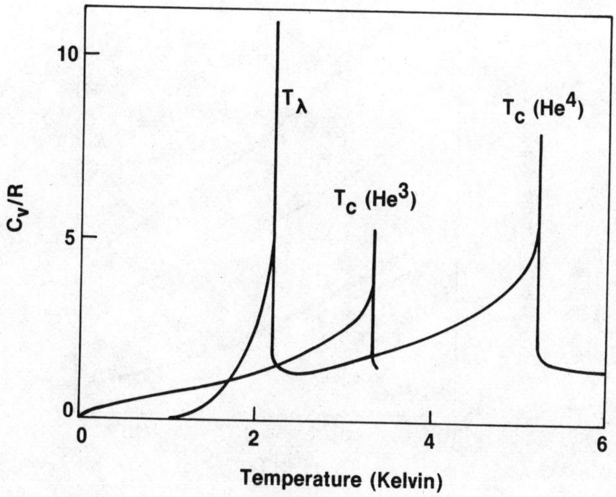

Fig. 6. The constant volume specific heat of He^3 and He^4 at their respective critical densities.

The most detailed studies of C_v in the critical region have been made for CO_2, C_2H_6, Ar, He^4, and He^3 (references 7-9, 10). For the latter three fluids, data are available on many isochores. For all five fluids, the original data are available for further analysis.

In Figure 6, C_v is sketched for He^3 and He^4 at their critical densities. (The λ-transition at 2.2 K in He^4 is discussed by Professor Ahlers elsewhere in this volume: it belongs to a different universality class than the liquid-vapor critical points that are considered here.) At the critical density, C_v is dominated by the critical point singularity. At temperatures only slightly above T_c, C_v declines to values close to the ideal gas value for He^3 and He^4, as well as for other fluids which have been studied.

In Figure 7, we reproduce a plot from Lipa[16] et al. of $C_v - C_v^{Ideal\ Gas}$ for CO_2 at ρ_c (Here the ideal gas contribution to C_v is taken as $(7/2)R$ or about 29 joule/mol deg). The data are compared with a three dimensional Ising model. The CO_2 data have been scaled by the factor ρ_c/ρ_{max} where ρ_{max} is an estimate for the liquid density at zero temperature. The agreement between the data and the model is remarkable, particularly when we recall that the data are taken along the path $\rho=\rho_c$ which does not correspond with the symmetry line of the fluid. It is interesting to note that below T_c, where the experimental path departs more strongly from the fluids symmetry line, the data fall further from the model.

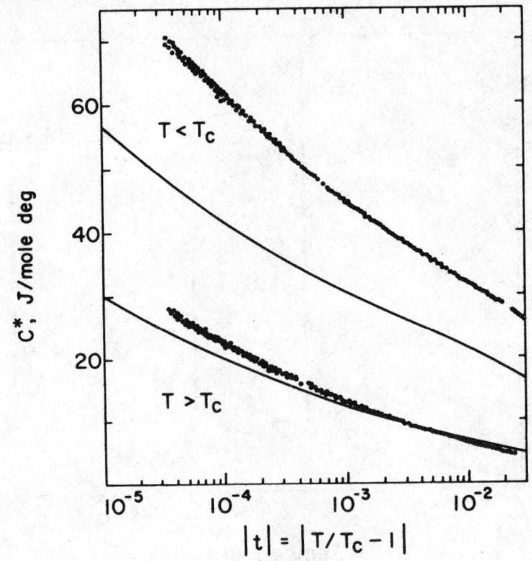

Fig. 7. The constant volume specific heat of CO_2 at its critical density is compared with a three dimensional Ising model (solid lines). (Ref. 16). The ideal gas heat capacity of CO_2, 7R/2, (29 J/mol deg) has been subtracted from the data.

Lipa et al. fit various equations to their data in an effort to determine the form of the leading singularity in C_v. An example of one of their results is:

$$f^{\pm} = \frac{C_v^{\pm}}{R} = \begin{cases} 5.694\ t^{-0.124} - 3.594 & : \quad (t>0) \\ 10.585(-t)^{-0.124} - 0.053 & : \quad (t<0) \end{cases} \qquad (7)$$

Note: This equation has two singularities in it. First there is a symmetrical divergence of C_v at T_c. Second there is a jump in the additive constant term at T_c. The jump is not consistent with the analytic structure of C_v expected from RG; however, it may be a good numerical approximation to a sum of singularities which are weaker than a divergence.

In this particular fit Lipa et al. used the constraint $\alpha=\alpha'$. The literal numbers in Eq. (7) all result from the fitting process. A plot of the deviations of the data from Eq. (7) is shown in Figure 8. The function obviously fits the data within their scatter (roughly ±1% of C_v). Lipa et al. were quite concerned with careful statistical measures of error in fitting the data. They were fully aware of how correlations among the various fitting parameters necessarily leads to larger errors in the estimates of parameters.

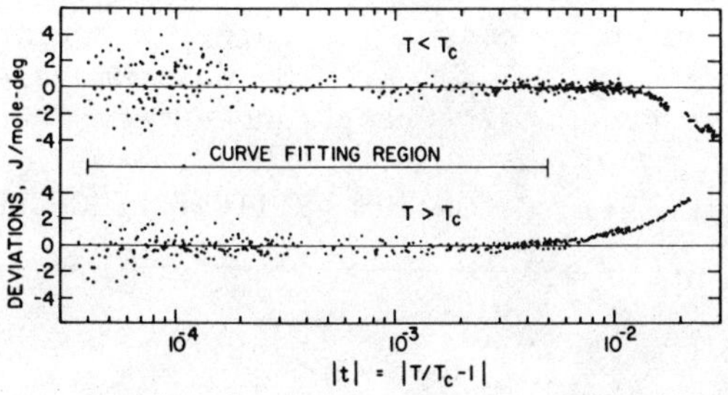

Fig. 8. Deviations of the CO_2 data[16] in Figure 7 from Eqs. (7) and (8).

In fact they provided graphical cross-sections of the chi-squared hypersurface to discuss such correlations. With these precautions in mind they state that there is a 95% probability that the true value of α is within ± 0.005 of the value 0.124, provided Eq. (7) is an appropriate model for the data.

As part of the preparation for this lecture, I refit these same data, which the authors thoughtfully published, using the same weighting of experimental errors that they did. I used functions of the form suggested by RG predictions[4,5]:

$$g^{\pm} = \frac{C_v^{\pm}}{R} = A^{\pm} |t|^{-\alpha} (1 + D^{\pm} |t|^{0.5}) + E \tag{8}$$

Using the same range of data designated on Figure 8, the best parameters for Eq. (8) are: $A^+=8.05$; $D^+=0.37$; $A^-=15.33$; $D^-=0.41$; $E=-6.95$; and finally $\alpha=0.105$. The value $\alpha=0.105$ falls well outside the range 0.124 ± 0.005. I do not need to draw another deviation plot to illustrate how well this new equation describes the data. Within the entire fitting range Eq. (7) and (8) are essentially identical:

$$\frac{|f^- - g^-|}{f^-} \leq 0.0013; \quad \frac{|f^+ - g^+|}{f^+} \leq 0.0033 \tag{9}$$

Thus this Eq. (7) cannot be distinguished from Eq. (8) by these data which have a scatter of approximately 1%. The point I have made in such detail is that error estimates must be viewed in the context of: What question are we asking the data? If we ask a different question a very different answer will appear.

In Table I we compare He^4 data[8] from 1969 with these CO_2 data[16] from 1977 using Eq. (8) with the correction-to-scaling exponent fixed at 0.5. Two ranges of data have been used to get a feel for the noise in the fitting process, not to seriously explore any possible range dependence of apparent exponents. The larger range is $4\times10^{-4}<|t|<0.01$ for He^4 and $4\times10^{-5}<|t|<0.01$ for CO_2. The maximum value of $|t|$ is 0.005 in the smaller range. In these ranges terms such as $t^{2\Delta}$ and t should be small.

TABLE I: C_v ON THE CRITICAL ISOCHORE

SYSTEM (RANGE)	α	A^+/A^-	D^+/D^-	E*	A^-	D^-	E
He^4 (larger)	0.116	0.48	0.78	0.74	3.06	0.61	-0.8
He^4 (smaller)	0.108	0.50	0.69	0.75	3.35	0.48	-1.0
CO_2 (larger)	0.101	0.53	1.40	0.69	16.2	0.29	-7.7
CO_2 (smaller)	0.103	0.53	0.99	0.68	15.6	0.40	-7.2
RG	0.110	0.48	1.26				
3d Ising series	0.125	0.51					

The tabulated results and some others not shown deserve several comments. First, if one asks an RG question, one gets very nearly an RG answer. These two very different sets of data were taken with very different experimental techniques by authors who were expecting to find $\alpha \approx 1/8$. Nevertheless, the quantities α, A^+/A^-, and D^+/D^- are very close to the RG values[2,20,21]. Second, if Eq. (8) is the model that we are testing, a value of α as large as 1/8 is inconsistent with the CO_2 data by the standard statistical tests. On the other hand, consistency with such a large value of α can be restored by adding to Eq. (8) theoretically reasonable terms such as ones proportional to t, $t^{2\Delta}$, or $t^{1-\alpha}$. (The latter term will be introduced when we account for the fact that the experimental path, $\rho=\rho_c$, is not the path y=0.) A value for Δ somewhat smaller than 1/2 would also permit α=1/8. Thus these C_v data are really not suited for distinguishing between the RG and HTS values for α without making fairly specific assumptions. Third, the tabulated values of D^+/D^- should not be taken very seriously. If additional terms are added to Eq. (8), the sign of D^+/D^- can easily be reversed. Coincidentally, the ε-expansion for D^+/D^- is very poorly behaved. (The value tabulated is a Pade approximation to the series in reference 21.) Fourth, I would like to call attention to the fact E* is well determined by these data and is nearly the same for both fluids. Here E* is defined by the relation:

$$E^* \equiv (E - C_v^{\text{Ideal Gas}})/A^- \tag{10}$$

where $C_v^{\text{Ideal Gas}}$ is 3R/2 for He^4 and 7R/2 for CO_2. In going from

He^4 to CO_2, E* changes about 10% while A^- changes 500%. E* would be essentially the same if it were defined using an estimate of the "real gas" heat capacity instead of $C_v^{\text{Ideal Gas}}$. (Such an estimate can be obtained from C_v at ρ_c at temperatures much higher than T_c.) Thus it appears that the constant term in Eq. (8) is actually a fluctuation induced contribution to C_v with the temperature dependence t^0, a result that I did not expect but one which is apparently not unreasonable. Voronel has remarked that away from the melting line, C_v for a fluid has an ideal gas contribution, a critical point divergence, and nothing else[22].

Finally, we remark that the non-universal amplitude, A^-, is much smaller for the heliums than for classical fluids (A^- is near 10 in CO_2, Ar, C_2H_6, Xe etc).

The analysis of C_v data can now be carried much further than when the data were first published. In a preliminary study, I have found that the C_v data for He^4 on all isochores obeys an appropriate scaling relation with a scaling function that is consistent with equation of state data. Details will be published in the future.

I will now briefly consider the limitations and prospects for future C_v measurements. The smallest value of t for which meaningful equilibrium measurements of C_v can be made is determined by the establishment of a density-vs-height profile within the calorimeter[23]. For small values of t, the fluid's weight leads to a sigmoid profile (see Figure 9) whose shape is strongly t-dependent. Under these circumstances, redistribution of fluid within the calorimeter makes the dominant contribution to entropy changes. C_v can no longer be measured. For a 1mm high sample the minimum value of t at which C_v can be measured is near 10^{-4} for He^4 and $2x10^{-5}$ for CO_2. Meaningful C_v measurements have been made closer to T_c in stirred samples where the density is not allowed to stratify. Virtually all measurements at consolute points in liquid mixtures are made in times short compared with the time required for stratification. Thus, in this global sense, many of the experiments discussed by Professors Sengers and Beysens in this volume have not been made in thermal equilibrium. The reduced gravity environment that will become available in Spacelab might make it possible to reduce the minimum value of t by a factor of 10^2-10^3. For values of t larger than the gravity limit, the precision with which an apparent exponent can be measured depends upon three things, the precision of the C_v measurement, the ratio of the sizes of successive terms in the model for the data, and finally the range of data available. With certain important reservations, the precision of the C_v measurements at small t is determined by the resolution of the primary thermometer. Substantial advances have occured in the technology for measuring small temperature changes in the past few years so that a resolution, δ_T, near 10^{-8} is attainable near 300 K. This implies C_v measurements could have a resolution approaching 0.1% for $|t|<10^{-5}$. Thus α can be determined with an accuracy of better than 1% if the correction-to-scaling exponent for fluids is not too much smaller than 0.5 and the correction coefficients are of order unity.

IV. STUDIES OF THE DENSITY VS HEIGHT PROFILE AND THE COEXISTING DENSITIES

In this section I will review measurements of the density vs height profile. These measurements have provided the best experimental information about the equation of state of fluids asymptotically close to the critical point. Some remarks concerning the techniques involved and their limitations will be made first; then the results will be described and interpreted.

Techniques

Figure 9 is a sketch of the density vs height profile of xenon at its critical temperature. This sketch is also a μ vs ρ isotherm because at equilibrium, in the earth's gravitational field, the potential energy per unit mass is a linear function of height:

$$\Delta\mu = \mu(\rho(z), T) - \mu(\rho(z_o),T) = -g(z-z_o) \qquad (11)$$

Here z_o is a reference height which will usually be the height of the inflection point or centrus in the μ vs ρ isotherm above T_c or the height of the liquid-vapor interface. Thus a measurement of the shape of the profile is a measurement of the equation of state. The unique feature of such a measurement is that it involves extraordinarily small changes in the chemical potential or in the pressure.

Upon going from the top to the bottom of a 1 cm tall cell, the dimensionless chemical potential, μ/k_BT_c, increases by 10^{-5} if the cell is filled with He^3, a fluid which is unusually responsive to gravity near its critical point[23]. The change is only $3x10^{-7}$ for steam near its critical point. These small changes in chemical potential are accompanied by extremely small changes in pressure: $3.5x10^{-5}P_c$ in the case of He^3 and $1.4x10^{-6}P_c$ in the case of steam. Such pressure changes can barely be resolved by direct pressure measurements.

All studies of the profile require very careful thermostat designs. At a typical liquid-vapor critical point a vertical temperature gradient as small as 1mK/cm will cause substantial convection totally upsetting the profile expected in equilibrium.

A variety of techniques have been used to study the profile. Th most direct is simply to use small floats[24]. Simple refractometers[25] as well as a Mach-Zender interferometer[26] have been used to directly measure the index of refraction as a function of $\Delta\mu$ and the

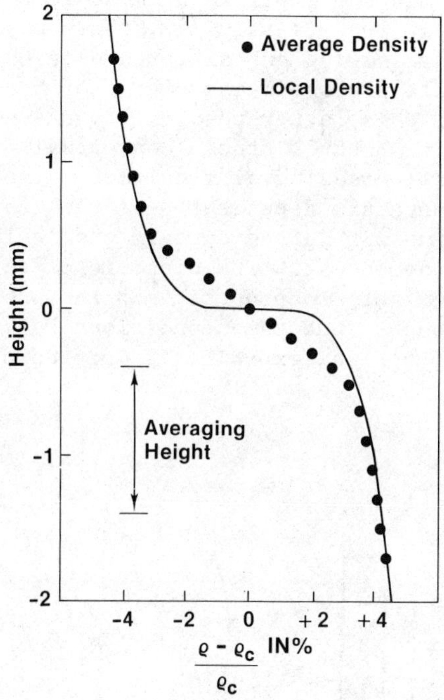

Fig. 9. The variation of the density of xenon with height at the critical temperature (solid curve). The points show the average density that would be measured by a 1mm high probe such as a magnetically levitated float 1mm high or a capacitor with plates separated by 1mm.

temperature. A pair of capacitors separated by a known height has been used[27] to study He^3. A column of many capacitors has been used[28,29] to study the dielectric constant as a function of $\Delta\mu$ in O_2 and SF_6. The data are interpreted by assuming the index of refraction or the dielectric constant is directly proportional to the density. (Searches for critical point deviations from the Clausius-Mossotti or Lorentz-Lorenz relations have not yielded convincing evidence of the expected weak anomaly[30]. Thus the amplitude of any such anomaly is small in the common, non-polar fluids. The anomaly would appear to first order in measurements of the symmetrical part of the equation of state such as $\rho_D(t)$; however the anomaly would cancel out to high order in measurements of the antisymmetric part of the equation of state such as $\Delta\rho$ as a function of $\Delta\mu$.)

At temperatures above T_c, say in the range $10^{-2}<t<10^{-4}$, these techniques for studying the profile are effective methods for measuring the isothermal compressibility, hence the exponent γ. For t greater than about 10^{-2}, the density difference across the height of a cell becomes too small to be measured reliably. At temperatures below T_c, say $10^{-4.5}<-t<10^{-0.5}$, these experiments are excellent methods of measuring the difference in the coexisting densities. They permit precise measurements of B and the correction exponent Δ; however, this latter possibility has not really been exploited to its limit. Near T_c these conceptually straightforward experiments encounter increasing difficulties. The profile's shape is most strongly temperature dependent near the centrus. The location of the centrus within the experimental cell also becomes strongly temperature dependent. The discrete capacitor arrays that have been used are not able to properly map the profile under these conditions. Under roughly the same conditions beam bending makes the interpretation of optical experiments more complicated.

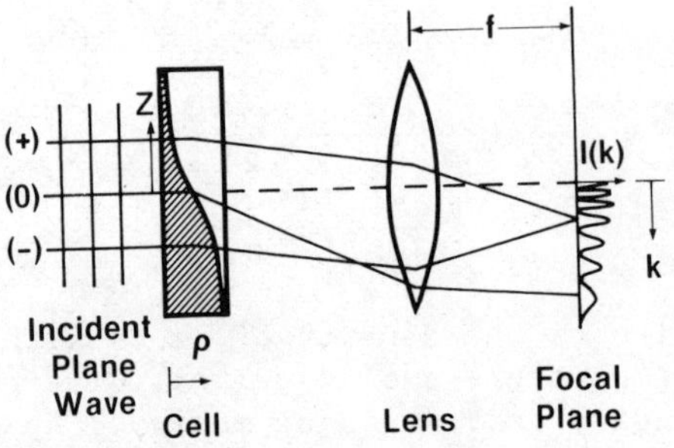

Fig. 10. The formation of a Franhofer diffraction pattern from a density vs height profile.

A somewhat more subtle optical technique has been applied to the study of the profile conditions[31,32]. In Figure 10, we illustrate this method which has the potential for extremely accurate studies of the profile at very small values of t and $\Delta\mu/k_BT_c$. The fluid sample is illuminated with a broad beam of coherent light. The beam is deflected downward because the optical path in the lower part of the cell is longer than that in the in the upper part of the cell. The deflection angle is proportional to the product of sample's thickness and the index of refraction <u>gradient</u>. (The gradient, dn/dz, is proportional to the isothermal suceptibility, $\chi_T = \partial\rho/\partial\mu$.) The most deflected ray ((o) in Figure 10) is the one passing through the inflection point in the μ-ρ isotherm. Other rays ((+) and (-) in Figure 10) are deflected less. The lens brings parallel rays to a focus where they interfere. There are two contributions to the phase difference between interfering rays. The lower ray (-) is advanced with respect to the upper ray (+) because it passes through a greater density of fluid. This contribution to the phase difference is proportional to $\Delta\rho$. The (+) ray is advanced because it travels over a longer geometrical path outside the cell. The length of this extra path is proportional to the product of the separation in height of the two rays ($\Delta z \sim \Delta\mu$) and the deflection angle ($k \sim dn/dz \sim d\rho/d\mu$). This product is proportional to the product of $\Delta\mu$ and χ_T. Thus, the diffraction pattern which is a map of optical phase as a function of deflection angle is essentially a representation of $\Delta\rho-\chi_T\Delta\mu$ as a function of χ_T. A photograph of the diffraction pattern made by Hocken and Moldover is shown in Figure 11. They used a microdensitometer to measure the angular locations of the maxima of the diffraction pattern with great precision. Because only the maxima were recorded the optical phase is, in effect, digitized in units of 2π.

We are now in a position to contrast the instrumental advantages and limitations of the Fraunhofer method of studying the profile with other methods. First, the Fraunhofer method is independent of the temperature dependence of the centrus height in the cell. All that is required is that the centrus remains within

Fig. 11. The Fraunhofer diffraction pattern from the density vs height profile. Light at various deflection angles is spread out vertically. Elapsed time is the horizontal coordinate. Initially the CO_2 sample is 2 mK above T_c ($t=6.6\times10^{-6}$). The thermostat temperature is then raised 2 mK to $t=13.1\times10^{-6}$. After some hours the profile has relaxed to a new one with smaller index gradients; therefore, the diffraction pattern is narrower. This record covers about 24 hours.

the cell. Second, the Fraunhofer method is free from many systematic errors. The deflection angles may be measured with respect to a portion of the incident beam which passes beside the fluid filled cell but otherwise travels through the same optical components as the main beam. This greatly reduces the requirements for mechanical stability of the entire apparatus. Thirdly, the Fraunhofer method can not be carried nearly as far above T_c as other methods. This limit arises from the fact that the dynamic range of angle measurements realized to date is less than that of measurements of electrica quantities such as capacitance. In practical terms, the experiments of Hocken and Moldover were limited to 1-1 1/2 decades in χ_T, namely: $6\times10^3 < k_T/k_T(\text{ideal gas}) < 10^5$. In a rough way this corresponds to the range $-1.5\times10^{-5} < t < 5\times10^{-5}$.

Results and Interpretation.

In Figure 12, the results of measurements[32] of the density profile are compared with RG and HTS calculations as well as with earlier data taken much further from the critical point. The profile data, which encompass only 1-1/2 decades in χ_T, yield apparent critical exponents and amplitude ratios which are in much better agreement with either of the theoretical ones than earlier experiment which study the equation of state further from the critical point.

It is remarkable that the exponents β and γ determined by the optical studies of the density profile can be combined with the exponent α determined from the isochoric C_v data to satisfy the scaling relation $\gamma+2\beta+\alpha=2$. This result is taken for granted today; however, it should be recognized as a triumph for the phenomenology

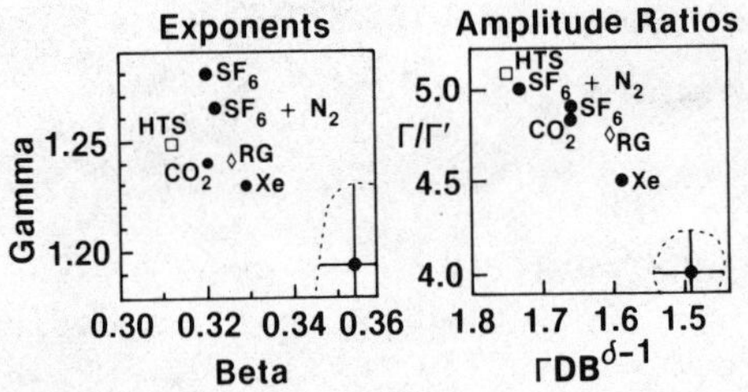

Fig. 12. Results of the density profile measurements[32] (solid dots) are compared with RG predictions[2] and HTS predictions[1]. The dashed ellipses indicate typical values of the same parameters from experiments further from the critical point which do not utilize the profile[47].

of scaling when such very different experimental results can be unified. This consistency is as important as the specific numerical value of any particular exponent.

Two questions arise in connection with these profile measurements. First, how accurate are these experimental numbers, or equivalently, are the various fluids studied really different from one another in the χ_T range studied; and second, what are the prospects for improving these measurements. On the issue of accuracy, Hocken and Moldover comment[32], that by standard statistical measures of error which allow for the correlations among parameters, the various fluids studied do indeed differ from one another. It is likely that some of the difference can be attributed to nonuniversal correction to scaling terms. (The analysis assumed simple scaling of the data in the variables $\Delta\mu$ and t. Both corrections to scaling terms and mixing of the variables $\Delta\mu$ and t are expected.) Another unknown portion of the differences between fluids may be attributed to the evolving nature of the experiment. The fluids were studied in the order: SF_6, SF_6+N_2, Xe, and CO_2. Each time the fluid sample was changed, improvements in technique were made. Unfortunately, the opportunity to repeat the initial experiments was not available.

The amplitude ratios indicated on the right hand side of Figure 12 are not independent of one another. They result from fitting the Wilcox-Estler equation of state[33] to the profile data. This equation has only one parameter which determines the scaling function, hence the experimental amplitude ratios all lie on a smooth trajectory in the figure. As similar phenomenon would occur if the "linear model" approximation to the scaling function were used[34]. Neither of these scaling functions is correct. Thus these amplitude ratios (and indeed the apparent critical exponents) are biased in a way which is hard to estimate and which is not inherent in the experimental method.

The amplitudes Γ and B obtained from optical profile studies have been combined[35] with the amplitudes of the correlation length divergence (ξ_o) in a test of the two-scale-factor universality hypothesis for CO_2,Xe, and SF_6. This test is satisfied within $\pm$ 8%, the accuracy with which ξ_o is known from measurements of the angular distribution of scattered light or from the turbidity. Again the results of very different kinds of experiments carried out in different laboratories have been combined in a way which lends strong support to the phenomenology upon which RG calculations are based. In this case the numerical value calculated for a particular amplitude ratio is also confirmed.

Optical studies of the profile are limited in their closeness of approach to the critical point by a trade-off between resolution and beam bending (or thick cell) errors. A thick cell is desirable to obtain both a large optical phase difference between the rays (+)

and (-) in Figure 10 and large deflection angles. On the other hand a ray deflected from the horizontal will pass through a range of heights within the cell which is proportional to the square of the thickness of the cell. Thus a thin cell is desirable so that each ray will carry information about the compressibility at a particular height. The analysis of Moldover et al.[23] indicates that the Hocken-Moldover experiment[32] was carried as close to the critical point as possible with a 3mm thick cell - perhaps even a bit too close.

The basic measurements in the Fraunhofer technique, those of the reduced temperature and the angles of the diffraction pattern's extrema, can be made with a precision exceeding 0.1% in the region $10^{-4}<r<10^{-5}$. The correction terms to the asymptotic equation of state in this range are expected to be at most a few percent here. Thus we expect that, given a suitably parameterized analytic model for the experiment, asymptotic critical exponents and a few amplitude ratios could be determined with a precision near 1%. In these experiments the analytic model is just as important as the data.

Other recent studies of the density vs height profile have yielded results which are consistent with the picture we have presented above. Of particular interest are the studies of the coexisting densities. As examples we mention the study of SF_6 by Balzarini and Ohrn[36] and one by Weiner et al[29]. Ethane has been studied by Balzarini and Burton[39] and He^3 by Horst Meyer and his collaborators[27].

The Balzarini papers are notable because they are among the first to show clearly that the apparent value of β is range-dependent or equivalently, the correction to scaling exponent, Δ, is substantially less than 1.0. In Figures 13 and 14 we reproduce two figures from Balzarini and Burton[37]. From Figure 14 it is clear than an effective exponent $\beta \simeq 0.35$ accurately describes the data in the range $-t>10^{-2}$, the range accessible to conventional equation of state measurements. A smaller value of β near 0.34 is needed to describe the data in the range $10^{-3}>-t>10^{-4}$. The data in Figures 13 and 14 span the range $0.04<\Delta\rho/\rho_c<0.52$. The lower value of $\Delta\rho/\rho_c$ corresponds to a value of the density which would occur within 1mm of the centrus at T_c. The minimum value of $-t$ is -1.5×10^{-5}, which is just about the maximum value of $-t$ attained in the Fraunhofer profile studies mentioned above. Thus these data just end where the Fraunhofer data begin. Unfortunately, the index of refraction data which led to Figures 13 and 14 and similar data for other fluids has not been published. Data of this sort could provide a nice experimental value for Δ.

The capacitance study of He^3 by Pittman[27] et al. is of interest because it resolved a dilemma of long duration, namely that the apparent exponents β and γ measured in He^3 (and to a lesser extent

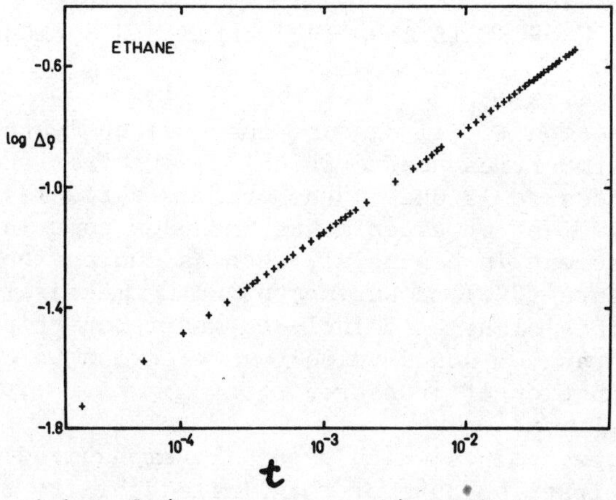

Fig. 13. Logarithm of $(\rho_{liquid}-\rho_{vapor})$ as a function of the logarithm of t.

He^4) were much further from the theoretical values than apparent exponents in other fluids. Pittman et al. find that their He^3 data are consistent with asymptotic exponents such as β=0.32 and γ=1.24 together with correction-to-scaling amplitudes that are comparable to those found in other fluids assuming Δ=0.5 in all cases. They also note that earlier data near the critical points of the helium isotopes did not extend to as small values of t as data on other fluids. It is likely that the descriptions of the earlier data included a value for T_c somewhat higher than the true value of T_c. A high value for T_c will simultaneously increase the apparent value for β and reduce the apparent value of γ, leading both exponents to be nearer to their mean field values.

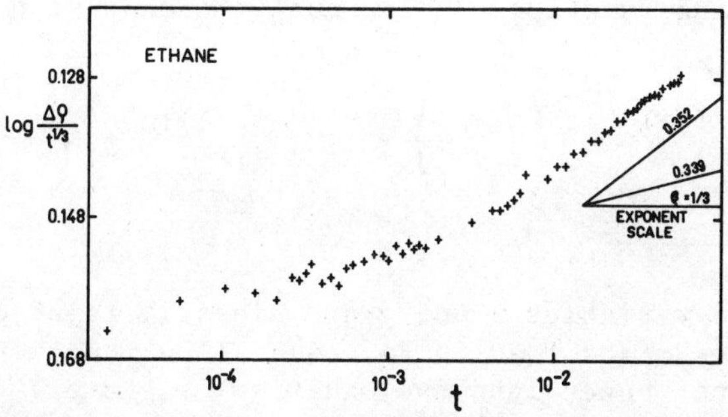

Fig. 14. Logarithm of $(\rho_{liquid}-\rho_{vapor})/(2\rho_c t^{1/3})$ as a function of t. If the asymptotic value of β were 1/3 and the experiment were perfect, the data would approach a horizontal straight line at small values of t.

V. THE EQUATION OF STATE AND OTHER EXPERIMENTS WHERE THE PRESSURE IS MEASURED

In this section I will discuss that can be learned about thermodynamic properties near critical points from those experiments in which the pressure is one of the primary variables which is measured. This class of experiments includes some in which the pressure measurement is essential, such as conventional pressure-volume-temperature (PVT) measurements and C_p measurements using flow calorimetry. This class also includes many properties (such as dielectric constant or sound velocity) which can be chosen to be studied as a function of pressure.

There are two points which should be emphasized about these measurements. First, because of the limited ability to resolve pressure changes, these measurements can tell us very little about details of asymptotic critical phenomena which are of theoretical concern today. Second, because of the great quantity of such data and the commercial importance of the systems studied, these measurements challenge us to find ways of applying the understanding we have gained about critical phenomena in the past decade.

To illustrate these points, I will describe two recent correlations of the equation of state of water and steam, one based on the Wegner expansion of the thermodynamic potential about the critical point and one which essentially ignores the existence of the critica point.

The equation of state based on the Wegner expansion can describ all the data in the small shaded region on Figure 5. We shall see this equation of state is awkward to write down and involves many parameters that are not universal. This equation was obtained by Balfour et al.[38] from a potential which was written as a sum of two singular scaling functions and an analytic expansion in two field variables:

$$\frac{P}{P_c} = |x|^{\beta(\delta+1)} \; g_0\left(\frac{y}{|x|^{\beta\delta}}\right) + \; |x|^{\beta(\delta+1)+\Delta} \; g_1\left(\frac{y}{|x|^{\beta\delta}}\right) + P_0(x,y) \tag{12}$$

Here the field variables x and y are linearly related to the usual field variables μ and T as in Eq. (4). (The function $\mu_0(T)$ in Eq. (4) does not enter into the equation of state.) The scaling functions g_0 and g_1 were chosed to reproduce the so called "linear model" equation of state. This choice of g_0 is justified by its correctness[39] to order ε^2 but not in higher order. The choice of g_1 was ad hoc. Today one might choose g_1 such that it is consistent with the amplitude ratios that have been calculated for the first subdominant scaling function[21]. In the linear model, the scaling

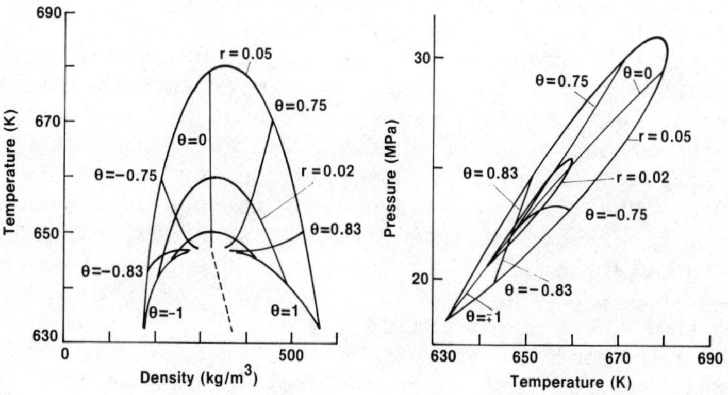

Fig. 15. The relationship between the parametric variables (r,θ) and the laboratory variables (T,ρ) and (P,T) near the critical point of steam. The "mixing of variables" parameter, q, and the parameter A_{11}, have been multiplied by 10 to exaggerate the weak singularity q produces in the θ=0 curve on the (ρ,T) plot.

function is defined in terms of two parametric variables r and θ. A map relating these variables to the laboratory ones using parameters appropriate for steam is shown in Figure 15. The actual equation of state used to fit data is:

$$P/P_c = ar^{\beta(\delta+1)}(k_0 p_0 + k_1 p_1 r^{\Delta}) + ar^{\beta\delta}\theta(1-\theta^2)(1-A_{11}t)$$

$$+ 1 - A_1 t - A_2 t^2 - A_3 t^3 \qquad (13)$$

$$\rho/\rho_c = r^{\beta}\theta(k_0 + k_1 r^{\Delta}) + qar^{1-\alpha}(k_0 s_0 + k_1 s_1 r^{\Delta}) + 1 - A_{11}t$$

$$T/T_c = t + 1 = 1 + r(1-b^2\theta^2) - qar^{\beta\delta}\theta(1-\theta^2)$$

Here p_0, p_1, s_0 and s_1 are the rational polynomials in θ that are consistent with the linear model. In this equation of state there are 11 parameters that are not expected to be universal: P_c, ρ_c, T_c, a, k_0, k_1, q, A_{11}, A_1, A_2, and A_3. In testing this equation against the steam data the critical exponents were fixed at the RG values. These 11 parameters together with b^2 could be adjusted to fit the data within their resolution in a region roughly bounded by $t>-10^{-3}$ and $r<0.07$. Reassuringly, the best value of b^2 turned out to be consistent with recent RG estimates[20] of the amplitude ratios A^+/A^-, Γ^+/Γ^- and $\Gamma^+DB^{\delta-1}$. Thus one can say: the equation of state of steam is consistent with a reasonable interpretation of the Wegner expansion for fluids. Now we consider the question: How strong a consistency check is a correlation such as this one?

In high-quality PVT measurements, the resolution of the pressure measurement is near that of a piston gauge, about 1 part in 10^5. To determine critical exponents from PVT data, the data must be differentiated, either twice with respect to temperature to examine the weak divergence of the vapor pressure or once with respect to density to obtain the strong divergence of the isothermal compressibility, k_T. If the density differentiation involves density changes of a few percent of ρ_c, the limited resolution of the pressure measurements implies that k_T can be measured with a 1% accuracy at $r=10^{-2}$ but only a 10% accuracy at $r=10^{-3}$. (Here r is the parametric variable in the equation of state, Eq. 13. Along the path $\theta=0$ we have $r \approx t$.) It follows that the compressibility exponent, γ, can be determined with an accuracy near 1% from $P(\rho,T)$ data in the region $r>10^{-2}$. Unfortunately, in this region several correction terms in the potential which vary as r^{Δ}, $r^{2\Delta}$, etc., will make substantial contributions to the apparent exponent γ. Thus it is not surprising that the same steam data are equally well described by totally ignoring the subdominant singularity (by setting $g_1=0$ in eq. 12 or $k_1=0$ in eq. (13)) and using exponents such as $\beta=0.34$ and $\gamma=1.22$ which differ from theoretical values. These apparent exponents will be determined with high precision by PVT in a region such as $10^{-2}<r<10^{-1}$. Thus Balfour et al. find $\beta = 0.3384 \pm 0.0002$ under one set of assumptions. (The small error estimate they provided is an extreme example of the misleading practice of quoting the diagonal element of an error matrix as an error estimate. This practice neglects the large correlations among the 12 parameters used.) If one is interested in measuring asymptotic exponents with a precision of say 3%, then the data after differentiation must have a precision on the order of 3% in the range $r<10^{-3}$. This is not possible when the pressure is one of the primary variables measured.

We have seen a correlation of PVT data for steam based on the critical region is consistent with the excellent data in the shaded region on Fig. 5. The equation of state is defined in terms of parametric variables and contains about 12 numerical parameters which enter nonlinearly in the equation of state and must be found by fitting data. These are serious obstacles to widespread adoption of this physically based equation; however, this equation of state is consistent with the delicate singularities in C_v and the speed of sound.

This correlation can be contrasted with the correlation of Haar et al.[19] which totally ignores the critical region and can describe almost all of the data outside of the shaded region of Fig. 5. It uses about 50 numerical parameters, each of 12 digits, which enter the fitting problem linearly. This correlation, which uses functions that are analytic in ρ and T at the critical point, fails in a characteristic way. The correlation's critical temperature is about 1°C higher than water's critical temperature. (I am told this discrepancy can be reduced to 0.1°C by adding a few more parameters.) The singularities in properties such as C_v and the speed of sound are not well represented near the critical point.

Now one of the challenges that faces anyone who would use these two correlations is that they don't quite overlap. In a region slightly below T_c near the saturated vapor pressure neither correlation is entirely satisfactory. I would expect that a similar problem would occur in a serious attempt to accurately represent all the numerical information that is available about Ising models. Does it? I hope that further theoretical advances will provide functional forms and numerical methods which are suitable for connecting the critical region to the rest of the thermodynamic surface.

VI CONCLUDING REMARKS CONCERNING MIXTURES

I have not discussed the thermodynamic data available near liquid-vapor critical lines in mixtures. Recent, detailed studies exist for the mixtures CO_2-C_2H_6 (ref. 40), C_2H_6-C_7H_{16} (ref. 41), and He^3-He^4 (ref. 42). Because the thermodynamic space of a binary mixture is a dimension higher than that of a pure fluid the density of the data is much lower and the precise specification of the experimental path with respect to the critical locus is much more difficult to obtain. Nevertheless, an excellent thermodynamic model for He^3-He^4 mixture data has been made by Leung and Griffiths. Their model does not make use of the most accurate asymptotic values of critical exponents and scaling functions which have been established in the past few years and are discussed throughout this volume. As I understand it, Leung and Griffiths do rely very heavily on a geometrical idea of the smoothness of the thermodynamic potential as a function of field variables. My naive interpretation of this idea is illustrated in Fig. 16. This figure is a sketch of a saddle shaped surface of coexisting densities near the critical line of mixtures of CO_2-C_2H_6. The densities are shown as a function of two thermodynamic field variables, T and ζ. Here we follow Leung and Griffiths in the definition:

$$\zeta = \frac{\exp(\mu_1/RT)}{\exp(\mu_1/RT) + \exp(\mu_2/RT)} \tag{14}$$

where μ_1 is the chemical potential for C_2H_6 and μ_2 is the chemical potential for CO_2. Thus ζ goes from 0 to 1 as the composition of the mixture is changed from pure CO_2 to pure C_2H_6. The smoothness idea is simply this: if ζ is a judicious choice for a mixture variable, each constant-ζ slice of the saddle in Fig. 16 (and each constant-ζ slice of the thermodynamic potential) will have the same shape as that of a pure fluid. If ζ is a poor choice, the potential of the

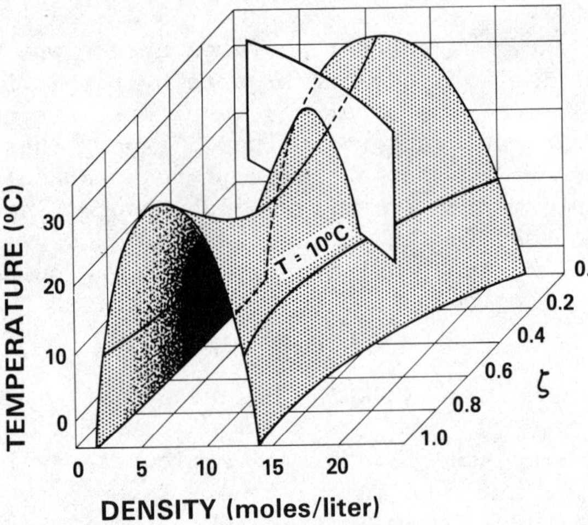

Fig. 16. The saddle shaped surface of coexisting liquid and vapor densities near the critical locus of CO_2-Ethane mixtures. The critical locus is the heavy curve at the top of the saddle. The curved sheet intersecting the saddle is a schematic representation of all of the thermodynamic states available to a mixture of fixed overall composition.

mixture will be similar to that of a pure component only asymptotically close to the critical point.

In most experiments, one works with mixtures of fixed composition. Typical experimental paths would appear in Fig. 16 as smooth curves confined to the sheet which cuts the saddle much as a curved knife would. The intersection of the saddle and the curved knife does not look at all like the inverted parabola found on constant ζ slices. Indeed, the critical point is not the highest temperature at which a mixture of a fixed composition will phase separate. This phenomenon is obvious in Fig. 16. It is known as "retrograde condensation" in engineering practice and is important in underground oil reservoirs.[44]

The idea of smoothness in field space can be applied quantitatively to the critical region mixtures which are more complicated than He^3-He^4. The first problem encountered in trying to do this is: How does one estimate the 10-12 parameters (now 10-12 smooth functions of ζ) needed to write down an equation of state such as Eq. (13) for a mixture? A partial solution to this problem would be extremely valuable.

Fig. 17 shows the results obtained by Moldover and Gallagher[45] in attempting to apply smoothness ideas to data for the liquid and vapor densities of mixtures of propane and octane - data taken from the chemical engineering literature[46]. The solid curves were obtained from a model that assumed that the critical locus is a smooth function of ζ and that all the other parameters in the potential are linear functions of ζ; thus they could be estimated by interpolation from the parameters of the pure components. Clearly this model reproduces many of the features the data exhibit; however, it seems likely more sophisticated ideas are needed. Ultimately, we need an understanding of the relationship between the nonuniversal parameters in Eq. (13) and the physics of fluids away from the critical point - parameters such as molecular core sizes, multipole moments, and virial coefficients.

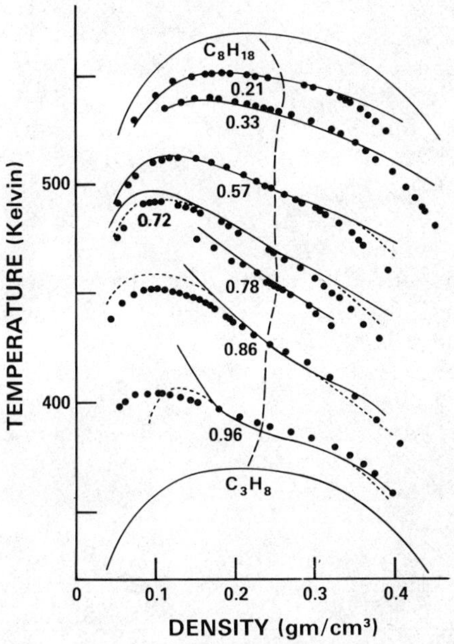

Fig. 17. Vapor liquid equlibria in propane-octane mixtures. The data of Kay et al.[46] (points) are compared with two different thermodynamic models which use the idea that the potential is a smooth function of ζ. The critical locus is near the vertical dashed curve.

ACKNOWLEDGEMENTS

I am grateful to D. A. Balzarini, J. A. Lipa and J. M. H. Levelt Sengers for use of their figures. I have profited greatly from discussions with J. V. Sengers, S. C. Greer, and especially J. F. Nicoll.

REFERENCES

1. C. Domb, in Phase Transitions and Critical Phenomena. Vol. 3, C. Domb and M. S. Green, eds. (Academic Press, New York, 1974), Ch. 6; See also B. G. Nickel, this volume.
2. G. A. Baker, B. G. Nickel, and D. I. Meiron, Phys. Rev. B17, 1365 (1978); J. C. LeGuillou and J. Zinn-Justin, Phys. Rev. B21, 3976 (1980).
3. J. J. Rehr and N. D. Mermin, Phys. Rev. A8, 472 (1973).
4. F. J. Wegner, Phys. Rev. B5, 4529 (1972); F. J. Wegner, Phys. Rev. B6, 1891 (1972).
5. M. Ley-Koo and M. S. Green, Phys. Rev. A16, 2483 (1977).
6. C. N. Yang and C. P. Yang, Phys. Rev. Lett. 13, 303 (1964).
7. G. R. Brown and H. Meyer, Phys. Rev. A6, 364 (1972).
8. M. R. Moldover, Phys. Rev. 182, 342 (1969).
9. A. V. Voronel, V. G. Gorbunova, V. A. Smirnov, N. G. Shmakov, and V. V. Shchekochikhina, Zh. Eksp. Teor. Fiz. 63, 964 (1972) [Eng. Transl. Sov. Phys. JETP 36, 505 (1973)]; M. A. Anisimov, A. T. Berestov, L. S. Veksler, B. A. Kovalchuk, and V. A. Smirnov, Zh. Eksp. Teor. Fiz. 66, 742 (1974) [Eng. Transl. Sov. Phys. JETP 39, 359 (1974)].
10. B. Widom, J. Chem. Phys. 43, 3898 (1965).
11. J. F. Nicoll, T. S. Chang, A. Hankey, and H. E. Stanley, Phys. Rev. B11, 1176 (1975).
12. R. B. Griffiths and J. C. Wheeler, Phys. Rev. A2, 1047 (1970).
13. R. B. Griffiths, J. Chem. Phys. 43, 1958 (1965).
14. H. A. Kierstead, Phys. Rev. A3, 329 (1971).
15. R. P. Behringer, T. Doiron, H. Meyer, J. Low Temp. Phys. 24, 315 (1976).
16. J. A. Lipa, C. Edwards, and M. J. Buckingham, Phys. Rev. A15, 778 (1977).
17. J. M. H. Levelt Sengers and W. T. Chen, J. Chem. Phys. 56, 595 (1972).
18. J. M. H. Levelt Sengers, Physica 73, 73 (1974).
19. L. Haar, J. Gallagher, and G. S. Kell in Water and Steam: Proc. 9th Intl. Conf. on Properties of Steam, J. Straub and K. Scheffler, eds. (Pergamon Press, New York, 1980) p. 69.
20. A. Aharony and P. C. Hohenberg, Phys. Rev. B13, 3081 (1976).
21. M. C. Chang and A. Houghton, Phys. Rev. B21, 1881 (1980).
22. A. V. Voronel in Phase Transitions and Critical Phenomena, Vol. 5B, C. Domb and M. S. Green, eds. (Academic Press, New York, 1976) p. 343-94.

23. M. R. Moldover, J. V. Sengers, R. W. Gammon, and R. J. Hocken, Rev. Mod. Physics 51, 79 (1979).
24. L. M. Artyukhovskaya, E. T. Shimanskaya, and Yu I. Shimanskii, Zh. Eksp. Teor. Fiz. 63 2159 (1972) [Eng. Transl: Sov. Physics JETP 36, 1140 (1973).
25. E. H. W. Schmidt in Critical Phenomena, M. S. Green and J. V. Sengers, eds. National Bureau of Standards Miscellaneous Publication 273 (U.S. Govt. Printing Office, Washington, D.C., 1966) p. 13.
26. D. A. Balzarini, Canadian J. Phys. 50, 2194 (1972).
27. C. Pittman, T. Doiron, H. Meyer, Phys. Rev. B20, 3678 (1979).
28. L. A. Weber, Phys. Rev. A2, 2379 (1970).
29. J. Weiner, K. H. Langley, N. C. Ford, Phys. Rev. Lett. 32, 879 (1974); J. Weiner, Ph.D. Thesis, Univ. of Massachusetts, 1974.
30. J. V. Sengers, D. Bedeaux, P. Mazur, and S. C. Greer, Physica, 104A, 573 (1980).
31. L. R. Wilcox and D. Balzarini, J. Chem. Phys. 48, 753 (1968); W. T. Estler, R. Hocken, T. Charlton, and L. R. Wilcox, Phys. Rev. A12, 2118 (1975).
32. R. Hocken and M. R. Moldover, Phys. Rev. Lett. 37, 29 (1976).
33. L. R. Wilcox and W. T. Estler, J. Phys. (Paris), Colloq. 32 C5A-175 (1971).
34. P. Schofield, Phys. Rev. Lett. 22, 606 (1969).
35. J. V. Sengers and M. R. Moldover, Phys. Lett. 66A, 44 (1978); see also J. V. Sengers in this volume.
36. D. Balzarini and K. Ohrn, Phys. Rev. Lett. 29, 840 (1972).
37. D. Balzarini and M. Burton, Canadian J. Phys. 57, 1516 (1979).
38. F. W. Balfour, J. V. Sengers, M. R. Moldover, and J. M. H. Levelt Sengers in Proc. 7th Symposium on Thermophysical Properties, A. Cezairliyan, ed. (American Society of Mechanical Engineers, New York, 1977) p. 786.
39. D. J. Wallace in Phase Transitions and Critical Phenomena, Vol. 6, C. Domb and M. S. Green, eds., (Academic Press, New York, 1976)
40. A. V. Voronel, V. G. Gorbunova, and N. G. Shmakov, Zh. Eksp. Teor. Fiz. Pis'ma Red. 9, 333 (1969) [Eng. Transl. Sov. Phys. JETP Lett. 9, 195 (1969)].
42. T. Doiron, R. P. Behringer, and H. Meyer, J. Low Temp. Phys. 24, 345 (1976); G. R. Brown and H. Meyer, Phys. Rev. A6, 1578 (1972).
43. S. S. Leung and R. B. Griffiths, Phys. Rev. A8, 2670 (1973); see also ref. 42.
44. K. E. Bett, J. S. Rowlinson, and G. Saville, Thermodynamics for Chemical Engineers (MIT Press, Cambridge, Mass., USA, 1975) p. 371.
45. M. R. Moldover and J. S. Gallagher, AIChE Journal 24, 267 (1978).
46. W. B. Kay, J. Genco, and D. A. Fichtner, J. Chem. Eng. Data 19, 275 (1974).

47. J. M. H. Levelt Sengers, W. L. Greer, and J. V. Sengers, J. Phys. Chem. Ref. Data, 5, 1 (1976).

UNIVERSALITY OF CRITICAL PHENOMENA IN CLASSICAL FLUIDS

J. V. Sengers

Institute for Physical Science and Technology
University of Maryland
College Park, MD 20742, U.S.A.

OUTLINE

I. INTRODUCTION

To characterize the asymptotic behavior of the physical properties of systems near a critical point, critical-point phase transitions are grouped into universality classes as first suggested by Kadanoff (1971). Systems with critical-point phase transitions assigned to the same universality class have identical critical exponents and identical scaling functions. Universality classes are distinguished by such general features as the dimensionality of the system, the number of components of the order parameter and whether the intermolecular forces are short- or long-range. Universality classes can be further distinguished with respect to the critical behavior of static and dynamic properties.

In these lectures we shall be concerned with fluids near the gas-liquid critical point and with binary liquids near the critical point of mixing. It is commonly believed that fluids near a critical point belong to the same static universality class as a uniaxial ferromagnet represented by the three-dimensional Ising model (Domb, 1974) or the Landau-Ginzburg-Wilson model with a one-component order parameter (Wilson and Kogut, 1974; Pfeuty and Toulouse, 1977; Patashinskii and Pokrovskii, 1979). Furthermore, fluids near the gas-liquid critical point and binary liquids near the critical mixing point are expected to belong to the same dynamic universality class for which theoretical models have been developed (Hohenberg and Halperin, 1977).

It is the purpose of these lectures to assess experimental evidence for the asymptotic behavior of static and dynamic properties of fluids near a critical point. I shall be guided primarily by experimental results with which I am most familiar, but a comparison with other experimental results will be included where appropriate. However, I do not claim to give a comprehensive review of the subject here.

The lectures are organized as follows: First we review the theoretical predictions for the asymptotic behavior of the thermodynamic properties of fluids in Section II. We then consider the experimental results for the thermodynamic critical exponents and amplitude ratios in Section III and the corresponding quantities for the correlation function in Section IV. The asymptotic critical behavior of dynamic properties is discussed in Section V. The main conclusions are summarized in Section VI.

II. THEORETICAL PREDICTIONS FOR THE THERMODYNAMIC BEHAVIOR OF FLUIDS NEAR THE CRITICAL POINT

2.1 Critical-Point Analogy with Uniaxial Ferromagnets

The oldest example of a critical point is the one associated with the gas-liquid phase transition in a fluid (Rowlinson, 1969). Below the critical temperature T_c, it is possible for the fluid to be in a state with two coexisting phases with different densities, a vapor phase and a liquid phase. In Fig. 1a we indicate schematically the densities of the vapor and liquid phases as a function of temperature. The shaded region corresponds to the region of two-

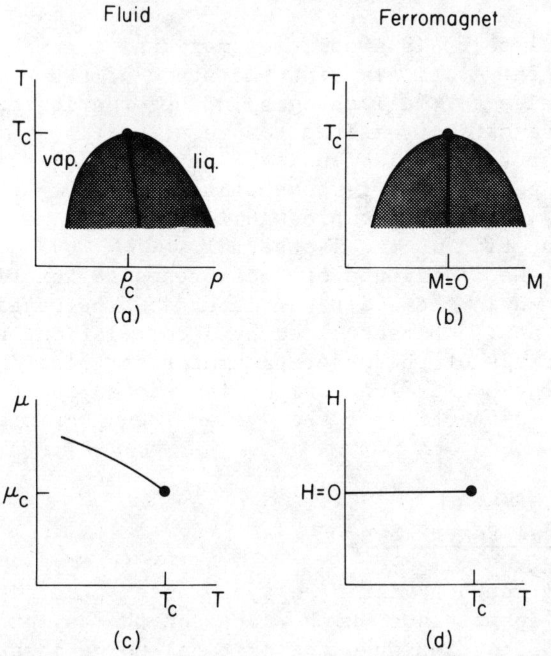

Fig. 1 Region of vapor-liquid equilibrium in a fluid near the critical point (Figs. 1a and 1c) and region of spontaneous magnetization in a ferromagnet near the Curie point (Figs. 1b and 1d).

phase equilibrium. In a ferromagnetic system, spontaneous magnetization develops when the temperature is lowered to a value below the Curie temperature T_c. The region of spontaneous magnetization in zero magnetic field is schematically indicated in Fig. 1b; the direction of this magnetization will depend on the direction of the magnetic field before it was turned off. With the phase transition one associates an order parameter s. In the ferromagnetic system the order parameter is the magnetization M; in the gas-liquid system the order parameter is the difference between the actual density ρ and the critical density ρ_c. The intensive thermodynamic variable conjugate to the order parameter s is the ordering field h. For the ferromagnetic system it is the magnetic field H; for the gas-liquid system the ordering field h at $T = T_c$ may be identified with $\mu - \mu_c$, where μ is the actual chemical potential and μ_c the chemical potential at the critical point. A more general relationship between the ordering field h and the physical variables will be discussed in Section 2.3. In terms of the intensive variables, the region of spontaneous magnetization corresponds to the line $H = 0$ for $T \leq T_c$ as indicated in Fig. 1d. For the gas-liquid system it is the curve of the saturation chemical potential as a function of temperature as indicated in Fig. 1c.

The critical point is a point of marginal thermodynamic stability and near the critical point the state of the system becomes extremely sensitive to small changes in the ordering field h. A measure of the sensitivity is the response function $\chi_T = (\partial s/\partial h)_T$ which diverges at the critical point. For a fluid near the gas-liquid critical point, this response function is $\chi_T = (\partial \rho/\partial \mu)_T = \rho^2 K_T$, where K_T is the isothermal compressibility. For a magnetic system, this response function is the isothermal susceptibility $\chi_T = (\partial M/\partial H)_T$. Associated with the divergence of these response functions are large-scale fluctuations in the order parameter s. The spatial extent of these fluctuations is characterized by a correlation length ξ which represents the range of the order-parameter correlation function G(r). These large-scale fluctuations lead to an anomalous behavior of many physical properties of the system when a critical point is approached.

2.2 Critical Power Laws

To describe the behavior of a system near a critical point one defines critical power laws which represent the asymptotic behavior of a number of properties when the critical point is approached along specific thermodynamic paths. In terms of the variables $\Delta T = T - T_c$ and $\Delta\rho = \rho - \rho_c$ or M, these paths are schematically indicated in Fig. 2; they are the coexistence curve or phase boundary (CXC), the critical isotherm $\Delta T = 0$ (CIT) and the line $\Delta\rho = 0$ or $M = 0$ in the one-phase region, referred to as critical isochore (CIC). The critical power laws defined along these paths are listed in Table I.

In this table we also include the definition of the exponent ν which determines the divergence of the correlation length and the exponent η which characterizes the long-range behavior of the order-parameter correlation function G(r) at the critical point. The critical exponents satisfy the scaling relations (Fisher, 1967)

$$2 - \alpha = \beta(\delta + 1) \quad , \tag{2.1}$$

$$\gamma = \beta(\delta - 1) \quad , \tag{2.2}$$

$$\gamma = (2 - \eta)\nu \quad . \tag{2.3}$$

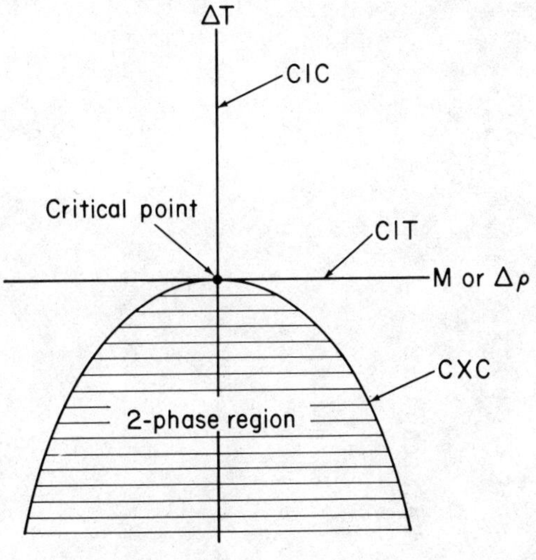

Fig. 2 Paths in the plane ΔT versus Δρ or M for the definition of critical power laws.

Table I. Critical power laws

Property	Fluid	Magnet	Path	Power law
Heat capacity	C_v/V	C_H	CIC, $\Delta T \geq 0$	$(A^+/\alpha)\lvert\Delta T\rvert^{-\alpha}$
Heat capacity	C_v/V	C_H	CIC, $\Delta T \leq 0$	$(A^-/\alpha)\lvert\Delta T\rvert^{-\alpha}$
Order parameter s	$\Delta\rho$	M	CXC	$\pm B\lvert\Delta T\rvert^{\beta}$
Response function	$(\partial\rho/\partial\mu)_T$	$(\partial M/\partial H)_T$	CIC, $\Delta T \geq 0$	$\Gamma^+\lvert\Delta T\rvert^{-\gamma}$
Response function	$(\partial\rho/\partial\mu)_T$	$(\partial M/\partial H)_T$	CXC	$\Gamma^-\lvert\Delta T\rvert^{-\gamma}$
Ordering field	$\mu-\mu_c$	H	CIT	$\pm D\lvert s\rvert^{\delta}$
Correlation length	ξ	ξ	CIC, $\Delta T \geq 0$	$\xi_o\lvert\Delta T\rvert^{-\nu}$
Correlation function	G(r)	G(r)	Critical point	$r^{-(1+\eta)}$

The values predicted theoretically for these critical exponents are summarized in Table II. In this table we also quote the value of the first correction-to-scaling exponent Δ_1 to be discussed below. The theoretical exponent values have been generally obtained in two ways. In the first column of Table II we give the values quoted for the three-dimensional Ising model as deduced from high-temperature series estimates (Fisher, 1967; Moore et al. 1969; Domb, 1974; Camp et al., 1976) and in the second column the values estimated for the Landau-Ginzburg-Wilson model using renormalization methods originally developed in quantum field theory (Baker et al., 1976, 1978; Le Guillou and Zinn-Justin, 1977, 1980).

The two methods have yielded similar exponent values, but some unresolved discrepancies do exist. In particular, the critical exponents that follow from the renormalization-group theory of critical phenomena satisfy the hyperscaling relation (Fisher, 1967)

$$d\nu = 2 - \alpha \quad , \tag{2.4}$$

Table II. Theoretical critical exponent values

Ising model[a]	Landau-Ginzburg-Wilson model[b]
$\alpha = 0.125 \pm 0.020$	$\alpha = 0.110 \pm 0.005$
$\beta = 0.312 \pm$ (02	$\beta = 0.325 \pm 0.002$
$\gamma = 1.250 \pm 0.003$	$\gamma = 1.241 \pm 0.002$
$\Delta_1 = 0.50 \pm 0.08$	$\Delta_1 = 0.50 \pm 0.02$
$\nu = 0.638 \pm 0.002$	$\nu = 0.630 \pm 0.002$
$\eta = 0.041 \pm 0.006$	$\eta = 0.031 \pm 0.004$

[a] Estimates from high-temperature series

[b] Estimates from renormalization-group methods

where d(=3) is the dimensionality of the system. On the other hand, the validity of the hyperscaling relationship for the 3-dimensional Ising model has been questioned by Baker (1977; this volume) and by Baker and Kincaid (1980). It has been suggested that this breakdown of hyperscaling is reflected in the differences between the exponent values obtained from high-temperature series and those obtained by the use of renormalization methods. However, it is not clear whether the numerical differences in the quoted exponent values are truly significant (Rehr, 1979, McKenzie, 1979; Oitmaa and Ho-Ting-Hung, 1979; Nickel and Sharpe, 1979, Zinn-Justin, 1979; Sheludyak and Rabinovich, 1979). For a more detailed discussion the reader is referred to the lecture notes of Gaunt and of Nickel in this volume.

2.3 Confluent Singularities

2.3.1 Wegner expansion

The power laws introduced above represent the behavior of a system asymptotically close to the critical point. In order to describe the properties of a system in a finite range around the critical point, one needs an estimate of the corrections to this asymptotic behavior. For a spin system represented by a Landau-Ginzburg-Hamiltonian, it was shown by Wegner (1972a) that the Gibbs free energy F (more precisely F/k_BT, where k_B is Boltzmann's constant) can be represented by an expansion of the form

$$F = F_o + |\tau|^{2-\alpha}\left[f_o\left(\frac{h}{|\tau|^{\beta\delta}}\right) + \sum_{i=1} c_i|\tau|^{\Delta_i} f_i\left(\frac{h}{|\tau|^{\beta\delta}}\right) + \frac{1}{2}\sum_{i=1}\sum_{j=1}|\tau|^{\Delta_i+\Delta_j} f_{ij}\left(\frac{h}{|\tau|^{\beta\delta}}\right) + \dots\right] . \tag{2.5}$$

Here τ and h are the two relevant scaling fields and c_i are irrelevant scaling fields. These scaling fields are analytic functions of the physical fields, temperature and magnetic field for a magnetic system (more precisely: T_c/T and H/k_BT). The relevant scaling fields τ and h vanish at the critical point; the irrelevant scaling fields c_i approach a finite value at the critical point. The function F_o is an analytic background function. The critical exponents β,δ, the correction-to-scaling exponents Δ_i and the scaling functions f_i are universal. Hence, apart from F_o, system-dependent quantities only appear in the relationship between the scaling fields and the physical fields.

In many cases, such as that of a ferromagnet, it is sufficient to restrict oneself to models that are symmetric in the order parameter. In that case all scaling functions $f_i(x)$ in (2.5) are even functions while odd scaling functions have vanishing coefficients; furthermore, the scaling field h reduces to H/k_BT, while F_o, τ and c_i depend on the temperature only (Wegner, 1972a). The first correction-to-scaling exponent $\Delta_1 = 0.5$ as quoted in Table II, while in the symmetric case the second correction-to-scaling exponent Δ_2 is approximately 0.9 (Rehr, 1979).

2.3.2 Application to fluids near the gas-liquid critical point

The analogy between critical phenomena in ferromagnets and fluids is illustrated theoretically by the analogy between Ising model and lattice gas (Lee and Yang, 1952; Fisher, 1967; Sengers and Levelt Sengers, 1978). The model can be generalized to decorated lattice-gas models that incorporate deviations from perfect gas-liquid symmetry (Mermin, 1971; Zollweg and Mulholland, 1972); these models were recently reviewed by Wheeler (1977).

A systematic procedure for applying the theoretical predictions of the Landau-Ginzburg-Wilson model to the behavior of fluids near the gas-liquid critical point was proposed by Ley-Koo and Green (Ley-Koo, 1976; Ley-Koo and Green, 1977). The procedure combines a revised scaling hypothesis, inspired by the decorated lattice gas models proposed by Rehr and Mermin (1973), with the "correction-to-scaling" or "extended scaling" terms of the Wegner expansion.

To apply the Wegner expansion (2.5) to fluids, the following observations are made. First, for fluids one needs to consider the full Wegner expansion including the terms generated by odd terms in the Landau-Ginzburg-Wilson-Hamiltonian. Thus scaling functions $f_i(x)$ in (2.5) that are odd must be retained, although they are not present in symmetric models such as the Ising model. Furthermore, the scaling functions, as well as the background function F_o, should now be allowed to depend on both the temperature and the variable analogous to the magnetic field. Finally, and most importantly, it is assumed that for a fluid near the gas-liquid critical point the pressure P is the potential analogous to the Gibbs free energy F of a spin system.

The pressure P as a function of its characteristic variable satisfies the differential relation

$$dP = sdT + \rho d\mu \quad , \tag{2.6}$$

where s is the entropy per unit volume. Strictly speaking the potential in (2.5) is the dimensionless free energy F/k_BT. Therefore, in our more recent work, we have used the modified variables (Balfour et al., 1978; Levelt Sengers and Sengers, 1980)

$$\tilde{P} = \frac{P}{T}, \quad \tilde{\mu} = \frac{\tilde{\mu}}{T}, \quad \tilde{T} = -\frac{1}{T} \quad , \tag{2.7}$$

with

$$d\tilde{P} = ud\tilde{T} + \rho d\tilde{\mu} \quad , \tag{2.8}$$

where u is the energy per unit volume.

Ley-Koo and Green postulate that this potential satisfies a Wegner expansion

$$\tilde{P} = \tilde{P}_o(\tilde{\mu},\tilde{T}) + |\tau|^{2-\alpha} f_o\left(\frac{h}{|\tau|^{\beta\delta}}\right) + c_1|\tau|^{\Delta_1} f_1\left(\frac{h}{|\tau|^{\beta\delta}}\right) + c_1'|\tau|^{\Delta_1'} f_1'\left(\frac{h}{|\tau|^{\beta\delta}}\right) + \ldots \quad , \tag{2.9}$$

where the term proportional to $|\tau|^{\Delta_1}$ is the first symmetric correction term discussed earlier and the term proportional to $|\tau|^{\Delta_1'}$ the first asymmetric correction term with an asymmetric scaling function $f_1'(x)$. In their original analysis Ley-Koo and Green assumed $\Delta_1' \simeq 1.5$ as estimated theoretically when calculated up to first order in $\varepsilon = 4-d$ (Wegner, 1972b). This exponent was recently calculated by Nicoll and Zia (1980) up to order ε^2; they find that the convergence of the expansion is poor and that Δ_1' could be anywhere between 0.5 and 1.5. In our analysis of experimental equation of state data we have assumed Δ_1' to be of order unity so that $\Delta_1' > \Delta_1$. Hence, we neglect terms of order $|\tau|^{\Delta_1'}$ as compared to terms of order $|\tau|^{\Delta_1}$. We do not know whether Δ_1' is larger or smaller than $2\Delta_1$ and Δ_2, exponents generated by the symmetric terms in the Hamiltonian as discussed in Section 2.3.1.

The background function $\tilde{P}_o(\tilde{\mu},\tilde{T})$ and the scaling fields τ, h, c_1, c_1' are analytic functions of $\tilde{\mu}$ and $\tilde{T}$ which can be expanded in a Taylor series around the critical point.† In first approximation (Levelt Sengers and Sengers, 1980)

$$\tilde{P}_0(\tilde{\mu},\tilde{T}) = \sum_{i=0} \tilde{P}_j(\Delta\tilde{T})^j + \rho_c\Delta\tilde{\mu} + \tilde{P}_{11}\Delta\tilde{\mu}\Delta\tilde{T} \quad , \tag{2.10}$$

$$\tau = c_{00}\Delta\tilde{T} + c_{01}\Delta\tilde{\mu} \quad , \tag{2.11}$$

$$h = c_{10}\Delta\tilde{T} + c_{11}\Delta\tilde{\mu} \quad , \tag{2.12}$$

† The assumption that the scaling fields are analytic functions of the physical field has been questioned by Mistura (1979). However, any deviations from analyticity affect terms of the expansions at higher order than those considered here.

with

$$\Delta\tilde{T} = \tilde{T} - \tilde{T}_c \quad , \quad \Delta\tilde{\mu} = \tilde{\mu} - \tilde{\mu}_c \quad . \tag{2.13}$$

Here we have retained only terms linear in the field $\Delta\tilde{\mu}$, while c_1 and c_1' in first approximation can be identified with their limiting values at the critical point. Since $\rho = (\partial\tilde{P}/\partial\tilde{\mu})_{\tilde{T}}$, the term proportional to $\Delta\tilde{\mu}$ accounts for the fact that ρ_c is finite, while the term proportional to $\Delta\tilde{\mu}\Delta\tilde{T}$ yields a linear contribution to the slope of the coexistence curve diameter as a function of temperature. At the phase boundary $h = 0$ in this approximation (Ley-Koo, 1976) and the ratio c_{10}/c_{11} is to be identified with the limiting slope $-d\tilde{\mu}/d\tilde{T}$ of the saturation chemical potential $\tilde{\mu}$ as a function of $\tilde{T}$:

$$h = c_{11}\left[\tilde{\mu} - \left(\tilde{\mu}_c + \frac{d\tilde{\mu}}{d\tilde{T}}\Delta\tilde{T}\right)\right] \quad . \tag{2.14}$$

This expression is the first approximation to the field $h = \tilde{\mu} - \tilde{\mu}_o(\tilde{T})$, where $\tilde{\mu}_o(\tilde{T})$ is an analytic function of temperature; this is the field originally proposed by Widom (1965) and retained in all subsequent formulations of the scaling laws (Sengers and Levelt Sengers, 1980). However, a new feature appears, equivalent to the revised scaling hypothesis of Rehr and Mermin (1973), as a result of a term proportional to $\Delta\tilde{\mu}$ in the temperature-like scaling field τ.

Recently, Nicoll (1980) made a detailed analysis of the consequences of the presence of cubic and quintic terms when added to the conventional Landau-Ginzburg-Wilson Hamiltonian for a spin system. This result supports the procedure of Ley-Koo and Green to the order considered. Specifically, the cubic and quintic terms considered together, cause linear mixing of the scaling fields, lead to a term in the background function linear in the field h and generate the term proportional to $|\tau|^{\Delta_1'}$ in (2.9).

The Wegner expansion (2.9) with the revised scaling fields leads to expansions generalizing the power laws introduced in Table I. An extensive survey of these expansions can be found in the thesis of Ley-Koo (1976). As an example, we quote here the first few terms of the expansions for the internal energy density u along the critical isochore above the critical temperature and for the densities ρ_+ and ρ_- along the two branches of the phase boundary below the critical temperature.

$$u = u_c\left[1 + u_o|\Delta\tilde{T}^*|^{1-\bar{\alpha}} + u_1\Delta\tilde{T}^* + u_2|\Delta\tilde{T}^*|^{1-\alpha+\Delta_1} + \ldots\right] \quad , \tag{2.15}$$

$$\frac{\rho_+ - \rho_-}{2} = \rho_c\left[B_o|\Delta\tilde{T}^*|^{\beta} + B_1|\Delta\tilde{T}^*|^{\beta+\Delta_1} + \ldots\right] \quad , \tag{2.16}$$

$$\frac{\rho_+ + \rho_-}{2} = \rho_c \left[1 + D_0 |\Delta\tilde{T}^*|^{1-\alpha} + D_1 \Delta\tilde{T}^* + D_2 |\Delta\tilde{T}^*|^{1-\alpha+\Delta_1} + \right. \\ \left. + D_3 |\Delta\tilde{T}^*|^{\beta+\Delta_1'} + \ldots \quad , \right] \tag{2.17}$$

where we have introduced the reduced variable

$$\Delta\tilde{T}^* = \tilde{T} + 1 = (T - T_c)/T \quad . \tag{2.18}$$

The coefficients u_i, B_i, D_i, can be readily related to the coefficients in the expansions (2.10), (2.11), (2.12). The term $D_0|\Delta\tilde{T}^*|^{1-\alpha}$ indicates a weak divergence of the slope of the coexistence curve diameter predicted by various authors (Widom and Rowlinson, 1970; Green et al., 1971; Mermin and Rehr, 1971). The term $D_2|\Delta\tilde{T}^*|^{\beta+\Delta_1'}$ represents the first correction term to the coexistence curve diameter due to asymmetric terms in the Hamiltonian. The work of Ley-Koo and Green (Ley-Koo, 1976) implies that this term is present *in addition* to the revised scaling terms proportional to $|\Delta\tilde{T}^*|^{1-\alpha}$ and $|\Delta\tilde{T}^*|^{1-\alpha+\Delta_1}$ in contrast to a suggestion of Vause and Sak (1980). However, Ley-Koo (1976) assumed $\Delta_1' = 1.5$ so that the term $|\Delta\tilde{T}^*|^{\beta+\Delta_1'}$ was thought to be of higher order than the term $D_1|\Delta\tilde{T}^*|^{1-\alpha+\Delta_1}$. As discussed earlier, the value of the exponent Δ_1' is not known accurately.

The expansions implied by the Wegner expansion with revised scaling fields are quite complicated and contain many terms that are not well separated. For instance, the expansion (2.16) for the coexistence curve contains terms proportional to $|\Delta\tilde{T}|^{\beta+2\Delta_1}$, $|\Delta\tilde{T}|^{\beta+1}$ and $|\Delta\tilde{T}|^{\beta+\Delta_2}$. Furthermore, there is considerable experimental (Levelt Sengers and Sengers, 1980) and theoretical (Nicoll, 1980) evidence that the so-called analytic corrections to the leading $|\Delta\tilde{T}|^{1-\alpha}$ behavior ($u_1\Delta\tilde{T}^*$ in (2.15) and $D_1\Delta\tilde{T}^*$ in (2.17)) contain fluctuation-induced contributions and cannot be identified with a "background" behavior deduced from data outside the critical region.

Most experimentalists have analyzed their data in terms of the more commonly used variable

$$\Delta T^* = (T - T_c)/T_c \quad , \tag{2.19}$$

rather than the modified variable $\Delta\tilde{T}^*$ defined in (2.18). The difference between ΔT^* and $\Delta\tilde{T}^*$ can be neglected in the expansions up to the order considered here.

2.3.3 Application to binary liquids near the critical mixing point

For a binary mixture the differential relation (2.8) for the potential $\tilde{P}$ is generalized to

$$d\tilde{P} = ud\tilde{T} + \rho_1 d\tilde{\mu}_1 + \rho_2 d\tilde{\mu}_2 \quad , \qquad (2.20)$$

with $\tilde{\mu}_i = \mu_i/T$. Here ρ_i and μ_i are the density and chemical potential of component i in the mixture. The potential $\tilde{P}$ is the appropriate potential to generalize the behavior of a one-component fluid point to the behavior of a binary mixture near the gas-liquid critical point (Saam, 1970; Leung and Griffiths, 1973).

There exists some uncertainty as to the most appropriate choice of the dependent potential to characterize the behavior of a binary liquid near the critical mixing point: it is commonly taken to be $\tilde{\mu}_1$ or $\tilde{\mu}_2$ (Griffiths and Wheeler, 1970; Scott, 1978; Sengers and Levelt Sengers, 1978). Dividing by the total density $\rho = \rho_1 + \rho_2$ of the mixture, we rewrite (2.20) as

$$d(-\tilde{\mu}_2) = Ud\tilde{T} + Xd\tilde{\Delta} - \rho^{-1}d\tilde{P} \quad , \qquad (2.21)$$

where $U = \rho^{-1}u$ is the energy per unit mass, X the concentration of component 1 and $\tilde{\Delta} = \tilde{\mu}_1 - \tilde{\mu}_2$. The procedure of Ley-Koo and Green can be generalized to a binary liquid near the critical mixing point by postulating that the potential $\tilde{\mu}_2$ satisfies the Wegner expansion (2.9) with background function and scaling fields that now depend analytically on the physical fields $\tilde{T}$, $\tilde{\Delta}$ and $\tilde{P}$ (Sengers et al., 1980).

A simple analogy with the behavior of a one-component fluid near the gas-liquid critical point is obtained if we consider experiments in binary liquids at constant $\tilde{P}$ so that $d(-\mu_2) = Ud\tilde{T} + Xd\tilde{\Delta}$. On comparing with (2.8) and (2.9) we note a one-to-one correspondence between the potential $\tilde{P}$ as a function of $\tilde{T}$ and $\tilde{\mu}$ for the one-component fluid and the potential $-\tilde{\mu}_2$ as a function of $\tilde{T}$ and $\tilde{\Delta}$ for the binary liquid. Table III gives a list of several thermodynamic properties that exhibit analogous critical behavior (Sengers and Levelt Sengers, 1978). In particular, the energy U per unit mass will satisfy the expansion (2.15) at the critical isochore $X = X_c$ with u_c replaced by U_c and the concentrations X_+ and X_- along the two branches of the phase boundary will satisfy the expansions (2.16) and (2.17) with ρ_c replaced by X_c. Experiments in binary liquids are often not performed at constant P or $\tilde{P}$, but at the vapor pressure of the binary liquid. Nevertheless, the changes in pressure are sufficiently small for the analogy to remain valid in practice (Greer, 1978).

Table III. Analogy between critical behavior of one-component fluids near the gas-liquid critical point and binary liquids near the critical mixing point

	One-component fluid near gas-liquid critical point	Binary liquid near critical mixing point (constant $\tilde{P}$)
Fundamental equation	$\tilde{P}(\tilde{T},\tilde{\mu})$	$\tilde{\mu}_2(\tilde{T},\tilde{\Delta})$
Ordering field	$\tilde{\mu}-\tilde{\mu}_c$	$\tilde{\Delta}-\tilde{\Delta}_c$
Order parameter	$\rho=(\partial\tilde{T}/\partial\tilde{\mu})_{\tilde{T}}$	$X=-(\partial\tilde{\mu}_2/\partial\tilde{\Delta})_{\tilde{T}}$
Coexistence curve	$\rho_{\pm}$	$X_{\pm}$
Equation of state	$\tilde{\mu}(\rho,\tilde{T})$	$\tilde{\Delta}(X,\tilde{T})$
Response function	$(\partial\rho/\partial\tilde{\mu})_{\tilde{T}}=(\partial^2\tilde{P}/\partial\tilde{\mu}^2)_{\tilde{T}}$	$(\partial X/\partial\tilde{\Delta})_{\tilde{T}}=-(\partial^2\tilde{\mu}_2/\partial\tilde{\Delta}^2)_{\tilde{T}}$
Specific heat (weakly divergent)	C_v per unit volume	$c_{\tilde{P},X}$ per unit mass

The binary liquid near the critical mixing point will belong to the same universality class independent of the actual value of $\tilde{P}$. Hence, only the constants in the relationship between the scaling fields (and background function) and the physical fields $\tilde{T}$ and $\tilde{\Delta}$ will depend (analytically) on $\tilde{P}$. In particular, the expressions (2.11) and (2.12) for the relevant scaling fields may now be written as (Sengers et al., 1980)

$$\tau = c_{00}(\tilde{P})\{\tilde{T} - \tilde{T}_c(\tilde{P})\} + c_{01}(\tilde{P})\{\tilde{\Delta} - \tilde{\Delta}_c(\tilde{P})\} \quad , \qquad (2.22)$$

$$h = c_{10}(\tilde{P})\{\tilde{T} - \tilde{T}_c(\tilde{P})\} + c_{11}(\tilde{P})\{\tilde{\Delta} - \tilde{\Delta}_c(\tilde{P})\} \quad , \qquad (2.23)$$

where the coefficients c_{ij} and the critical parameters $\tilde{T}_c$ and $\tilde{\Delta}_c$ depend on $\tilde{P}$. It is convenient to consider $\tilde{\mu}_2$ as a function of τ, h, $\tilde{P}$ rather than as a function of T, $\tilde{\Delta}$, $\tilde{P}$. From (2.21) it follows that the specific volume $V = \rho^{-1}$ is given by

$$V = \rho^{-1} = \left(\frac{\partial\tilde{\mu}_2}{\partial\tilde{P}}\right)_{\tau,h} + U\left(\frac{\partial\tilde{T}}{\partial\tilde{P}}\right)_{\tau,h} + X\left(\frac{\partial\tilde{\Delta}}{\partial\tilde{P}}\right)_{\tau,h} \quad . \qquad (2.24)$$

The coefficients $(\partial\tilde{T}/\partial\tilde{P})_{\tau,h}$ and $(\partial\tilde{\Delta}/\partial\tilde{P})_{\tau,h}$ are analytic functions of the field variables which may be obtained by inverting (2.22) and (2.23). In first approximation

$$\lim_{\substack{\tau\to 0\\ h\to 0}}\left(\frac{\partial\tilde{T}}{\partial\tilde{P}}\right)_{\tau,h} = \frac{d\tilde{T}_c}{d\tilde{P}}, \quad \lim_{\substack{\tau\to 0\\ h\to 0}}\left(\frac{\partial\tilde{\Delta}}{\partial\tilde{P}}\right)_{\tau,h} = \frac{d\tilde{\Delta}_c}{d\tilde{P}} \quad . \qquad (2.25)$$

From (2.24) we conclude that the specific volume will exhibit a singular behavior induced by the singular behavior of the energy U and the concentration X. For instance, at the critical concentration $X = X_c$ we obtain from (2.15) and (2.24)

$$\rho^{-1} = \rho_c^{-1}\left[1 + R_0|\Delta\tilde{T}^*|^{1-\alpha} + R_1\Delta\tilde{T}^* + R_2|\Delta\tilde{T}^*|^{1-\alpha+\Delta_1} + \ldots\right] \quad , \qquad (2.26)$$

with

$$\rho_c^{-1}R_0 = U_c u_0 d\tilde{T}_c/d\tilde{P} \quad . \qquad (2.27)$$

The term $(\partial\tilde{\mu}_2/\partial\tilde{P})_{\tau,h}$ yields additional singular contributions due to the fact that the irrelevant scaling fields c_i depend on $\tilde{P}$, the leading singular contribution being proportional to $|\tau|^{2-\alpha+\Delta 1}$. The term $R_0|\Delta\tilde{T}|^{1-\alpha}$ represents a weak divergence in the thermal expansion coefficient predicted by Griffiths and Wheeler (1970) and supported by some experimental observations (Greer and Hocken, 1975; Morrison and Knobler, 1976; Scheibner et al., 1978).

III. EQUATION OF STATE AND THERMODYNAMIC PROPERTIES

3.1 Thermodynamic Exponents

The early analysis of experimental equation of state data of fluids near the gas-liquid critical point covering a range of approximately $5 \times 10^{-4} \leq \Delta T^* \leq 10^{-2}$ yielded exponent values $\beta \simeq 0.355$ and $\gamma \simeq 1.19$ in disagreement with the at that time available theoretical estimates from the high-temperature series expansions for the 3-dimensional Ising model (Levelt Sengers, 1974; Levelt Sengers and Sengers, 1975; Levelt Sengers et al., 1976). The values deduced for the exponent γ from light scattering measurements tended to be slightly larger ($\gamma \simeq 1.23$), in retrospect because light scattering data are usually taken closer to the critical temperature than equation of state data (Chu, 1972; Levelt Sengers, 1974; Scott, 1978). For binary liquids near the critical mixing the exponent β was generally found to be about 0.33 or 0.34, again larger than the theoretical estimate $\beta = 0.312$ from high-temperature series expansions (Stein and Allen, 1974; Scott, 1978).

A change occurred when numerical exponent estimates based on the renormalization-group approach became available. The new estimates indicated that the theoretical exponent value for β could be slightly larger than 0.312 and, even more importantly, indicated the presence of a confluent singularity with an exponent Δ_1 as small as 0.5. When data are taken with a precision of about 1%, corrections to a simple asymptotic power law behavior may already enter at temperatures corresponding to $\Delta T^* > 10^{-4}$. Hence, the attention focused at producing data closer to the critical temperature.

Close to the critical point experiments are hampered by a number of difficulties such as long relaxation times, thermal gradients effects and gravity effects (Levelt Sengers, 1975). In particular, gravity effects become very severe in fluids near the gas-liquid critical point. The use of a probe of finite height is always necessary for measuring fluid properties. It can be shown that data taken with a probe with a height of only 1 mm are subject to gravitational rounding effects at $\Delta T^* \leq 10^{-4}$ (Moldover et al., 1979). Hence, it is impossible to investigate the range of simple asymptotic power-law behavior with traditional thermodynamic experiments.

To eliminate such gravity effects, one has in principle the following options:

a) To reduce the experimental averaging height to heights well below 1 mm. In practice, this option requires optical detection techniques.

b) To prepare the fluid in a spatially homogeneous state and to complete the measurements before the gravitationally-induced gradients can establish themselves.

c) To consider data in a larger temperature range, but include the correction-to-scaling terms in the analysis.

d) To perform critical phenomena experiments in an orbiting laboratory at reduced gravitational levels.

Option (b) is the option commonly adopted in measuring the properties of binary liquids near the critical mixing point, where the relaxation time for the establishment of gravitationally-induced concentration gradients is often very large. This option has also been considered for experiments near the gas-liquid critical point, but thus far with limited success only (Moldover et al., 1979). Option (d) is under consideration, but has not yet materialized (Sengers and Moldover, 1978a).

An interferometric method for measuring the local compressibility as a function of the local density was pioneered by Wilcox and coworkers (Wilcox and Balzarini, 1968; Estler et al, 1975); this method enables one to approach the critical point to about $\Delta T^* \simeq 10^{-5}$ (Moldover et al., 1979). In studying the coexistence curve of SF_6, Balzarini and Ohrn (1972) noticed that the exponent β decreased when the temperature range was reduced. For the binary liquid cyclohexane + aniline near the critical mixing point Balzarini (1974) found $\beta = 0.328$; this result seemed puzzling at the time since it did not agree with the then only available theoretical prediction $\beta = 0.312$ nor with the values generally found from other experiments. Hocken and Moldover (1976) used the method to make a detailed study of the equation of state of Xe, SF_6 and CO_2 in the one-phase region and found values for β ranging from 0.321 to 0.329 and for γ ranging from 1.23 and 1.28; these values are consistent with the theoretical values quoted in Table II and significantly different from the effective exponents earlier deduced from conventional PVT data. A detailed discussion of these experiments is given in the lecture notes of Moldover in this volume. We shall return to these experiments when considering the issue of two-scale factor universality in Section 4.2.

In order to judge whether one has reached the range of truly asymptotic behavior an estimate of the magnitude of the correction terms to the asymptotic behavior is necessary. As an example, we consider coexistence curve data when analyzed in terms of the Wegner expansion (2.16)

$$\frac{\rho_+ - \rho_-}{2} = \rho_c \left[B_0 |\Delta T^*|^{\beta} + B_1 |\Delta T^*|^{\beta+\Delta_1} \right] . \tag{3.1}$$

Here ρ is to be identified with the density in fluids near the gas-liquid critical point and with the concentration in binary mixtures near the critical mixing point. Following Ley-Koo and Green (1977), one has in practice also considered higher-order terms in (3.1) proportional to $|\Delta T^*|^{\beta+2\Delta_1} \simeq |\Delta T^*|^{\beta+1}$ but omitting terms proportional to $|\Delta T^*|^{\beta+\Delta_2}$ (Greer, 1976). Some values found for the exponent β by this procedure are listed in Table IV. In these lecture notes errors quoted for the experimental critical exponents generally correspond to two standard deviations; they do not include the effect of possible systematic errors either in the experimental data or induced by the analysis. Moreover, in the case of binary liquids the results depend somewhat on the choice of concentration units. Generally, the best results are obtained when the concentration is expressed in terms of volume fractions (Greer, 1976; 1978). In addition, the values $\beta = 0.326 \pm 0.006$ and $\beta = 0.318 \pm 0.012$ have been found for methanol + cyclohexane (Jacobs et al., 1977) and for nitroethane + isooctane (Beysens, 1979), respectively, without the inclusion of a correction-to-scaling term. The amplitude of this correction term will depend on the choice of fluid; whether the correction terms indeed vary appreciably for different systems requires further study. A recent investigation of the dielectric constant of SF_6 close to the critical point yielded $\beta = 0.325 \pm 0.010$

Table IV. Values of the exponent β from data analyzed in terms of the Wegner expansion (3.1)

Substance	Value of β	Authors
Sulfurhexafluoride	0.327 ± 0.006	Ley-Koo and Green (1977)
^{3}Helium	0.321 ± 0.006	Pitman et al. (1979)
isobutyric acid + water	0.324 ± 0.012	Greer (1976)
carbon disulfide + nitromethane	0.316 ± 0.008	Greer (1976)
polystyrene + cyclohexane	0.327 ± 0.008	Nakata et al. (1978)

(Thijsse, 1980). In summary, the experimental values are in good agreement with the theoretical value $\beta = 0.325 \pm 0.002$ derived from the renormalization-group approach. There is a tendency for the values to be slightly larger than the theoretical value $\beta = 0.312 \pm 0.002$ proposed on the basis of high-temperature series expansions for the Ising model. An exception is the exponent value reported by Hayes and Carr (1977) for Xe, but the data were obtained under highly nonequilibrium conditions.

It is difficult to deduce the "best" exponent value from experiments, since much depends on the assumptions in the analysis of the data. A more realistic goal is to answer the question whether the experimental data are consistent with a given theory. Obviously, the data are consistent with the predictions from the renormalization-group theory. It is somewhat more difficult to demonstrate consistency with the predictions from high-temperature series expansions. The reason is that effective exponents, when the data were analyzed in terms of a single power law, gave values for the exponent β that were too high. Hence, fewer correction terms are needed to reduce the apparent exponent value to 0.325 than to 0.312. It has been shown that a value $\beta = 0.325$ is also consistent with coexistence curve data for n-heptane + acetic anhydride (Nagarajan et al., 1980) and with equation of state data for steam (Balfour et al., 1978) and ethylene (Hastings et al., 1980).

The more detailed studies of the exponent γ have all been made by measuring the intensity of critical fluctuations with light-scattering. We shall return to this subject in Section 4.1 when discussing measurements of the critical correlation function.

It follows from (2.15) that the specific heat c_V per unit volume of a fluid near the gas-liquid critical point will diverge at the critical isochore as

$$c_V = \frac{P_c}{T_c}\left[\frac{A_0}{\alpha}|\Delta\tilde{T}^*|^{-\alpha} + A_1 + A_2|\Delta\tilde{T}^*|^{-\alpha+\Delta_1} + \ldots\right] , \tag{3.2}$$

with

$$A_0 = \alpha(1-\alpha)u_o u_c/P_c ,\; A_1 = u_1 u_c/P_c ,\; A_2 = (1-\alpha+\Delta_1)u_2 u_c/P_c \quad . \tag{3.3}$$

The same expansion applies to $c_{\tilde{P}X}$ per unit mass of a binary liquid at the critical concentration.

An accurate experimental determination of the exponent α is difficult because of the weak character of the singular behavior. The analysis is further complicated by the fact that the constant term A_1 in (3.2) cannot be identified with the physically observed background specific heat away from the critical temperature. In some cases A_1 seems even to be negative (Moldover, 1969; Lipa et al., 1977). Furthermore, specific heat measurements near the gas-liquid critical point are severely affected by gravity effects (Schmidt, 1971; Hohenberg and Barmatz, 1972; Moldover et al., 1979).

Lipa and coworkers (1977) measured the specific heat of CO_2 near the critical point and reported $\alpha = 0.124 \pm 0.005$ in apparent good agreement with the value predicted on the basis of high-temperature series expansions. However, this value was obtained from a fit to the data in terms of the simple power law $(A_0/\alpha)|\Delta T^*|^{-\alpha}+A_1$; we now know that the temperature range $4\times10^{-5}<\Delta T^*<3\times10^{-2}$ of the experimental data is too large for a simple power law to be adequate. Very recently, Bloemen and coworkers (1980) reported a detailed study of the specific heat of triethylamine + water near the critical mixing point. They verified that the correction-to-scaling term could be neglected at temperatures within $\Delta T^* \leq 6\times10^{-4}$ and concluded $\alpha = 0.107 \pm 0.002$. Beysens and Bourgou (1979) had earlier found $\alpha = 0.110 \pm 0.006$ for the same system; the latter value was deduced from refractive index measurements, but there are some complications with the data analysis to be discussed in Section 4.2. The above information does indicate that the asymptotic behavior of the specific heat is consistent with the theoretical predictions, but I doubt whether the experimental data allow us to discriminate firmly between the theoretical values of 0.110 and 0.125. This issue is also discussed by Moldover in this volume.

3.2 Thermodynamic Scaling Function and Thermodynamic Amplitude Ratios

Not only the critical exponents, but also the scaling functions $f_i(x)$ in (2.9) are predicted to be universal. Universality of the asymptotic scaling function $f_0(x)$ in (2.9) implies universal relationships between the amplitudes of the asymptotic power laws defined in Table I. Theoretical estimates for the universal amplitude ratios have been obtained from series expansions for the Ising model and from an ε-expansion of the renormalization-group equations (Aharony and Hohenberg, 1976).

The scaling functions can be deduced experimentally from equation-of-state data, i.e. pressures as a function of both temperature and density. However, to make the analysis practical one needs a functional representation of the scaling function. For this purpose, one commonly introduces a coordinate transformation expressing the scaling fields in terms of parametric variables. A popular realization of a parametric scaled equation of state has been the so-called linear model. For a survey the reader is referred to a previous review (Sengers and Levelt Sengers, 1978). The model, although not exact, is theoretically correct up to order ε^2 (Wallace and Zia, 1974) and is expected to give a reasonable approximation to the asymptotic scaling function. Balfour and coworkers (1977, 1978) formulated a revised and extended linear model which includes the first correction-to-scaling term in (2.9) and allows for mixing of the scaling fields discussed in Section 2.3.2. A list of equations for various thermodynamic properties implied by this model can be found elsewhere (Levelt Sengers and Sengers, 1980). The equation was fitted to classical PVT data of steam (Balfour et al, 1977, 1978, 1980) and ethylene (Hastings et al., 1980). The system-dependent constants in the expansions of the background $\tilde{P}_o(\tilde{\mu},\tilde{T})$ and the scaling fields were treated as adjustable parameters, but the critical exponents were assumed to have the theoretical values $\beta = 0.325$, $\gamma = 1.241$ and $\Delta_1 = 0.50$. This choice was motivated by our experience that data for fluids can be more easily accommodated with the exponent values predicted by the renormalization-group theory than with the values estimated from high-temperature series expansions. The results yield values for the various thermodynamic amplitude ratios of steam and ethylene; they are, however, not independent but interrelated by the linear model approximation. The values found for the amplitude ratios are compared with the theoretical estimates in Table V taken from a recent review (Levelt Sengers and Sengers, 1980). In this table we have also included the specific heat amplitude ratio recently determined by Bloemen et al. (1980) for the triethylamine + water system near the critical mixing point. The values are in satisfactory agreement given the accuracy with which these amplitude ratios are currently known theoretically and experimentally.

Universality of the correction-to-scaling function $f_1(x)$ implies similar universal relationships between the various correction-to-scaling amplitudes. Attempts to calculate theoretical values for the correction-to-scaling amplitude ratios have been initiated (Chang and Houghton, 1980), but they are not yet known accurately. Some consequences have been considered by Aharony and Ahlers (1980) as further discussed by Ahlers in this volume. Past experience has shown the necessity of incorporating the effect of the first correction terms to deduce reliable experimental values for the asymptotic critical exponents and amplitudes. Similarly, an analysis of the effect of higher order terms in the Wegner expansion will be needed before one can deduce accurate experimental values for the amplitudes of the first correction-to-scaling terms.

IV. STATIC CORRELATION FUNCTION

4.1 Correlation Function Exponents

The critical fluctuations can be investigated experimentally by light scattering, X-ray scattering and neutron scattering.

Table V. Some thermodynamic amplitude ratios.

	Theory[a]	Theory[b]	Steam[c]	Ethylene[d]	Triethylamine + H_2O[e]
A^+/A^-	0.51	0.55	0.53	0.50	0.57
Γ^+/Γ^-	5.07	4.80	4.89	5.31	
$\Gamma^+ DB^{\delta-1}$	1.75	1.6	1.6	1.76	
$A^+\Gamma^+/B^2$	0.059	0.066	0.057	0.051	

[a] From series expansions for Ising model (Aharony and Hohenberg, 1976).

[b] From ε-expansion of renormalization-group equations (Aharony and Hohenberg, 1976).

[c] From analysis by Balfour et al. (1980).

[d] From measurements and analysis by Hastings et al. (1980).

[e] From measurements and analysis by Bloemen et al. (1980).

Measurements of the scattering intensity as a function of scattering angle yields the structure factor as a function of the wave number k. A definitive analysis of the structure factor as a function of both the temperature and density (or concentration in binary liquids) is currently not available and we restrict our discussion to the behavior of the structure factor as a function of temperature and wave number at the critical isochore in the one-phase region.

The structure factor at the critical isochore is predicted to vary asymptotically as (Fisher, 1964, 1967)

$$\chi(k) = \Gamma(\Delta T^*)^{-\gamma} g(k\xi) \quad , \tag{4.1}$$

where ξ is the correlation length which diverges as $\xi = \xi_0(\Delta T^*)^{-\nu}$. The correlation scaling function g(y) satisfies the boundary conditions

$$g(y) = \frac{1}{1+y^2} \qquad \text{for } y \ll 1 \quad , \tag{4.2}$$

$$g(y) = \frac{C_1}{y^{2-\eta}}\left[1 + \frac{C_2}{y^{(1-\alpha)/\nu}} + \frac{C_3}{y^{1/\nu}}\right] \qquad \text{for } y \gg 1 \quad . \tag{4.3}$$

Thus for small values of the scaling variable y, the correlation function scaling function approaches the Ornstein-Zernike form (4.1), while the expansion (4.3) for large y is often referred to as the Fisher-Langer expansion.

The asymptotic behavior (4.1) should be approached for temperatures sufficiently close to the critical temperature, i.e. ξ sufficiently large, and for wave numbers k sufficiently small, i.e. for wavelengths much larger than the distances between the molecules. Just as in the case of the thermodynamic properties, confluent singularities need to be accounted for when the structure factor is studied over larger ranges of temperatures and wave numbers. When one includes the first Wegner correction only, the asymptotic expansion for the structure factor becomes (Wegner, 1975)

$$\chi(k) = \Gamma(\Delta T^*)^{-\gamma} g_0(y) + \Gamma_1(\Delta T^*)^{-\gamma+\Delta_1} g_1(y) \tag{4.4}$$

where the scaling variable y continues to be defined as

$$y = k\xi_0(\Delta T^*)^{-\nu} \quad . \tag{4.5}$$

The scaling function $g_1(y)$ satisfies boundary conditions analogous to the conditions (4.2) and (4.3) for $g_0(y)$ (Wegner, 1975; Chang et al., 1979).

Most light scattering experiments have probed a range of small y, yielding experimental values for the exponents γ and ν via (4.2). The early light scattering measurements were reviewed by Chu (1972) who concluded $\gamma = 1.23 \pm 0.02$ and $\nu = 0.63 \pm 0.02$ consistent with the hypothesis of critical-point universality. Since light scattering measurements can be obtained closer to the critical temperature than conventional thermodynamic measurements, the effects of correction-to-scaling terms appear somewhat less severe.

Nevertheless, the subject remained controversial, because of the discrepancies between the predictions from series expansions and renormalization-group theory for Ising-like systems and because of the difficulties associated with obtaining a reliable determination of the critical exponent η. The difficulties are compounded by the fact that the exponent determined experimentally is $2-\eta$ rather than η itself as is evident from (4.3). A modified procedure, alleviatin this difficulty, was recently proposed by Gürmen et al. (1980).

One method of deducing η is obtaining it from the Fisher relation $2 - \eta = \gamma/\nu$. However, the result is sensitive to small errors in γ and ν and, hence, to any effects of confluent singularities in determining γ and ν. Another method is to determine it from the Fisher-Langer expansion (4.3) by making $k\xi$ large. For ligh scattering experiments, which correspond to small k values, it means that data must be taken very close to the critical temperature, wher problems associated with multiple scattering usually become very severe.

Warkulwiz et al. (1974) attempted to determine η from neutron scattering measurements, probing much larger values of k, and reported $\eta = 0.11 \pm 0.03$ for neon near the gas-liquid critical point. This result seemed to agree with the value $\eta = 0.10 \pm 0.05$ reported by Lin and Schmidt (1974) from X-ray scattering measurements for argon, but was in definite disagreement with the value predicted theoretically. It should be emphasized that in order for the asymptotic behavior to be present, not only $k\xi$ should be large, but k should be sufficiently small. There is some doubt whether this condition is realized in neutron scattering experiments.

Chang et al. (1976, 1979) made light scattering measurements in the binary liquid 3-methylpentane + nitroethane. This system has the advantage that the refractive index difference between the two components on the one hand is sufficiently small so that multiple scattering effects remain small and on the other hand the difference is sufficiently large so that the temperature dependence of the observed structure factor can be attributed to the temperature dependence of the order-parameter fluctuations (Chang et al., 1979). The experiment covered the range $0.18 < y < 25$ in $k\xi$ and $10^{-6} < \Delta T^* < 2.7 \times 10^{-3}$ in temperature. To analyze the data one needs a correlation scaling function approximant. Approximants interpolating between the Ornstein-Zernike form for small η and the asymptotic $y^{\eta-2}$ behavior for large y have been proposed by Fisher and Burford (1974) and Ferrell and Scalapino (1975). A refined approximant incorporating the full Fisher-Langer expansion (4.3) was proposed by Bray (1976). In Table VI we list the exponent values deduced when the data were analyzed in terms of the available approximants (Chang et al., 1979; Burstyn, 1979). The values are in agreement with the theoretical estimates predicted by the renormalization-group theory independent of the choice of approximant and whether a correction-to-scaling term is included or not. The values obtained for ν and η have a tendency of favoring the predictions from the renormalization-group theory over those from series expansions (Chang et al., 1979). Of course, one can always argue that more correction terms are needed than the one incorporated in (4.4).

Schmidt and coworkers have reanalyzed their X-ray measurements for argon either by allowing for an analytic background (Bale et al., 1977) or by incorporating the first correction-to-scaling term (Gürman et al., 1980) producing a revised value $\eta = 0.03 \pm 0.02$.

Very recently, Schneider et al. (1980) made neutron scattering measurements of the system isobutyric acid + D_2O near the critical mixing point and reported $\eta = 0.09 \pm 0.03$, in agreement with the earlier result of Warkulwiz et al. (1974). It remains my conjecture that correction-to-scaling effects must be incorporated in the analysis of neutron scattering experiments, but the discrepancy between neutron scattering results and light scattering results has clearly not yet been resolved.

Accurate light scattering measurements for SF_6 near the gas-liquid critical point were earlier reported by Cannell (1975) who concluded that the data could be represented by a simple power law behavior with $\gamma = 1.223$ and $\nu = 0.621$ for $\Delta T^* \leq 3 \times 10^{-2}$. Nevertheless, this range seems too large for a simple power law to be applicable. It is noted that he was able to measure the correlation length ξ at temperatures very close to T_c by combining light scattering measurements with turbidity measurements and using a special quenching technique. I reanalyzed his turbidity data, restricted to the range $4 \times 10^{-6} < \Delta T^* < 6 \times 10^{-4}$, and found $\nu = 0.622 \pm 0.008$. Hence, a case can be made that these data are also consistent with the predictions from the renormalization-group theory. Very recently, Cannell and Güttinger (1980) obtained accurate light-scattering and turbidity measurements for xenon; they detected definite deviations from a simple power-law behavior and concluded that the asymptotic behavior was indeed consistent with $\gamma = 1.241$.

We conclude that all recent accurate light-scattering measurements yield correlation exponent values that are in agreement with the predictions from the renormalization-group theory.

Table VI. Correlation function exponents obtained for the 3-methylpentane + nitroethane system (Chang et al., 1979)

Correlation scaling function	γ	ν	η
Approximant of Bray (1976)	1.240 ± 0.017	0.625 ± 0.006	0.017 ± 0.015
Approximant of Fisher and Burford (1974)	1.235 ± 0.016	0.625 ± 0.006	0.024 ± 0.022
Approximant of Ferrell and Scalapino (1975) with $\omega = 1$	1.238 ± 0.018	0.625 ± 0.006	0.020 ± 0.017
Approximant of Ferrell and Scalapino (1975) with $\omega = 0$	1.232 ± 0.020	0.626 ± 0.006	0.030 ± 0.025
Approximant of Bray (1976) plus first correction-to-scaling term	1.246 ± 0.033	0.628 ± 0.013	0.016 ± 0.016

4.2 Hyperscaling and Two-Scale-Factor Universality

The renormalization-group theory of critical phenomena predicts a relationship between the correlation length exponent ν and the specific heat exponent α, referred to as hyperscaling and a corresponding amplitude relationship, known as two-scale-factor universality (Stauffer et al., 1972; Hohenberg et al., 1976). These relationships follow from the statements that the Helmholtz free energy in a volume ξ^d should be finite and universal. Since the free energy varies asymptotically as $|\Delta T^*|^{2-\alpha}$, while ξ^d diverges as $|\Delta T^*|^{-d\nu}$, the statement implies the hyperscaling relation

$$d\nu = 2 - \alpha \quad , \qquad (4.6)$$

further discussed by Nickel and Baker in this volume. The universal relationship between the corresponding amplitudes implies the ratio

$$R_\xi = (A^+/k_B)^{1/d}\,\xi_o \quad , \qquad (4.7)$$

to be universal, where A^+ is the amplitude of the asymptotic power law for the specific heat as defined in Table I. The most recent theoretical estimates for this ratio are $R_\xi = 0.2547 \pm 0.007$ from high-temperature series expansions and $R_\xi = 0.2699 \pm 0.0008$ from the renormalization-group theory as quoted by Bervillier and Godrèche (1980). It should be noted that there exists a strong correlation between the values adopted for the critical exponents and the values for the corresponding amplitudes.

In the previous sections we have argued that the experimental data for fluids are in agreement with the exponent values predicted by the renormalization-group theory and which, in turn, satisfy the hyperscaling relation (4.6) exactly. However, the presumed possible violations of the hyperscaling relation as considered by Baker (1977) for the Ising model are small and within the accuracy with which the exponent values for fluids are firmly established experimentally.

Assuming the hyperscaling relation to be valid for fluids, we turn our attention to experimental studies of the validity of two-scale-factor universality for fluids. An early analysis for binary liquids near the critical mixing point was made by Woermann and coworkers, but the experimental errors were generally too large to arrive at a definitive conclusion (Klein and Woermann, 1976; Pelger et al., 1977). An analysis of the data for a number of fluids near the gas-liquid critical point provided support for the validity of two-scale-factor universality, but the analysis was clouded by the use of effective critical exponent values for fluids (Sengers and Levelt Sengers, 1978).

More recent studies of two-scale-factor universality were made by Bloemen et al. (1980) and by Beysens et al. (1979) for binary liquids near the critical mixing point and by Sengers and Moldover (1978b) for fluids near the gas-liquid critical point.

Thoen et al. (1978) measured the specific heat of triethylamine + water and reported evidence in support of the validity of two-scale-factor universality. However, the analysis was clouded by the observation that the specific heat data seemed to prefer a value for the exponent α distinctly larger than predicted by the theory. Recently, they reanalyzed the data for triethylamine + H_2O, complemented with additional data for triethylamine + D_2O, and found $R_\xi = 0.27 \pm 0.02$ and $R_\xi = 0.26 \pm 0.03$ for these two systems in good agreement with the theoretical predictions mentioned above (Bloemen et al., 1980).

Beysens and coworkers (1979) reported evidence in support of the validity of two-scale-factor universality for a number of binary liquids. These investigators did not measure the specific heat directly, but considered instead the expansion (2.26) for the density relating the amplitude R_o of the term proportional to $|\Delta T^*|^{1-\alpha}$ to the specific heat amplitude via (2.27) and (3.3) as further discussed by Beysens in this volume. However, some complications associated with the procedure should be noted. First, the method requires separation of the term proportional to $|\Delta T^*|^{1-\alpha}$ from the term proportional to ΔT^* which is notoriously difficult. In the analysis of Beysens and Bourgou (1979) an apparent high precision is obtained, because the term $R_1 \Delta T^*$ is identified with the linear term deduced from data away from the critical temperature. This assumption is not justified because of the presence of fluctuation-induced contributions to the linear term as discussed in Section 2.3.2. Furthermore, the quantity measured experimentally is not the density, but the refractive index which is assumed to be an analytic function of the density ρ. However, there exists strong theoretical evidence that the refractive index contains an additional singular contribution whose amplitude is proportional to the rate of change of T_c with the application of an electric field (Sengers et al., 1980). A significant discrepancy exists between the specific heat amplitude deduced by Beysens and Bourgou (1979) from refractive index measurements and the amplitude deduced by Bloemen et al. (1980) from direct specific heat data, apparently compensated by different values adopted for the correlation length amplitude ξ_o.

To avoid the difficulties associated with an accurate determination of the amplitude of a weakly divergent quantity such as the specific heat, in fluids near the gas-liquid critical point further complicated by gravity effects, Sengers and Moldover (1978b) reformulated two-scale-factor universality in terms of a ratio of amplitudes corresponding to more strongly singular quantities:

$$R'_\xi = (B^2/k_B\Gamma^+)^{1/d}\,\xi_o \quad , \tag{4.8}$$

where B and Γ^+ are again amplitudes defined in Table I. Universality of the ratios $(A^+/k_B)^{1/d}\xi_o$ and $A^+\Gamma^+/B^2$ implies that also the ratio R'_ξ, defined in (4.8) should be universal. Theoretical estimates from series expansions yield $R'_\xi = 0.65$ and from the ε-expansion of the renormalization-group equations $R'_\xi = 0.67$ (Hohenberg et al., 1976).

To test this prediction we reconsidered the interferometric data of Hocken and Moldover (1976) for SF_6, CO_2 and Xe, mentioned in Section 3.1, but this time imposing the theoretical exponent estimates $\beta = 0.325$ and $\gamma = 1.24$ and then deducing the corresponding optimum values for the amplitudes. The correlation length amplitude ξ_o for these fluids can be deduced from light-scattering data of Cannell (1975) for SF_6, of Lunacek and Cannell (1971) for CO_2 and of Cannell and Güttinger (1980) for Xe. For this purpose we assume the hyperscaling relation for the critical exponents which implies $\nu = 0.63$. The results found for R'_ξ are given in Table VII. The experimental value $R'_\xi = 0.69 \pm 0.04$ is in agreement with the theoretical estimate $R'_\xi = 0.67$ from the renormalization-group theory, thus providing good evidence for the validity of two-scale-factor universality (Sengers and Moldover, 1978b).

V. DYNAMIC CRITICAL PHENOMENA

5.1 Dynamic Critical Exponents

The principle of dynamic scaling, formulated originally by Ferrell et al. (1968) and by Halperin and Hohenberg (1969) as a phenomenological postulate and subsequently confirmed by the renormalization-group theory of critical dynamics (Hohenberg and Halperin, 1977; Mazenko, 1978), asserts that the decay rate Γ of the order-parameter fluctuations sufficiently close to the critical point will assume the form

$$\Gamma = k^z \Omega(k\xi) \quad . \tag{5.1}$$

In addition it is predicted that the shear viscosity η_s will diverge as (Hohenberg and Halperin, 1977)

$$\eta_s \propto \xi^\phi \quad . \tag{5.2}$$

Table VII. Test of two-scale-factor universality

	$(B^2/k_B\Gamma^+)^{1/3}$ nm^{-1}	ξ_o nm	R'_ξ
SF_6	3.7 ± 0.1[a]	0.186 ± 0.006[b]	0.68 ± 0.04
CO_2	4.6 ± 0.1[a]	0.150 ± 0.009[c]	0.69 ± 0.06
Xe	3.7 ± 0.1[a]	0.186 ± 0.006[d]	0.69 ± 0.04

[a] From data of Hocken and Moldover (1976) with $\beta = 0.325$ and $\gamma = 1.24$.

[b] From data of Cannell (1975) with $\nu = 0.63$.

[c] From data of Lunacek and Cannell (1971) with $\nu = 0.63$.

[d] From data of Cannell and Güttinger (1980) with $\nu = 0.63$.

The dynamic scaling function $\Omega(y)$ satisfies the boundary conditions

$$\lim_{y\to 0} \Omega(y) = C_o y^{-(1+\phi)} \quad , \tag{5.3}$$

$$\lim_{y\to\infty} \Omega(y) = C_\infty \tag{5.4}$$

where C_o and C_∞ are constants. The dynamic critical exponents z and ϕ satisfy the scaling relation

$$z = 3 + \phi \quad . \tag{5.5}$$

At a given temperature, i.e. at a given value for ξ, Γ will vary as k^2 for $k\xi << 1$ in agreement with the laws of hydrodynamics. On the other hand, Γ will vary as k^z in the critical regime $k\xi >> 1$.

To characterize dynamic critical phenomena, systems can be grouped in dynamic universality classes. Systems within the same dynamic universality class have identical dynamic critical exponents as well as the same dynamical scaling function $\Omega(y)$ when properly normalized as discussed in Section 5.2.

The dynamic critical exponents have been calculated from the dynamic renormalization-group theory (Siggia et al., 1976) and from the mode-coupling theory of critical dynamics (Gunton, 1979). Using a perturbation expansion up to order ε^2 one finds the theoretical estimates

$$\phi = 0.065, \qquad z = 3.065 \quad . \tag{5.6}$$

As discussed earlier, the thermodynamic properties of fluids approach a simple asymptotic power law behavior provided the critical temperature is approached to within 10^{-4} or 10^{-3}. Hence, one must expect a priori that the validity of the asymptotic dynamic behavior is restricted to a similar small range in temperature. The restrictions in dealing with dynamic critical phenomena, however, are even more severe. In deriving the asymptotic equations one assumes implicitly that the critical point is approached sufficiently closely so that the singular contribution to the shear viscosity is large compared with the normal shear viscosity $\bar{\eta}_s$ in the absence of critical fluctuations (Garisto and Kapral, 1976; Hohenberg and Halperin, 1977). This condition is almost never satisfied in experiments.

To account for this difficulty, one derives expressions for the decay rate and the shear viscosity from an approximate solution to the mode-coupling equations which then reduce asymptotically to the behavior predicted by (5.1) and (5.2). These considerations lead to the postulate that the shear viscosity exhibits a multiplicative anomaly, i.e. the ratio $\eta_s/\bar{\eta}_s$, rather than η_s itself, should diverge as ξ^ϕ (Ohta, 1977), so that

$$\eta_s/\bar{\eta}_s = (q\xi)^\phi \quad , \tag{5.7}$$

where q is a system-dependent constant. Indeed, when experimental viscosity data of binary liquids near the critical mixing point are analyzed in terms of (5.7) one often finds a value of the exponent ϕ close to the value predicted theoretically (Calmettes, 1977; Lee 1978).

As an example, we consider the viscosity data of Tsai and McIntyre (1974) for 3-methylpentane + nitroethane. In Fig. 3 the

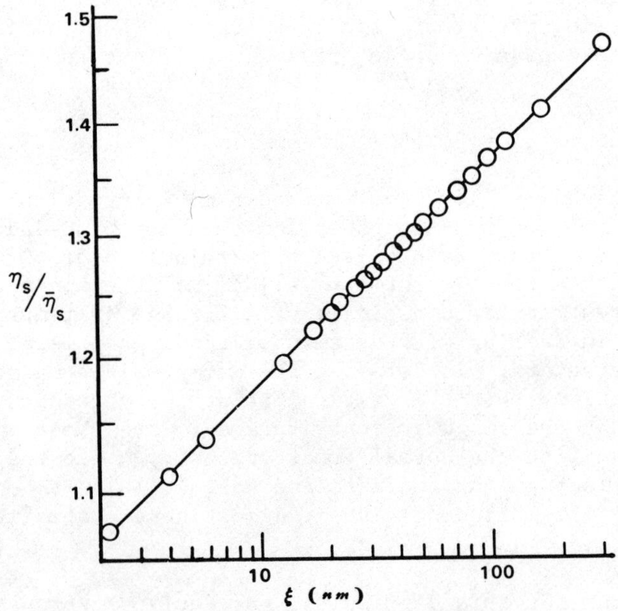

Fig. 3 Log-log plot of the viscosity ratio $\eta_s/\bar{\eta}_s$ of 3-methylpentane nitroethane as a function of ξ (Burstyn, 1979).

ratio $\eta_s/\bar{\eta}$ is plotted as a function of ξ on a double-logarithmic scale (Burstyn, 1979). The data can be represented by a straight line with slope $\phi = 0.0635 \pm 0.0004$ in good agreement with the theoretical prediction. It is noted that when experimental viscosity data for fluids near the gas-liquid critical point are analyzed in terms of (5.7), slightly smaller values for the exponent ϕ are frequently found (Lee, 1978; Basu et al., 1980) and more correction terms need to be taken into account.

The decay rate Γ of the critical fluctuations can be determined from light-scattering measurements. To determine the dynamic scaling exponent z one needs to probe a range where $k\xi$ becomes large. Again such experiments are difficult for fluids near the gas-liquid critical point because of multiple scattering effects (Güttinger and Cannell, 1980). Several investigators have attempted to determine the exponent z experimentally for liquid mixtures. An early investigation of Chang et al. (1971, 1972) yielded $z = 2.99 \pm 0.05$ for 3-methylpentane + nitroethane in apparent good agreement with the prediction of $z = 3.00$ available at that time. A value close to 3.00 has also been reported for other mixtures (Chu and Lin, 1974; Harker and Schmidt, 1977). More recently, Sorensen et al. (1978) reported $z = 2.992 \pm 0.014$ for 3-methylpentane + nitroethane in agreement with the earlier result of Chang et al., but in definite disagreement with the now available theoretical prediction.

It is noted that the hypothesis of dynamical scaling does not presuppose exponential decay of the time-dependent correlation function, but applies to any suitably defined decay rate Γ (Hankey and Stanley, 1972; Hohenberg and Halperin, 1977). Very close to the critical temperature small deviations from exponential decay have indeed been observed (Burstyn, 1979; Burstyn et al., 1980). The effect appears to agree with a frequency dependence of the viscosity close to the critical point predicted theoretically (Bhattacharjee and Ferrell, 1980). To check the dynamic scaling prediction, Burstyn and coworkers defined Γ as the effective decay rate when the experimental time-dependent correlation function was fitted to an exponential decay law over a time interval corresponding to one relaxation time. At a given temperature, an effective dynamic scaling exponent z_{eff} may be defined as (Burstyn, 1979; Burstyn and Sengers, 1980):

$$\Gamma(T) \propto k^{z_{eff}(T)} \quad . \tag{5.8}$$

From (5.3) it follows that

$$\lim_{T \to T_c} z_{eff}(T) = z \quad . \tag{5.9}$$

Fig. 4 shows this effective exponent z for 3-methylpentane + nitroethane as a function of $T - T_c$ in a range of 10 millidegrees from the critical temperature (Burstyn and Sengers, 1980). From these data it is concluded that in the limit $T \to T_c$ $z_{eff}(T) \to 3.063 \pm 0.024$, in good agreement with the theoretical value predicted for this exponent.

5.2 Dynamic Critical Amplitude Ratio

The decay rate Γ of the order-parameter fluctuations is related to the diffusion coefficient D by

$$D = \lim_{k \to 0} \Gamma/k^2 \quad . \tag{5.10}$$

For fluids near the gas-liquid critical point D is to be identified with the thermal diffusivity and for liquid mixtures near the critical mixing point with the binary diffusion coefficient. The theory of dynamic critical phenomena predicts that this diffusion coefficient sufficiently close to the critical point should assume the form (Hohenberg and Halperin, 1977)

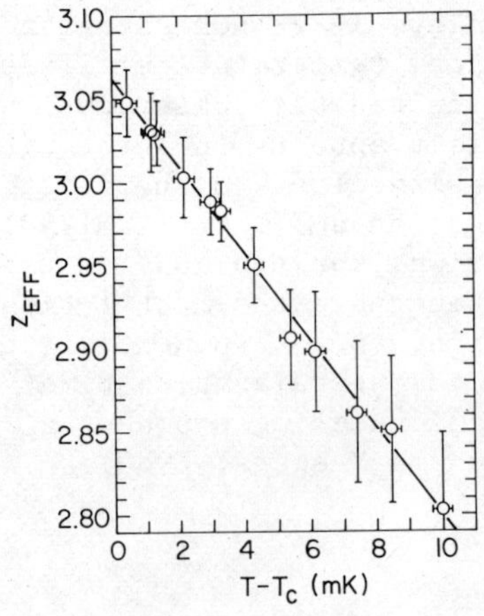

Fig. 4 The effective dynamic critical exponent z_{eff} for 3-methylpentane + nitroethane as a function of $T - T_c$ close to the critical temperature (Burstyn and Sengers, 1980).

$$D = \frac{R k_B T}{6\pi \eta_s \xi} , \tag{5.11}$$

where R is a universal constant. This constant R may be interpreted as the dynamic critical amplitude ratio associated with the dynamic scaling relation (5.5).

In the original derivations of (5.11) the constant R was found to be equal to unity (Kawasaki, 1970; Ferrell, 1970). Siggia et al. (1976) evaluated R to first order in ε from the dynamic renormalization-group equations and proposed as their best theoretical estimate $R = 1.20$. Experimental evidence in support of the latter value was reported by Chen et al. (1978) for n-hexane + nitrobenzene and by Sorensen et al. (1978) for 3-methylpentane + nitroethane. This result has been interpreted as indicating $R/6\pi \simeq 1/5\pi$ which corresponds to Stokes' law for a spherical droplet moving in a medium with the same viscosity as that of the liquid in the droplet. It has led Beysens et al. (1979) to assume $R/6\pi = 1/5\pi$ and then to use (5.11) for determining the correlation length ξ from experimental decay rate data.

However, a number of complications should be noted. As was repeatedly stressed before in the analysis of other fluid properties, in order for the asymptotic behavior (5.11) to be valid the critical temperature should be approached sufficiently close. It has been well established that a noncritical background contribution should be taken into account in the analysis of thermal diffusivity data of fluids near the gas-liquid critical point (Sengers and Keyes, 1971; Sengers, 1972; Swinney and Henry, 1973). A similar noncritical background contribution should be expected in the experimental diffusion coefficient data of binary liquid mixtures near the critical mixing point (Chang et al., 1971, 1972; Oxtoby and Gelbart, 1974), but it has been ignored by most investigators. Furthermore. the experimental value of R is affected by the combined errors in ξ, η_s and D or Γ and it appears to be difficult to determine R accurately. A detailed study of these quantities for 3-methylpentane + nitroethane was recently reported by Burstyn et al. (1980), who found $R = 1.02 \pm 0.06$. The same quantity was investigated by Güttinger and Cannell (1980) for xenon. After correcting for a noncritical background they found $R = 1.01 \pm 0.06$. The close agreement between the values thus found for a binary liquid near the critical mixing point and for a gas near the gas-liquid critical point provides good evidence in support of the universality of the dynamic critical amplitude ratio R. However, the experimental values appear to be smaller than the theoretical value 1.20 suggested earlier and it would be desirable to have a more accurate theoretical calculation of this amplitude ratio.

VI. CONCLUSIONS

The experimental data for fluids near the gas-liquid critical point and for binary liquids near the critical mixing point are consistent with the asymptotic behavior predicted for the one-component Landau-Ginzburg-Wilson model on the basis of the renormalization-group equations. In particular, we have considered the critical exponents that characterize the asymptotic behavior of the thermodynamic properties, the correlation function and the decay rate of the critical fluctuations together with the amplitude ratios implied by static scaling, hyperscaling and dynamic scaling. The experimental data have a tendency of favoring the estimates predicted from the renormalization-group theory over those commonly deduced from high-temperature series expansions. All critical exponents and amplitudes ratios agree within error with the values predicted from the renormalization-group theory of static and dynamic critical phenomena except for the dynamic critical amplitude ratio which appears slightly smaller than the most recent theoretical estimate.

ACKNOWLEDGMENTS

In preparing this review the author has benefitted from a close collaboration with F. W. Balfour, H. C. Burstyn, R. F. Chang, J. M. H. Levelt Sengers and M. R. Moldover and from many stimulating discussions with J. K. Bhattacharjee, D. S. Cannell, R. A. Ferrell, S. C. Greer and J. F. Nicoll. The research was supported by the National Science Foundation under grant DMR 79-10819.

Appendix

SOME RECENT REVIEW PAPERS DEALING WITH CRITICAL PHENOMENA IN FLUIDS

"Critical-point universality and fluids", A. Levelt Sengers, R. Hocken and J. V. Sengers, Physics Today 30 (12), pp. 42-51 (1977).

"Critical exponents for binary fluid mixtures", R. L. Scott, in Chemical Thermodynamics, Vol. 2, Specialist Periodical Report, M. L. McGlashan, ed. (Chemical Society, London, 1978), pp. 238-274.

"Critical phenomena in classical fluids", J. V. Sengers and J. M. H. Levelt Sengers, in Progress in Liquid Physics, C. A. Croxton, ed. (Wiley, New York, 1978), pp. 103-174.

"Liquid-liquid critical phenomena", S. C. Greer, Acc. Chem. Res. 11, pp. 427-432 (1978).

"Critical phenomena in fluids", P. C. Hohenberg in Microscopic Structure and Dynamics of Liquids, J. Dupuy and A. J. Dianoux, eds. (Plenum Publ. Corp., New York, 1978), pp. 333-366.

"Gravity effects in fluids near the gas-liquid critical point", M. R. Moldover, J. V. Sengers, R. W. Gammon and R. J. Hocken, Rev. Mod. Phys. 51, pp. 79-99 (1979).

"How close is "close to the critical point"?", J. M. H. Levelt Sengers and J. V. Sengers, in Perspectives in Statistical Physics, H. J. Raveché, ed. (North-Holland Publ. Comp., Amsterdam, in press).

"Light scattering investigations of the critical region in fluids W. I. Goldburg, in Light Scattering Near Phase Transitions, H. Z. Cummins and A. P. Levanyuk, eds. (North-Holland Publ. Comp., Amsterdam, in press).

"Critical-point phenomena in binary liquid systems", A. Kumar, H. R. Krishnamurthy and E. S. R. Gopal, to be published.

REFERENCES

Aharony, A., and G. Ahlers (1980), Phys. Rev. Lett. 44, 782.

Aharony, A. and P. C. Hohenberg (1976), Phys. Rev. B 13 3081.

Baker, G. A. (1977), Phys. Rev. B 15 1552.

Baker, G. A. and J. M. Kincaid (1980), J. Stat. Phys. (in press).

Baker, G. A., B. G. Nickel, M. S. Green and D. I. Meiron (1976), Phys. Rev. Lett. 36, 1351.

Baker, G. A., B. G. Nickel and D. I. Meiron (1978), Phys. Rev. B 17, 1365.

Bale, H. D., J. S. Lin, D. A. Dolejsi, J. L. Casteel, O. A. Pringle and P. W. Schmidt (1977), Phys. Rev. A 15, 2513.

Balfour, F. W., J. V. Sengers, M. R. Moldover and J. M. H. Levelt Sengers (1977), in Proceedings 7th Symposium on Thermophysical Properties, A. Cezairliyan, ed. (American Society of Mechanical Engineers, New York), p. 786.

Balfour, F. W., J. V. Sengers, M. R. Moldover and J. M. H. Levelt Sengers (1978), Phys. Lett. 65A, 223.

Balfour, F. W., J. V. Sengers, and J. M. H. Levelt Sengers (1980), in Water and Steam: Their Properties and Current Applications, J. Straub and K. Scheffler, eds. (Pergamon Press, Oxford and New York), p. 128.

Balzarini, D. A. (1974), Canad. J. Phys. 52, 499.

Balzarini, D. A. and K. Ohrn (1972), Phys Rev. Lett. 29, 840.

Basu, R. S., J. V. Sengers and J. T. R. Watson (1980), Intern. J. Thermophys. 1, 33.

Bervillier, C. and C. Godrèche (1980), Phys. Rev. B 21, 5427.

Beysens, D. (1979), J. Chem. Phys. 71, 2557.

Beysens, D. and A. Bourgou (1979), Phys. Rev. A 19 2407.

Beysens, D., R. Tufeu and Y. Garrabos (1979), J. Physique 40, L-623.

Bhattacharjee, J. K. and R. A. Ferrell (1980), Phys. Rev. A. (in press).

Bloemen, E., J. Thoen and W. Van Dael (1980), J. Chem. Phys. (in press).

Bray, A. J. (1976), Phys. Rev. Lett. 36, 285; Phys. Rev. B 14, 1248.

Burstyn, H. C. (1979), "An examination of the structure factor, near the consolute point, in the binary liquid 3-methylpentane+ nitroethane", thesis (Department of Physics and Astronomy, University of Maryland, College Park, MD).
Burstyn, H. C., R. F. Chang and J. V. Sengers (1980), Phys. Rev. Lett. 44, 410.
Burstyn, H. C. and J. V. Sengers (1980), Phys. Rev. Lett. 45, 259.
Burstyn, H. C., J. V. Sengers and P. Esfandiari (1980), Phys. Rev. A 22, 282.
Calmettes, P. (1977), Phys. Rev. Lett. 39, 1151.
Camp, W. J., D. M. Saul, J. P. Van Dyke and M. Wortis (1976), Phys. Rev. B 14, 3990.
Cannell, D. S. (1975), Phys. Rev. A 12, 225.
Cannell, D. S. and H. Güttinger (1980), private communication.
Chang, M. C. and A. Houghton (1980), Phys. Rev. Lett. 44, 785; Phys. Rev. B 21, 1881.
Chang, R. F., H. Burstyn, J. V. Sengers and A. J. Bray (1976), Phys. Rev. Lett. 37, 1481.
Chang, R. F., H. Burstyn and J. V. Sengers (1979), Phys. Rev. A 19, 866.
Chang, R. F., P. H. Keyes, J. V. Sengers and C. O. Alley (1971), Phys. Rev. Lett. 27, 1706.
Chang, R. F., P. H. Keyes, J. V. Sengers and C. O. Alley (1972), Ber. Bunsenges. physik. Chemie 76, 260.
Chen, S. H., C. C. Lai and J. Rouch (1978), J. Chem. Phys. 68, 1994.
Chu, B. (1972), J. Stat. Phys. 6, 173; Ber. Bunsenges. physik. Chemie 76, 202.
Chu, B. and F. L. Lin (1974), J. Chem. Phys. 61, 5132.
Domb, C. (1974), in Phase Transitions and Critical Phenomena, Vol 3, C. Domb and M. S. Green, eds. (Academic Press, New York), Ch. 6.
Estler, W. T., R. Hocken, T. Charlton and L. R. Wilcox (1975), Phys. Rev. A 12, 2118.
Ferrell, R. A. (1970), Phys. Rev. Lett. 24, 1169.
Ferrell, R. A., N. Menyhárd. H. Schmidt, F. Schwabl and P. Szépfalusy (1968), Ann. Phys. (N.Y.) 47, 565.
Ferrell, R. A. and D. J. Scalapino (1975), Phys. Rev. Lett. 34, 200.
Fisher, M. E. (1964), J. Math. Phys. 5, 944.
Fisher, M. E. (1967), Reports Progr. Phys. 30 (II), 615.
Fisher, M. E. and R. J. Burford (1974), Phys. Rev. B 10, 2818.
Garisto, F. and R. Kapral (1976), Phys. Rev. A 14, 884.
Green, M. S., M. J. Cooper and J. M. H. Levelt Sengers (1971). Phys. Rev. Lett. 26, 492, 941.
Greer, S. C. (1976), Phys. Rev. A 14, 1770.
Greer, S. C. (1978), Acc. Chem. Res. 11, 427.
Greer, S. C. and R. Hocken (1975), J. Chem. Phys. 63, 5067.
Griffiths, R. B. and J. C. Wheeler (1970), Phys. Rev. A 2, 1047.
Gürmen, E., M. Chandrasekhar, P. E. Chumbley, H. D. Bale, D. A. Dolejsi, J. S. Lin and P. W. Schmidt (1980), Phys. Rev. A 22, 170.

Güttinger, H. and D. S. Cannell (1980), Phys. Rev. A 22, 285.
Gunton, J. D. (1979), in Dynamical Critical Phenomena and Related Topics, C. P. Enz, ed. (Springer-Verlag, Berlin, Heidelberg, New York), p. 1.
Halperin, B. I. and P. C. Hohenberg (1969), Phys. Rev. 177, 952.
Hankey, A. and H. E. Stanley (1972), Phys. Rev. B 6, 3515.
Harker, G. G. and R. L. Schmidt (1977), J. Chem. Phys. 67, 332.
Hastings, J. R., J. M. H. Levelt Sengers and F. W. Balfour (1980), J. Chem. Thermod. (in press).
Hayes, C. E. and H. Y. Carr (1977), Phys. Rev. Lett. 24, 1558.
Hocken, R. and M. R. Moldover (1976), Phys. Rev. Lett. 37, 29.
Hohenberg, P. C. and M. Barmatz (1972), Phys. Rev. A 6, 289.
Hohenberg, P. C., A. Aharony, B. I. Halperin and E. D. Siggia (1976), Phys. Rev. B 13, 2986.
Hohenberg, P. C. and B. I. Halperin (1977), Rev. Mod. Phys. 49, 435.
Jacobs, D. T., D. J. Anthony, R. C. Mockler and W. J. O'Sullivan (1977), Chem. Phys. 20, 219.
Kadanoff, L. P. (1971), in Critical Phenomena, Proc. Intern. School of Physics, "Enrico Fermi", Course LI, M. S. Green, ed. (Academic Press, New York), p. 100.
Kawasaki, K. (1970), Ann. Phys. (N.Y.) 61, 1.
Klein, H. and D. Woermann (1976), J. Chem. Phys. 64, 5316.
Lee, S. P. (1978), Chem. Phys. Lett. 57, 611.
Lee, T. and C. N. Yang (1952), Phys. Rev. 87, 410.
Le Guillou, J. C. and J. Zinn-Justin (1977), Phys. Rev. Lett. 39, 95.
Le Guillou, J. C. and J. Zinn-Justin (1980), Phys. Rev. B 21, 3976.
Leung, S. S. and R. B. Griffiths (1973), Phys. Rev. A 8, 2670.
Levelt Sengers, J. M. H. (1974), Physica 73, 73.
Levelt Sengers, J. M. H. (1975), in Experimental Thermodynamics Vol. II, B. Le Neindre and B. Vodar, eds. (Butterworth, London) p. 657.
Levelt Sengers, J. M. H., W. L. Greer and J. V. Sengers (1976), J. Phys. Chem. Ref. Data 5, 1.
Levelt Sengers, J. M. H. and J. V. Sengers (1975), Phys. Rev. A 12, 2622.
Levelt Sengers, J. M. H. and J. V. Sengers (1980), in Perspectives in Statistical Physics, H. J. Raveché, ed. (North-Holland Publ. Comp, Amsterdam, in press).
Ley-Koo, M. (1976), "Consequences of the renormalization group for the thermodynamics of fluids near the critical point", thesis (Department of Physics, Temple University, Philadelphia, PA).
Ley-Koo, M. and M. S. Green (1977), Phys. Rev. A 16, 2483.
Lin, J. S. and P. W. Schmidt (1974), Phys. Rev. A 10, 2290.
Lipa, J. A., C. Edwards and M. S. Buckingham (1977), Phys. Rev. A 15, 778.
Lunacek, J. H. and D. S. Cannell (1971), Phys. Rev. Lett. 27, 841.
Mazenko, G. (1978), in Correlation Functions and Quasiparticle Interactions, J. W. Halley, ed. (Plenum Publ. Corp., New York) p. 1.

McKenzie, S. (1979), J. Phys. A 12, L185.
Mermin, N. D. (1971), Phys. Rev. Lett. 26, 957.
Mermin, N. D. and J. J. Rehr (1971), Phys. Rev. Lett. 26, 1155.
Mistura, L. (1979), Nuovo Cim. 52B, 277.
Moldover, M. R. (1969), Phys. Rev. 182, 342.
Moldover, M. R., J. V. Sengers, R. W. Gammon and R. J. Hocken (1979), Rev. Mod. Phys. 51, 79.
Moore, M. A., D. Jasnow and M. Wortis (1969), Phys. Rev. Lett. 22, 940.
Morrison, G. and C. M. Knobler (1976), J. Chem. Phys. 65, 5507.
Nagarajan, N., A. Kumar, E. S. R. Gopal and S. C. Greer (1980), J. Phys. Chem. 84, 2883.
Nakata, M., T. Dobashi, N. Kuwara, M. Kaneko and B. Chu (1978), Phys. Rev. A 18, 2683.
Nickel, B. G. and B. Sharpe (1979), J. Phys. A 12, 1819.
Nicoll, J. F. (1980), Phys. Rev. A (in press).
Nicoll, J. F. and R. K. P. Zia (1980), Phys. Rev. B (in press).
Ohta, T. (1977), J. Phys. C 10, 791.
Oitmaa, J. and J. Ho-Ting-Hun (1979), J. Phys. A 12, L281.
Oxtoby, D. W. and W. M. Gelbart (1974), J. Chem. Phys. 61, 2957.
Patashinkii, A. Z. and V. I. Pokrovskii (1979), Fluctuation Theory of Phase Transitions (Pergamon Press, New York).
Pelger, M., H. Klein and D. Woermann (1977), J. Chem. Phys. 67, 5362.
Pfeuty, P. and G. Toulouse (1977), Introduction to the Renormalization Group and to Critical Phenomena (Wiley, New York).
Pittman, C., T. Doiron and H. Meyer (1979), Phys. Rev. B 20, 3678.
Rehr, J. J. (1979), J. Phys. A 12, L179.
Rehr, J. J. and N. D. Mermin (1973), Phys. Rev. A 8, 472.
Rowlinson, J. S. (1969), Nature 224, 541.
Saam, W. F. (1970), Phys. Rev. A 2, 1461.
Scheibner, B. A., C. M. Sorensen, D. T. Jacobs, R. C. Mockler and W. J. O'Sullivan (1978), Chem. Phys. 31, 209.
Schmidt, H. H. (1971), J. Chem. Phys. 54, 3610.
Schneider, R., L. Belkoura, J. Schelten, D. Woermann and B. Chu (1980), Phys. Rev. A (in press).
Scott, R. L. (1978), in Chemical Thermodynamics, Vol. 2, Specialist Periodical Report, M. L. McGlashnan, ed. (Chemical Society, London), p. 238.
Sengers, J. V. (1972), Ber. Bunsenges. physik. Chemie 76, 234.
Sengers, J. V., D. Bedeaux, P. Mazur and S. C. Greer (1980), Physica A (in press).
Sengers, J. V. and P. H. Keyes (1971), Phys. Rev. Lett. 26, 70.
Sengers, J. V. and J. M. H. Levelt Sengers (1978), in Progress in Liquid Physics, C. A. Croxton, ed. (Wiley, New York, 1978), Ch. 4.
Sengers, J. V. and M. R. Moldover (1978a), in Cospar: Space Research, Vol. XVIII, M. J. Rycroft and A. C. Strickland, eds. (Pergamon Press, N.Y.), p. 495; Z. Flugwiss. Weltraumforsch. 2, 371.

Sengers, J. V. and M. R. Moldover (1978b), Phys. Lett. 66A, 44.
Sheludyak, Yu. E. and V. A. Rabinovich (1979), High Temp. 17, 40.
Siggia, E. D., B. I. Halperin, and P. C. Hohenberg (1976), Phys. Rev. B 13, 2110.
Sorensen, C. M., R. C. Mockley and W. J. O'Sullivan (1978), Phys. Rev. Lett. 40, 777.
Stauffer, D., M. Ferer and M. Wortis (1972), Phys. Rev. Lett. 29, 345.
Stein, A. and G. F. Allen (1974), J. Phys. Chem. Ref. Data 2, 443.
Swinney, H. L. and D. L. Henry (1973), Phys. Rev. A 8, 2586.
Thijsse, B. J. (1980), to be published.
Thoen, J., E. Bloemen and W. Van Dael (1978), J. Chem. Phys. 68, 735.
Tsai, B. C. and D. McIntyre (1974), J. Chem. Phys. 60, 937.
Vause, C. and J. Sak (1980), Phys. Lett. 77A, 191; Phys. Rev. A 21, 2099.
Wallace, D. J. and R. K. P. Zia (1974), J. Phys. C 7, 3480.
Warkulwiz, V. P., B. Mozer and M. S. Green (1974), Phys. Rev. Lett. 32, 1410.
Wegner, F. J. (1972a), Phys. Rev. B 5, 4529.
Wegner, F. J. (1972b), Phys. Rev. B 6, 1891.
Wegner, F. J. (1975), J. Phys. A 8, 710.
Wheeler, J. C. (1977), Ann. Rev. Phys. Chem. 28, 411.
Widom, B. (1965), J. Chem. Phys. 43, 3898.
Widom, B. and J. S. Rowlinson (1970), J. Chem. Phys. 52, 1670.
Wilcox, L. R. and D. Balzarini (1968), J. Chem. Phys. 48, 753.
Wilson, K. G. and J. Kogut (1974), Physics Reports 12C, 75.
Zinn-Justin, J. (1979), J. Physique 40, 63.
Zollweg, J. A. and G. W. Mulholland (1972), J. Chem. Phys. 57, 1021.

AN ANALYSIS OF THE CONTINUOUS-SPIN, ISING MODEL*

George A. Baker, Jr.
Theoretical Division
Los Alamos Scientific Laboratory
University of California
Los Alamos, N.M. 87545

Since the time when the study of relations between the various critical indices was systemitized,[1] these indices have been classed into groups. First, I remind you of some notation. If χ is the magnetic susceptibility, M the magnetization, C_H the specific heat at constant magnetic field, and ξ the correlation length, then near the critical point, temperature, $T = T_c$, and magnetic field, $H = 0$, for an Ising model on a d-dimensional, rigid, regular space-lattice we expect, $T > T_c$, $H = 0$,

$$\chi \simeq A_+(T-T_c)^{-\nu}, \ \xi \simeq D_+(T-T_c)^{-\nu},$$

$$-\frac{\partial^2\chi}{\partial H^2} \simeq B_+(T-T_c)^{-\gamma-2\Delta}, \ C_H \propto (T-T_c)^{-\alpha}, \tag{1}$$

$T = T_c$,

$$M \propto H^{1/\delta}, \quad \langle\sigma_o\sigma_{\vec{r}}\rangle\Big|_{H=0} \propto r^{-d+2-\eta} \tag{2}$$

where $\langle\sigma_o\,\sigma_{\vec{r}}\rangle\Big|_{H=0}$ is the spin-spin correlation function between a spin σ at the origin and one at $\vec{r}$ in zero magnetic field.

$T < T_c$, $H = 0$,

---

*Work supported in part by the U.S. D.O.E.

$$\chi \simeq A_-(T_c-T)^{-\gamma'}, \quad \xi \simeq D_-(T_c-T)^{-\nu'}$$

$$-\frac{\partial^2\chi}{\partial H^2} \simeq B_-(T_c-T)^{-\gamma'-2\Delta'}, \quad C_H \propto (T_c-T)^{-\alpha'}$$

$$M \propto (T_c-T)^{\beta}. \tag{3}$$

In terms of this notation, a selection of the relations between the critical indices (α, γ, δ, etc.) would be:

single temperature region,

$$\alpha' + 2 \quad + \gamma' = 2; \tag{4}$$

critical isotherm plus a single temperature region,

$$\alpha' + \beta(1+\delta) = 2,$$

$$\delta = \Delta/(\Delta-\gamma); \tag{5}$$

two temperature regions

$$\gamma = \gamma', \quad \alpha = \alpha',$$

$$\Delta = \Delta'; \tag{6}$$

relations involving correlation exponents,

$$\gamma = (2-\eta)\nu,$$

$$\gamma' = (2-\eta)\nu'; \tag{7}$$

and relations involving the spatial dimension or hyperscaling,

$$d\nu = 2 - \alpha,$$

$$2 - \eta = d(\delta-1)/(\delta+1),$$

$$2\Delta = d\nu + \gamma. \tag{8}$$

On the numerical evidence, the hyperscaling relations (8) were the least well supported and those of (6) suffered initially from the weakness of the accuracy in the $T < T_c$ numerical results. Many of these relations have been proven to be rigorous inequalities, e.g.,[2-6]

$$\alpha' + 2\beta + \gamma' \geq 2, \qquad \gamma \geq (2-\eta)\nu,$$

$$d\nu + \gamma \geq 2\Delta, \qquad \delta \geq \Delta/(\Delta-\gamma),$$

$$\alpha' + \beta(1+\delta) \geq 2. \tag{9}$$

In order to understand what was going on, and to gain a deeper understanding of these exponent relations, general ideas that related them to scaling properties were put forth.[7,8] These ideas were further developed and extended by the use of field theoretic methods[9,10] to yield the renormalization group theory of critical phenomena, which rests on the renormalization group hypothesis[11,12] and has all the index relations (4-8) as a consequence for $d \leq 4$.

Now the trouble starts when one compares the results of the renormalization group theory of critical phenomena with those of the high-temperature series numerical computations. These high temperature series results yield, for example[13-15]

$$\gamma = 1.250 \pm 0.003,$$

$$\nu = 0.638 \begin{array}{l} + 0.002 \\ - 0.001, \end{array}$$

$$\Delta = 1.563 \pm 0.003, \tag{10}$$

and[16] for the renormalization group equality (8),

$$2\Delta - d\nu - \gamma = - 0.028 \pm 0.003, \; d = 3,$$

$$= - 0.302 \pm 0.038, \; d = 4. \tag{11}$$

These results show small but persistent deviations[17] from the expected renormalization group results,[14,18] in three dimensions,

$$\gamma = 1.241 \pm 0.004, \quad \nu = 0.630 \pm 0.002,$$

$$2\Delta - d\nu - \gamma \equiv 0. \tag{12}$$

Other analyses of the high-temperature series coefficients are to be found in this volume.

To study this discrepancy in detail, I prefer to put it in a broader context[11,12,19] and consider the continuous-spin Ising model which has both the spin-½, Ising model and Euclidean, Boson, quantum field theory as limiting cases.

The partition function for this model is

$$Z(H) = M^{-1} \int_{-\infty}^{+\infty}\cdots\int \prod_{\vec{i}=1}^{N} d\phi_{\vec{i}} \exp\left[-\tfrac{1}{2}v \sum_{\vec{i}=1}^{N} \left\{ \frac{2d}{q} \sum_{\{\vec{\delta}\}} \frac{(\phi_{\vec{i}} - \phi_{\vec{i}+\vec{\delta}})^2}{a^2} + m_o^2 :\phi_{\vec{i}}^{\,2}: + \frac{2}{4!} g_o :\phi_{\vec{i}}^{\,4}: \right\} + \sum_{\vec{i}} H_{\vec{i}} \phi_{\vec{i}} \right], \tag{13}$$

where a is the lattice spacing, $v \propto a^d$ is the specific volume per lattice site, q is the lattice coordination number, $\{\vec{\delta}\}$ is one-half the set of nearest neighbor sites, and $H_{\vec{i}}$ is the magnetic field at site $\vec{i}$. This model looks like a lattice-cutoff model field theory. If we perform the usual amplitude (Z_3) and mass renormalizations ($m_o^2 = m^2 + \delta m^2$), then we can rewrite (13) as

$$Z(\tilde{H}) = \tilde{M}^{-1} \int_{-\infty}^{+\infty}\cdots\int \prod_{\vec{i}=1}^{N} d\sigma_{\vec{i}} \exp\left[\sum_{\vec{i}} \left\{ K \sum_{\{\vec{\delta}\}} \sigma_{\vec{i}} \sigma_{\vec{i}+\vec{\delta}} - \tilde{g}_o \sigma_{\vec{i}}^{\,4} - \tilde{A} \sigma_{\vec{i}}^{\,2} + \tilde{H} \sigma_{\vec{i}} \right\} \right], \tag{14}$$

where the relation between the field theory language of (13) and the statistical mechanical language of (14) is

$$\tilde{g}_o = g_o K^2 q^2 a^4/(96 d^2 v) \propto g_o a^{4-d},$$

$$\tilde{A} = qK(2d + m^2 a^2 + \delta m^2 a^2 - \tfrac{1}{2} C a^2 g_o)/4d,$$

$$\tilde{H} = H_i \left[qKa^2/(2dZ_3 v) \right]^{\frac{1}{2}}. \tag{15}$$

Note that we have added a free parameter, K, and imposed a normalization condition,

$$\langle \sigma^2 \rangle_{H=K=0} = 1 = \frac{\int_{-\infty}^{+\infty} dx\, x^2 \exp(-\tilde{g}_o x^4 - \tilde{A} x^2)}{\int_{-\infty}^{+\infty} dx\, \exp(-\tilde{g}_o x^4 - \tilde{A} x^2)}, \tag{16}$$

which fixes $\tilde{A}$ as a function of $\tilde{g}_0$. Further note that C is the usual $[\phi^-,\phi^+]$ commutator which diverges as a goes to zero for $d \geq 2$. As usual, the renormalization conditions imposed on the two-point function,

$$\Gamma_R^{(2)}(p,-p) = \left\{ v \sum_{\vec{j}=0}^{N-1} \left.\frac{\partial^2 \ell n Z(H)}{\partial H_o \partial H_j}\right|_{H=0} \exp\left[-2\pi i \vec{p}\cdot\vec{j}a\right] \right\}^{-1} \tag{17}$$

determine the renormalization constants Z_3 and δm^2. These renormalization conditions are

$$\begin{aligned} \Gamma_R^{(2)}(p,-p) &\simeq m^2 + 4\pi^2 p^2 + \cdots, \text{ as } p \to 0, \\ &= \frac{2dZ_3}{qKa^2}\chi^{-1}(1 + (2\pi)^2\xi^2 a^2 p^2 + \cdots, \end{aligned} \tag{18}$$

in terms of

$$\begin{aligned} \chi &= \sum_{\vec{j}=0}^{N-1} [\langle\sigma_o\sigma_{\vec{j}}\rangle - \langle\sigma_o\rangle^2] , \\ \xi^2 &= \frac{\sum_{\vec{j}=0}^{N-1} j^2(\langle\sigma_o\sigma_{\vec{j}}\rangle - \langle\sigma_o\rangle^2)}{2d\chi} , \end{aligned} \tag{19}$$

where the expectation values are determined by the partition function (14). These conditions lead to the relations,

$$\begin{aligned} &m^2\xi^2 a^2 = 1, \\ &Z_3 = (\chi/\xi^2)(qK/2d). \end{aligned} \tag{20}$$

The object to be studied is the dimensionless, renormalized, coupling constant

$$g = g_R m^{d-4} = \frac{-v}{a^d}\,\frac{\frac{\partial^2 K}{\partial H^2}}{\chi^2\xi^d} \simeq (1 - T_c/T)^{\gamma+d\nu-2\Delta} \tag{21}$$

This quantity is bounded as $T \to T_c$ by Schrader's[4] inequality. If it goes to zero, then hyperscaling fails (8) and the corresponding field theory is trivial.[20] If it is finite, then hyperscaling holds.

The conventional wisdom for the behavior of $g(g_0,a)$ is that there is a limiting curve which is smoothly approached as $a \to 0$. By eq. (20) for a fixed, renormalized mass, this limit is equivalent to $\xi \to \infty$ with fixed lattice spacing, i.e., the temperature approaches the critical temperature. This limiting curve is conventionally thought to rise monotonically from zero for $g_0 = 0$ to a finite limit g^* for $g_0 = \infty$. Specifically, the renormalization group hypothesis[11,12,19] is that there exists a unique, non-zero limit as $g_0 \to \infty$ and $a \to 0$ independent of the manner of approach. From this hypothesis, as a statistical-mechanical problem corresponds to $\tilde{g}_0$ fixed, and by eq. (15) $\tilde{g}_0 \propto g_0 a^{4-d}$, we must have $g_0 \to \infty$ as $a \to 0$ for $d < 4$ and so $g \to g^*$. As everything is thought to depend on g, we must, based on this hypothesis, get the same result, i.e., universality, for any $\tilde{g}_0$-fixed, statistical-mechanical model. The hypothesized smoothness and differentiability of the approach to the limit yields the critical index relations.

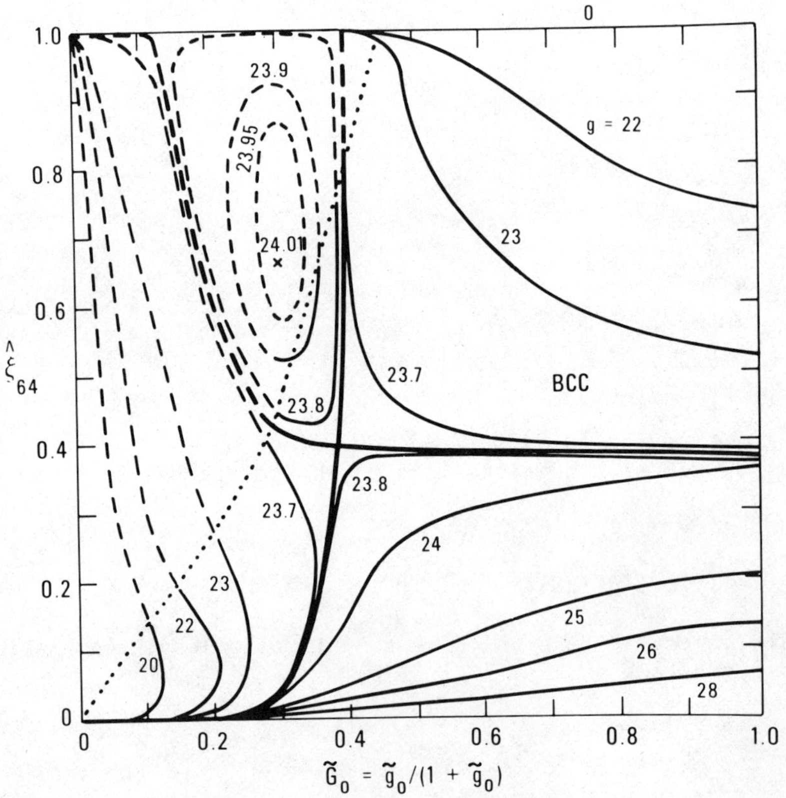

Fig. 1. Contours of the renormalized coupling constant, g, in the $\hat{\xi}_{64}$, $\tilde{G}_0$ plane for the body-centered-cubic lattice. Here $\hat{\xi}_{64} = \xi^2/(64+\xi^2)$ and $\tilde{G}_0 = \tilde{g}_0/(1+\tilde{g}_0)$. The boldface curve represents $g^* = 23.78$.

Baker and Kincaid[11,19] have made a detailed investigation using high-temperature series methods and concluded on numerical evidence that the renormalization group hypothesis holds for d = 1, 2 (known previously[21]) but fails in d = 3 and 4 dimensions. The results in three dimensions are particularly interesting as Fig. 1 illustrates. A much richer structure in the g-contour map is found than had been anticipated. The top edge of the figure is a spread-out version of the $g_0 = \infty$, $a = 0$ point. They[19] found that $g = g^* = 23.78$ alone did not appear to represent this point and that the g^* contour also extended into the interior and possessed a saddle point. We remark in passing that such a saddle point reconciles these numerical results with Schrader's[22] rigorous results.

Where can we look, theoretically, for the breakdown of hyperscaling in three and four dimensions? Looking back at eq. (8) we note that the occurence of the spatial dimension d appears in association with the relation of a single index such as ν or η for a microscopic property to a thermodynamic index such as α, γ, δ, etc. It is therefore interesting to introduce a thermodynamic coupling constant which replaces the dependence on ξ^d in (21) by a combination of thermodynamic variables. The most obvious move is to use the terms from Sokal's[23] proof of the Josephson inequality to make the replacement

$$\xi^d \to \left(\frac{\partial \chi}{\partial K}\right)^2 \Big/ (C_H\, \chi^2), \tag{22}$$

however, as $C_H \propto \ln(K_c - K)$ for d = 2, this replacement would lead to an infinite thermodynamic coupling constant in two dimensions. I prefer to make the replacement

$$\xi^d \to (\chi)^{(\delta+1)/(\delta-1)} \tag{23}$$

One finds directly by use of Fisher's[3] results

$$\gamma \le (2-\eta)\nu, \quad (2-\eta) \le d(\delta-1)/(\delta+1) \tag{24}$$

that

$$d\nu \ge \gamma(\delta+1)/(\delta-1) \tag{25}$$

Hence if we select

$$g_T = \frac{\dfrac{\partial^2 \chi}{\partial H^2}}{\chi^2\, \chi^{(\delta+1)/(\delta-1)}} \tag{26}$$

Then, including a dimension and lattice dependent constant, Ω, related[3] to the amplitude of the decay of the two-spin, correlation function with distance for $T = T_c$ in zero magnetic field, we may conclude

$$\Omega g_T \geq g, \quad (T \to T_c). \tag{27}$$

Since g is bounded from above[4] and goes to zero if hyperscaling fails, and, as we shall see below, since $A_+ \neq 0$ and $B_+ \neq 0$, g_T is not zero, although it could become infinite, we conclude that it is sufficient for (25) to be a strict inequality for

$$2\Delta < d\nu + \gamma. \tag{28}$$

That is to say, if one of the hyperscaling relations (8) fails [here (28)] then necessarily the others [here we will only see (25)] fail as well. Certainly this result is expected,[24] if the non-hyperscaling relations continue to hold. We remark that numerically g_T is finite for the cases tested (e.g., d = 2,3,∞) within error.

Now, to show that (26) does not go to zero as $T \to T_c$, consider

$$F(\tau) = \frac{M-\tau}{(1-\tau^2)} = (\chi-1)\,\tau + \left(\frac{\partial^2\chi}{\partial H^2} + \frac{4}{3}\,\chi-1\right)\tau^3 + O(\tau^5) \tag{29}$$

where $\tau = \tanh H$. Baker[25] has shown that the Yang-Lee theorem implies that

$$F(\tau) = \int_0^\infty \frac{\tau\, d\psi(\omega)}{1+\tau^2\omega} = f_1\tau - f_2\tau^3 + \cdots, \quad d\psi \geq 0, \tag{30}$$

i.e., $F(\tau)$ is τ times a series of Stieltjes. By standard theory[26], we must have

$$F(\tau) \geq \frac{f_1\tau}{1+f_3\tau^2/f_1}, \quad 0 \leq \tau \leq \infty. \tag{31}$$

If we choose,

$$\tau_s{}^2 = f_1/f_3, \tag{32}$$

since $\frac{\partial^2\chi}{\partial H^2}$ is negative and dominates[27] χ, then we have, as $\tau_s \to 0$ for $\delta > 1$, by monotonicity of the magnetization in temperature,[6]

$$M(\tau_s) \simeq \mu\tau_s{}^{1/\delta} \geq F(\tau_s) \geq \frac{1}{2}\,\chi^{3/2} \Big/ \left(\frac{\partial^2\chi}{\partial H^2}\right)^{\frac{1}{2}}, \tag{33}$$

which becomes,

$$\mu(T-T_c)^{\Delta/\delta} \geq \left[B_+^{3/2}/(2A_+^{\frac{1}{2}})\right](T-T_c)^{\Delta-\gamma} \tag{34}$$

or

$$\delta \geq \Delta/(\Delta-\gamma) \ . \tag{35}$$

This result is slightly stronger than the corresponding result of Gaunt and Baker[5] because their result is for Δ_∞, and this one is for Δ_4. The subscripts refer to the order of the derivative with respect to H involved in the definition. The result with Δ_4 is stronger than that with Δ_∞ as[25] $\Delta_{2m+2} \geq \Delta_{2m}$.

We have reduced the theoretical study of the apparent failure of hyperscaling in d = 3, 4 dimensions to a study of (25) which is defined in terms of only one and two-point correlations rather than (28) which also involves 4-point correlations. Presumably one could as well study the single-temperature relation, which involves only one and two-point correlations

$$2 - \eta \leq d\ (\delta-1)/(\delta+1) \tag{36}$$

which is equivalent to (25) if eq. (7) holds, but we have not proven this further simplification.

The failure of critical index, relations between correlation functions involving a different number of points is expected[28] to introduce, minimally, an anomalous dimension of the vacuum, i.e., replace d by $d-\omega^*$ in (8), and suggests that the genesis of the breakdown of hyperscaling comes in local properties at spin separations $r \ll \xi$, rather than sums over the whole lattice.

REFERENCES

1. M. E. Fisher, The Theory of Equilibrium Critical Phenomena, Repts. Prog. Phys. 30:615 (1967).
2. G. S. Rushbrooke, On the Thermodynamics of the Critical Region for the Ising Problem, J. Chem. Phys. 39:842 (1963).
3. M. E. Fisher, Rigorous Inequalities for Critical-Point Correlation Exponents, Phys. Rev. 180:594 (1969).
4. R. Schrader, New Rigorous Inequality for Critical Exponents in the Ising Model, Phys. Rev. B 14:172 (1976).
5. D. S. Gaunt and G. A. Baker, Jr., Low-Temperature Critical Exponents from High-Temperature Series: The Ising Model, Phys. Rev. B 1:1184 (1970).
6. R. B. Griffiths, Ferromagnets and Simple Fluids near the Critical Point: Some Thermodynamic Inequalities, J. Chem. Phys. 43:1958 (1965).
7. B. Widom, Equation of State in the Neighborhood of the Critical Point, J. Chem. Phys. 43:3898 (1965).

8. L. P. Kadanoff, Scaling Laws for Ising Models near T_c, Physics 2:263 (1966).
9. K. G. Wilson, Renormalization Group and Critical Phenomena, I. Renormalization Group and the Kadanoff Scaling Picture, Phys. Rev. B 4:3174 (1971); II. Phase-Space Cell Analysis of Critical Behavior, Phys. Rev. B 4:3184 (1971).
10. E. Brezin, J. C. LeGuillou, and J. Zinn-Justin, Field Theoretical Approach to Critical Phenomena in "Phase Transitions and Critical Phenomena, Vol. 6," C. Domb and M. S. Green, eds., Academic Press, London (1976).
11. G. A. Baker, Jr., and J. M. Kincaid, Continuous-Spin Ising Model and $\lambda{:}\phi^4{:}_d$ Field Theory, Phys. Rev. Lett. 42:1431 (1979).
12. G. A. Baker, Jr., The Continuous-Spin, Ising Model of Field Theory and the Renormalization Group in "Bifurcation Phenomena in Mathematical Physics and Related Topics," C. Bardos and D. Bessis, eds., D. Reidel Pub. Co., Dordrecht, Netherlands (1980).
13. C. Domb, Ising Model in "Phase Transitions and Critical Phenomena, Vol. 3," C. Domb and M. S. Green, eds., Academic Press, London (1974).
14. J. C. LeGuillou and J. Zinn-Justin, Critical Exponents from Field Theory, Phys. Rev. B 21:3976 (1980).
15. J. W. Essam and D. L. Hunter, Critical Behavior of the Ising Model Above and Below the Critical Temperature, J. Phys. C 1:392 (1968).
16. G. A. Baker, Jr., Analysis of Hyperscaling in the Ising Model by the High-Temperature Series Method, Phys. Rev. B 15:1552 (1977).
17. J. Zinn-Justin, Analysis of Ising Model Critical Exponents from High Temperature Series Expansion, J. de Phys. 40:63 (1979).
18. G. A. Baker, Jr., B. G. Nickel, M. S. Green, and D. I. Meiron, Ising-Model Critical Indices in Three Dimensions from the Callan-Symanzik Equation, Phys. Rev. Lett. 36:1351 (1976); J. C. LeGuillou and J. Zinn-Justin, Critical Exponents for the n-Vector Model in Three Dimensions from Field Theory, Phys. Rev. Lett. 39:95 (1977); G. A. Baker, Jr., B. G. Nickel, and D. I. Meiron, Critical Indices from Perturbation Analysis of the Callan-Symanzik Equation, Phys. Rev. B 17:1365 (1978).
19. G. A. Baker, Jr. and J. M. Kincaid, The Continuous-Spin Ising Model, $g_0{:}\phi^4{:}_d$ Field Theory, and the Renormalization Group, J. Stat. Phys. 24:471 (1981).
20. J. Glimm and A. Jaffee, The Coupling Constant in a ϕ^4 Field Theory, The Rockefeller University preprint (1979).
21. D. R. Nelson and M. E. Fisher, Ann. Phys. 91:226 (1975); L. P. Kadanoff, Correlations along a Line in the Two-Dimensional Ising Model, Phys. Rev. 188:859 (1969); J. Stephenson, Ising Model Spin Correlations on the Triangular Lattice, J. Math. Phys. 5:1009 (1964); B. M. McCoy and T. T. Wu, "The Two-Dimensional Ising Model" pp. 186-199, Harvard Univ. Press,

Cambridge (1973).

22. R. Schrader, A Possible Constructive Approach to ϕ_4^4 III, Ann. Inst. H. Poincaré 26:295 (1977).
23. A. Sokal, Rigorous Proof of the High-Temperature Josephson Inequality for Critical Exponents, Princeton University preprint (1980).
24. G. Stell, Weak Scaling, in "Proceedings of the International School of Physics 'Enrico Fermi', Critical Phenomena, Course LI," M. S. Green, ed., pg. 188, Academic Press, New York, 1971; M. E. Fisher, General Scaling Theory for Critical Points in "Proceedings of the Twenty-Fourth Nobel Symposium on Collective Properties of Physical Systems, Aspenäsgården, Sweden, 1973," B. Lundquist and S. Lundquist, eds., pg. 16, Academic Press, New York (1973).
25. G. A. Baker, Jr., Some Rigorous Inequalities Satisfied by the Ferromagnetic Ising Model in a Magnetic Field, Phys. Rev. Lett. 20:990 (1968).
26. G. A. Baker, Jr., "Essentials of Padé Approximants," Academic Press, New York (1975).
27. A. Sokal, "More Inequalities for Critical Exponents," Princeton University preprint (1980).
28. G. A. Baker, Jr. and S. Krinsky, Renormalization Group Structure for Translationally Invariant Ferromagnets, J. Math. Phys. 18:590 (1977).

WHAT IS SERIES EXTRAPOLATION ABOUT?*

George A. Baker, Jr.

Theoretical Division
Los Alamos Scientific Laboratory
University of California
Los Alamos, N. M. 87545

Much of the theoretical discussion in this volume revolves around the extraction of information about the value of a function from its Taylor series expansion. For example, the problem of finding a critical index can be set out mathematically as, given the Taylor series

$$f(x) = \sum_{n=0}^{\infty} f_n x^n \tag{1}$$

find for

$$F(z) = \frac{f(z)}{f'(z)} \simeq \gamma^{-1}(z-z_0) \tag{2}$$

the physically interesting zero, z_0, of F(z) and the slope, γ^{-1}, of F there. Many methods of obtaining estimates of these quantities are known and they are usually presented as formulas which relate the nth estimate to the first n coefficients f_n of the Taylor expansion of f(x). This mode of presentation might lead one to the hasty conclusion that series extrapolation is just a transformation of "input series data" into "output parameter estimates" without regard for intermediate aspects. The insight of the nineteenth century analysts was that the answer to this question is one of analytic continuation from information given at the point z = 0 to the point z_0 and is equivalent, in principle, to supplying F(z) on some path from 0 to z_0. Of course, for appropriately established methods, this work has already been done for

*Work supported in part by the U.S.D.O.E.

us. One well known example is the partial sums of the Taylor series inside its radius of convergence. However, not all methods have had the necessary work done, nor in some cases has it even been contemplated.

In order to illustrate the point, consider the example of Hardy.[1] Suppose

$$f(x) = 1 + 2x^2 + 2x^4 + 2x^6 + 2x^8 + \cdots\;, \tag{3}$$

which can be summed exactly for $|x| < 1$ to yield

$$f(x) = \frac{1 + x^2}{1 - x^2}\;. \tag{4}$$

On the other hand eq. (3) can be rearranged as

$$f(x) \overset{?}{=} 1 + \frac{1}{2}\left(\frac{2x}{1+x^2}\right)^2 + \frac{1}{2}\cdot\frac{3}{4}\left(\frac{2x}{1+x^2}\right)^4 + \frac{1\cdot3\cdot5}{2\cdot4\cdot6}\left(\frac{2x}{1+x^2}\right)^6 + \cdots\;, \tag{5}$$

which can be verified for $|x| < 1$ to be an exact rearrangement of (3) by expanding each term in (5) and comparing order-by-order in x. The new series plainly converges for small x, and indeed it gives the same answer there as do eqs. (3) and (4). However, (5) also plainly converges for very large x, which series (3) does not. For example, for x = 1000, series (5) yields 1.0000020 ····, whereas the analytic continuation of (4) yields -1.0000020 ···· which is plainly different!

What has happened? One should be careful when resumming a divergent series, e.g. series (3) for $|x| > 1$. Thus let us look at

$$1 + \frac{1}{2}\left(\frac{2x}{1+x^2}\right)^2\lambda + \frac{1}{2}\cdot\frac{3}{4}\left(\frac{2x}{1+x^2}\right)^4\lambda^2 + \frac{1\cdot3\cdot5}{2\cdot4\cdot6}\left(\frac{2x}{1+x^2}\right)^6\lambda^3 + \cdots\;, \tag{6}$$

which for sufficiently small λ can be summed to

$$\left[1 - \lambda\left(\frac{2x}{1+x^2}\right)^2\right]^{-\frac{1}{2}} = \left[\frac{1+2(1-2\lambda)x^2+x^2}{(1+x^2)^2}\right]^{-\frac{1}{2}}\;. \tag{7}$$

Now to go from $|x| < 1$ to $|x| > 1$ one must cross $|x| = 1$. If $|x| = 1$, then we may write $x = e^{i\theta}$, and

$$\frac{2x}{1+x^2} = \sec\theta \tag{8}$$

so for $|x| = 1$

$$1 \leqslant \left(\frac{2x}{1+x^2}\right)^2 \leqslant \infty \ . \tag{9}$$

For $\lambda = 1$, therefore the passage for $|x| < 1$ to $|x| > 1$ must cross the branch-cut of the square root in (7). Hence, we have the explanation of the failure of eq. (5) for $x = 1000$, in spite of the rapad convergence of the series there. There is no convergence of (5) on that point of the path from 0 to 1000 for which $|x| = 1$.

The warning of this example is that any method of series extrapolation, no matter how aesthetically appealing, must be thoroughly analyzed before it can be relied on. Padé methods[2] directly produce an approximate analytic continuation which can be cross-checked by, for example, the table of values method of Baker and Hunter.[3] For the ratio methods[3] in simple cases where

$$f(x) = A(x)\ (1 - \mu x)^{-\gamma} + B(x), \tag{10}$$

one uses the $n^{\underline{th}}$ order estimated parameters $\hat{A}_n$, $\hat{\mu}_n$ and $\hat{\gamma}_n$ to construct the orthodox, but rarely used, approximate analytic continuation

$$f_n(x) = \hat{A}_n(1-\hat{\mu}_n x)^{-\hat{\gamma}_n} + \sum_{j=0}^{n} \left[f_j - \hat{A}_n \binom{-\gamma}{j} (-\hat{\mu}_n)^j \right] x^j . \tag{11}$$

The convergence of the $f_n(x)$ can then be checked along a suitable path from $x = 0$ to $x = \hat{\mu}_n^{-1}$, and thus the convergence of the ratio method explored.

The example of Camp[4] serves well in the context of this volume as an illustration of the inadequacy of merely looking at the convergence of the "output parameter estimates." He shows how, using the first 10 terms of the very smooth looking series to some functions of the form

$$[A + Bx + (C + Dx)(1 - \mu x)^{\frac{1}{2}}](1 - \mu x)^{-5/4}, \tag{12}$$

he obtains by the "n-shifted" ratio method very good seeming convergence to a $\hat{\gamma} \simeq 1.23$ instead of the true value of 1.25. Had the corresponding $f_n(x)$ been computed and examined [here ${}_2F_1(1,\hat{\gamma}_n + \Delta n;\ 1 + \Delta n;\ -\hat{\mu}_n x)$ plays the role of $(1 - \hat{\mu}_n x)^{-\hat{\gamma}_n}$ in eq. (11)], a different conclusion could be foreseen. Likewise, the need for the use of orderly procedures of series analysis in the problems of actual physical interest, cannot be too strongly emphasized.

REFERENCES

1. G. H. Hardy, "Divergent Series", Oxford University Press, London (1956).
2. G. A. Baker, Jr., "Essentials of Padé Approximants", Academic Press, New York (1975).
3. D. L. Hunter and G. A. Baker, Jr., Methods of Series Analysis. I. Comparison of Current Methods Used in the Theory of Critical Phenomena, Phys. Rev. B 7:3346 (1973).
4. W. Camp, this volume.

HIGH TEMPERATURE SERIES, UNIVERSALITY AND SCALING CORRECTIONS+

William J. Camp

Sandia National Laboratories

Albuquerque, New Mexico 87185 U.S.A.

ABSTRACT

The evidence for universality of critical behavior is reviewed from a series expansion viewpoint. The necessity of incorporating non-analytic corrections to dominant scaling-theory singularities is pointed out, and the outlook for the verification of hyperscaling is discussed.

The history of critical phenomena contains perhaps the greatest successes of statistical physics. Indeed the cooperative behavior of large numbers ($\sim 10^{23}$) of interacting particles is remarkably well understood today. The path from Landau's self-consistent field theory for phase transitions[1] to Wilson's unified scaling theory[2] is marked by several remarkable mathematical triumphs, by inituitive leaps, and by--especially--the curious combination of rigorous mathematics and analogical reasoning that is series expansions.

Of the mathematical tours de force most significant was Onsager's solution of the two-dimensional Ising model,[3] which demonstrated the failure of Landau theory--at least in two dimensions. Perhaps the most important intuitive leap was that of Kadanoff[4] who at once postulated universality classes and scaling theory, based mainly on the notions that the only length of importance in the critical region is the correlation length, and that the symmetry of the interactions, but not their details, determines the critical behavior.

It is important to realize that during the period from Onsager to Wilson there did not exist a theory for phase transitions which at once predicted the structure of the phenomena and provided a calculus with which to quantify predictions--as had, albeit in a very crude way, Landau theory. This is not to overlook the work of Yang and Lee[5] who provided a powerful theoretical framework which demonstrated firmly that statistical mechanics provided an adequate mathematical foundation for phase transitions. Unfortunately, however, the Yang-Lee theory has provided neither a detailed phenomenology, nor a calculus for critical phenomena. Rather, during this period, most of our information came from series expansion studies, led notably by Domb and coworkers[6] at King's college.

By analogy with Onsager's solution of the two-dimensional Ising model, singularities of the form

$$\chi \sim |T-T_c|^{-\gamma} \tag{1}$$

i.e., branch cuts, were anticipated in the various thermodynamic and microscopic quantities at the critical point. Powerful statistical-mechanical computational methods were developed which allowed workers to obtain quite long Taylor series as functions of inverse temperature in two and three dimensions, for the Ising model especially.[6,7] Assuming that the nearest singularities to the origin are of this branch-out form, one could utilize series extrapolation techniques to analytically continue the functions of interest up to the critical point, and to estimate values for the critical temperature T_c and the critical exponent, γ. This method, when tested against the exact two-dimensional results, was found to be very successful.[8] It then was employed to analyze the critical behavior of three dimensional models, notably the spin-half Ising and Heisenberg models.[6,7,8,9] For the Ising model in both two and three dimensions, this work indicated that the Ising critical exponents depend on dimensionality but not on lattice structure.

The advent of more powerful computers, and the development of sophisticated series generation methods led to series analysis of models with more structure than the spin-half Ising model. Perhaps the spin-S Ising models,[10,11,12] the continuous-spin Ising models,[13,14] the n-vector models[15] and the classical anisotropic Heisenberg model[16,17] have been most studied. The longest, apparently best behaved and most studied series are still those of the spin-half Ising model[18]. Nonetheless, series for the spin-S and continuous-spin Ising models have been considerably lengthened[11,12,13,14] as have those for the anisotropic Heisenberg model.[16,19] Finally, very long series for Ising models on the bcc lattice have just been obtained by Nickel.[20]

The early work[10] on the spin-S Ising model, defined by its Hamiltonian

$$H = -(J/S^2) \sum_{n.n.} S_i S_j \tag{2}$$

where S_i takes values $-S$, $1-S$, ... $S-1$, S, indicated that the exponent γ, which characterizes the strength of the susceptibility divergence, was roughly independent of spin.[10] The anisotropic classical Heisenberg model with Hamiltonian

$$H = -J \sum_{\mu} \sum_{n.n.} g_\mu S_i^\mu S_j^\mu \qquad (\mu=x,y,z) \tag{3}$$

was first studied in detail by Jasnow and Wortis.[16] In this model three-dimensional unit spins interact via an anisotropic exchange force. Three levels of symmetry may be identified: $g_x > g_y \geqslant g_z$ which is Ising or n=1 symmetry (H is invariant only under $S^\mu \longrightarrow -S^\mu$); $g_x = g_y > g_z$ which is x-y or n=2 symmetry (H is invariant under rotation in the x-y spin plane as well as under inversion in S^μ); and finally $g_x = g_y = g_z$ which is Heisenberg or n=3 symmetry (H has full rotational symmetry in spin space). To good accuracy these authors found that the critical exponent γ was independent of g_μ within a given n-class.

The above observations, together with numerous experimental results, led very naturally to the Kadanoff universality hypothesis.

In addition to series studies of thermodynamic functions, microscopic correlation functions were attacked using series methods, mainly by Fisher and coworkers.[21] This work showed clearly that the classical Ornstein-Zernike theory of correlations breaks down near the critical point. This breakdown was characterized by introduction of Fisher's exponent η which measures the deviation of critical point correlations $G(R) \sim R^{-(d-2+\eta)}$ from classical predictions $G(R) \sim R^{-(d-2)}$. With the introduction of η which is 1/4 in two dimensions and ~0.03--0.06 in three dimensions, the destruction of the entire classical framework for critical phenomena was completed. Although the calculations of η (which invariably involves small increments on large backgrounds) proved very difficult within the series method, rather convincing evidence that η is greater than zero was obtained. Incidentally, it is in the work of Fisher and Burford[21] that the first indications of nonanalytic scaling corrections were seen in series results (for the true correlation range) at least five years before they were predicted theoretically!

Thus prior to the advent of modern renormalization group theory, there existed accurate estimates for many critical properties based on series expansions. These estimates had demonstrated the complete failure of classical theories of critical phenomena. In addition, a phenomenological theory of scaling and universality had grown, in part, out of this body of knowledge. Nonetheless, the situation was highly unsatisfactory because in a very real sense series analysis methods are only well founded when the analytic structure of the function under study is "sufficiently well" characterized. In contrast, the branch-cut structure of critical point singularities was only assumed by analogy with Onsager's results. Furthermore, even if the leading singularity is of the branch cut type, subdominant corrections can wreak havoc with detailed series estimates (see below).

In the early seventies, Wilson[2] and others[22] developed the renormalization group theory (RGT) of phase transitions. In addition to providing a detailed physical picture of scaling theory the renormalization approach has provided a number of useful calculi for critical phenomena. With the use of full-blown field-theoretic methods results comparable in accuracy with series results--and in very good agreement with those results--have been obtained.[23]

Since the development of RGT, in order to study its foundations and verify its conclusions, series studies of three dimensional models have concentrated four areas: incorporation of subdominant scaling corrections; more detailed verification of universality, studies of the so-called hyperscaling relations, and investigation of multicritical-point scaling. We shall only be concerned with the first three in this lecture.

According to RGT the two-point correlation function may be written as

$$\Gamma(R,t,H,\{h_k\}) \approx R^{-(d-2+\eta)} \mathscr{D}\left(Rt^{\nu},\ H/t^{\Delta},\ \left\{h_k/t^{\nu\lambda_k}\right\}\right) \tag{4}$$

where R is the spatial separation, $t = (T-T_c)/T_c$ is the temperature-like distance from the critical point, H is the ordering field, $\{h_k\}$ are other relevant ($\lambda_k > 0$) or irrelevant ($\lambda_k < 0$) fields and $\mathscr{D}(x,y,\ldots)$ is the scaling function. By standard moment relations, we then obtain

$$\mu_n \equiv \int_{\Omega} d\vec{R}\,\Gamma(\vec{R})\ |R|^n \approx A_n\left(H/\tau^{\Delta},\ \left\{h_k/t^{\nu\lambda_k}\right\}\right) t^{-(\gamma+n\nu)}\ , \tag{5}$$

where $\gamma = (2-\eta)\nu$ defines the susceptibility exponent γ through the zeroth-moment relation:

$$\chi = \mu_0 = \int d\vec{R}\,\Gamma(\vec{R}). \tag{6}$$

Unless all relevant scaling fields are set to zero, eqs. (4) and (5) are divergent at the critical point (signifying that we have chosen the wrong fixed point). Assuming that all relevant fields are zero, we can find the effect of irrelevant fields by expanding the scaling functions $\mathscr{D}(x,y,\ldots)$ or $A_n(y,\ldots)$:

$$\mu_n \approx A_n^{(0)} t^{-(\gamma+n\nu)} \left\{1 + B_{n,1}t^{\Delta_1} + B_{n,2}t^{\Delta_2} + C_{n,1}t^{2\Delta_1} + \ldots\right\}\ , \tag{7}$$

where $\Delta_j = -\nu\lambda_j$ is positive by assumption, and where $\Delta_1 < \Delta_2 < \ldots$.

Thus the expected effect of the irrelevant variables is to add subdominant, confluent, branch-cut singularities to the dominant behavior.

ISING MODELS

Although early series studies of short spin-S Ising series indicated that exponents (e.g., γ) were independent of spin and lattice structure,[10] as longer series became available[11,12] ratio and Padé analysis (appropriate for the simple branch cut case) indicated a persistent "spin effect" in exponent estimates. For example, the susceptibility exponents varied smoothly from $\gamma = 1.23$ at $S = \infty$ to 1.25 at $S = 1/2$.[11,12] In an unpublished talk at the Newport Beach conference, Wortis[24] suggested that the large-S estimates were affected by a persistent curvature due to a square-root confluent correction term:

$$\chi \approx \chi_0 t^{-\gamma} \{1 + \chi_1 \sqrt{t}\} \quad . \tag{8}$$

Wortis suggested that allowance for such "confluent corrections" would make the series results consistent with spin universality. Later Wegner[25] was to predict just such corrections through RGT analysis.

A number of series analysis methods have been developed which are appropriate to functions with confluent corrections to dominant singular behavior. These techniques have been applied in detail to the spin-S Ising model by Wortis and coworkers[11,12] and by Camp and Van Dyke.[12] Four principle methods have been employed: two versions of the method of n-fits,[11,12] a non-linear transformation of series introduced by Baker and Hunter,[26] and the "model series" method.

To understand n-fits, recognize that a scaling function

$$\chi = \chi_0 t^{-\gamma} \left[1 + \chi_{11} t^{\Delta_1} + \chi_{21} t^{\Delta_2} + \chi_{12} t^{2\Delta_1} + \ldots\right] \tag{9}$$

is specified by values for the exponents γ, Δ_1, Δ_2, ..., the amplitudes χ_0, χ_{11}, χ_{21}, χ_{12} ..., and the critical point, T_c. If we specialize to the dominant singularity and the leading correction five quantities χ_0, χ_{11}, γ, Δ_1 and T_c are to be specified. One can utilize the high temperature series for χ

$$\chi = \sum_{n=0}^{\infty} \chi_n K^n \tag{10}$$

($K = J/k_B T$) to obtain estimates for the parameters χ_0, χ_{11}, γ, Δ_1 and T_c. This is done by formally expanding the assumed functional form in powers of K. Then we can obtain m-th order estimates for

χ_0, χ_{11}, γ, Δ_1 and T_c by equating the m-th, (m-1)-th, (m-2)-th, (m-3)-th and (m-4)-th coefficients of the two series. The equations obtained are non-linear and "poorly behaved." In fact if all five parameters are treated as unknowns the method becomes the "exponential fitting problem" discussed by Nickel[20] in his lectures. In that case, useful estimates are not obtained. In fact, it is doubtful if one can even find unique estimates in the range of physical significance.[11,12] If we specify either T_c or (preferrably) γ, the equations for the remaining four parameters become quite well behaved. This is one version of the method of four fits. Another method arises from the recognition that the ratio of successive coefficients in the high temperature series, $R_n = X_n/X_{n-1}$, should behave as[12]

$$R_n \approx R_\infty \left\{ 1 + \frac{\gamma - 1}{n} + \frac{a}{n^{1+\Delta_1}} + \frac{b}{n^2} + \cdots \right\} \tag{11}$$

Here R_∞ is the critical temperature (K_c^{-1}). Again one can solve a set of non-linear equations to obtain m-th order parameter estimates. The method is essentially similar in terms of behavior of solutions to that based on the coefficients directly.

The Baker-Hunter non-linear series transformation[26] converts the set of confluent branch cut singularities to an equivalent set of non-confluent simple poles. Consider a function,

$$f(x) = \sum_{i=1}^{N} A_i/(1 - yx)^{\gamma_i} \quad , \tag{12}$$

composed of a sequence of confluent branch cuts. Introduce a variable $e^{\xi} = (1 - yx)$. Then we have

$$f(x(\xi)) = F(\xi) = \sum_{i=1}^{N} A_i \, e^{-\gamma_i \xi} \quad . \tag{13}$$

This function has a power series in ξ given by

$$F(\xi) = \sum_{n=0}^{\infty} \frac{\xi^n}{n!} \left\{ \sum_{i=1}^{N} A_i (-\gamma_i)^n \right\} \tag{14}$$

We can form an auxiliary function $\mathscr{F}(\xi)$ with series coefficients

$$\mathscr{F}_n = n! \; F_n$$

where $\{F_n\}$ are the series coefficients of $F(\xi)$ in Eq. (14). Then, the auxiliary function becomes

$$\mathscr{F}(\xi) = \sum_{i=1}^{N} A_i \sum_{n=0}^{\infty} \xi^n(-\gamma_i)^n = \sum_{i=1}^{N} A_i/(1-\gamma_i\xi) \qquad (15)$$

Thus, to use the Baker-Hunter transformation one provides a "good" estimate for the critical point, y, reverts the series to one in ξ and forms the auxiliary series, $\mathscr{F}(\xi)$. Then estimates for the exponents and amplitudes are found by straightforward Padé analysis of $\mathscr{F}(\xi)$.[12,25] To the extent that a good estimate for y is available, the Baker-Hunter method obviates the exponential fitting problem.

The final method for confluent series analysis to be discussed is the "model series" method introduced by Camp and Van Dyke.[12] This method in a sense treats the available series coefficients as experimental data. Given an assumed functional form, the series coefficients are least-squares fit to the assumed form. A weighted fitting procedure is used which weights the series coefficients according to the amount of configurational information they contain (i.e., according to their sizes). A model series is then obtained which is the best fit to the series "data" with the assumed functional form. The quality of the fit of course depends on how well the series is represented by the assumed form.

In using model series the series analyst has a significant advantage over his experimental colleague. Basically the experimentalist has the problem that the data most representative of the asymptotic critical behavior is that taken deepest in the critical region. However, that data is also the noisiest. The experimenter is then faced with a difficult optimization problem. He may end up with little data that both high quality and representative of the asymptotic behavior. It is harder for the series analyst to get into the critical region. If N is the maximum order in his series, his data is limited to $t > t_o$ where $\xi(t_o)/a \doteq N$. (Here $t = (T-T_c)/T_c$), $\xi(t)$ is the correlation length, and a is the lattice spacing). However, the theorist has no experimental noise in his series (unless he has made a mistake in developing the series). In practice, model series seems to characterize critical behavior quite nicely in many cases.

To date, the convergence of various confluent-singularity analysis techniques has not been analyzed in sufficient detail. Two kinds of error problems have to be considered. First, one must assess the sensitivity of the methods to errors in the assumed critical point or the assumed dominant exponent. Second, the kinds of errors caused by correction terms other than the confluent branch cuts must be assessed. These include additive and multiplicative analytic terms and non-confluent singularities (such as antiferromagnetic singularities). For the Baker-Hunter transformation, errors of the first type have been considered by Baker and Hunter.[25] Camp and Van Dyke[12] performed some heuristic response-surface analysis for errors of the second kind in the case of the Baker-Hunter transformation. For four-fits and model series only limited analysis of either error type has been carried out.[11,12]

The use of confluent-singularity analysis techniques has been limited to the various Ising models[11,12,13,27] in two and three dimensions and to the classical Heisenberg model at and near isotropy.[28] For the spin-S Ising model, analysis of susceptibility and second-moment series has been carried out on the fcc and triangular lattices.[11,12] Attempts to use these methods on loose-packed lattices have failed completely. This is because on such lattices the non-confluent antiferromagnetic singularities are at $-K_c$ on the radius of convergence. These antiferromagnetic singularities are then the dominant corrections to scaling, and as such wreak havoc with any attempt to extract secondary, confluent contributions.[29] For the close packed lattices, the antiferromagnetic singularities are in the complex plane, removed beyond the radius of convergence. However, they still affect the analysis to a significant degree on the fcc lattice (see below); and on the triangular net they are strong enough that they vitiate the analysis of all but the spin-half case for which T_c is known exactly, and for which long series are available.[12] Consider the spin-half case: γ is exactly 1.75 and Δ_1 is 1.0. However, employing the <u>exact</u> critical point $\tanh(K_c) = 2 - \sqrt{3}$ in the Baker-Hunter transformation, we do not obtain γ correct to three places (1.75) until we examine estimates involving the eighteenth order in K.[12] Furthermore, even at this order Δ_1 is underestimated by some 4% What has happened is that complex poles--presumably due to the anti-ferromagnetic singularities--badly mar Padé analysis of the transformed series. Although such effects should be a priori weaker on the fcc lattice--and are found a posteriori to be more difficult to see--the triangular-net results serve notice of the inherent difficulties of the method.

On the fcc net analysis of the spin-S Ising susceptibility was carried out first, using tenth-order[12] and twelfth-order[11,12] series. Recall that for confluent singularity analysis to work either K_c or γ has to be specified. In two dimensions, where both are known exactly, either will do. However, in three dimensions, neither K_c nor γ is known exactly. Historically two values of γ have appeared naturally. For the spin-half model, standard ratio and Padé methods appear to converge nicely to $\gamma \simeq 1.25$. However, for large S (characterized best by $S = \infty$), the series exhibit strong "curvature" when subjected to standard ratio methods. Nonetheless, convergence to $\gamma = 1.23$ is apparently very good. So, it made sense to test these two values of γ using four fits.

When the estimate $\gamma = 1.23$ is employed in the four-fit procedure, no reasonable <u>single</u> value for Δ_1 is found as a function of S.[12] (Tenth order estimates for Δ_1 vary from 1.9 at $S = 1$ to 0.89 at $S = \infty$). Furthermore, order-to-order scatter in the estimates is very large. By contrast when $\gamma = 1.25$ is used in four-fits a single universal value of Δ_1 near 0.58--0.60 is found through tenth order for all S. Furthermore, the sequence of estimates for a given S are quite well behaved.[11,12]

The Baker-Hunter transformation provides an alternate approach to the problem. Here estimates for K_c in a range near best estimates from ratio and Padé analyses are input to the transformation. A pragmatic procedure (which works well on test functions) is employed whereby the value of K_c is chosen such that apparent convergence of estimates for γ and Δ_1 is optimized. On this basis, Camp and VanDyke[12] found "best" estimates for γ and Δ_1 in the range $1.240 \leq \gamma \leq 1.255$, $0.50 \leq \Delta_1 \leq 0.70$. Very good apparent convergence of the spin-half and spin-infinity cases to $\gamma \simeq 1.250$ was found.

Finally, it is interesting to use the model series method[12] to show how well the series data through twelfth order are characterized by Eq. (9) with $\gamma = 1.25$ and $\Delta_1 = (1/2)\Delta_2 = 0.50$. Consider the model function

$$\chi_1 = A_1(1 - yK)^{-5/4} + A_2(1 - yK)^{-1/4} + B_1(1 - yK)^{-3/4} + B_2(1 - yK)^{1/4} \; . \tag{16}$$

By a weighted least-squares fit to the susceptibility coefficients through N-th order, we obtain an N-th order model function $\chi_1^{(N)}$. In Table I such a sequence of model functions is displayed for the S = ∞ Ising model on the fcc net. The apparent convergence of the model function parameters is spectacular, indicating that Eq. (16) provides a particularly good representation of $\chi(S = \infty)$. To show by direct comparison how faithful the representation is, we compare in Table II the end-shifted ratio analysis of $\chi_1^{(10)}$ to that of $\chi(S = \infty)$. The end-shift method is a standard ratio method which allows for curvature due to the arbitrariness of the N = 0 point by introduction of a shift δ in N. That is the ratios are fitted to

$$R_n = R_\infty\left(1 + \frac{\gamma - 1}{n + \delta}\right). \tag{17}$$

Note from Table II that all significant differences between $\chi(S = \infty)$ and $\chi_1^{(10)}$ disappear by seventh order!

By use of the confluent singularity techniques, Camp and VanDyke[12] and Wortis and coworkers,[11] thus concluded that (i) the apparent nonuniversality of $\gamma(S)$ in the Ising model arose from corrections to scaling; (ii) of the two choices $\gamma \simeq 1.25$ and $\gamma \simeq 1.23$ only the former was consistent with universality; and (iii) if the assignment $\gamma = 1.25$ were made, the amplitude of the confluent correction term apparently decreased smoothly to near zero at S = 1/2 (consistent with the apparent lack of curvature in S = 1/2 ratio analyses[6]).

Table 1. Parameters of the Least-Square Fit to Eq. (16) for S = ∞.

N	Y	A_1	B_1	A_2	B_2
6	3.50748	0.25876	0.06746	0.02232	-0.003
7	3.50734	0.25894	0.06747	0.01753	-0.009
8	3.50734	0.25894	0.06747	0.01742	-0.009
9	3.50733	0.25894	0.06748	0.01744	-0.009
10	3.50733	0.25894	0.06748	0.01747	-0.009
11	3.50733	0.25894	0.06748	0.01748	-0.008
12	3.50733	0.25894	0.06748	0.01744	-0.009

Table II. End-Shifted Ratio Analysis of $\chi_1^{(10)}$ and $\chi(S = \infty)$

	χ_1					$\chi(S = \infty)$
N	y(1)	γ(1)	δ(1)	y(∞)	γ(∞)	δ(∞)
4	3.4550	1.391	1.99	3.4831	1.300	1.30
5	3.5017	1.253	0.96	3.5053	1.241	0.82
6	3.5066	1.238	0.80	3.5082	1.232	0.72
7	3.5078	1.233	0.74	3.5079	1.233	0.74
8	3.5081	1.232	0.72	3.5081	1.232	0.72
9	3.5082	1.232	0.71	3.5081	1.232	0.72
10	3.5082	1.232	0.71	3.5081	1.232	0.72

A similar analysis of the second moment series for the spin-S Ising model on the fcc net[12] led to gratifyingly similar conclusions; namely, the correlation length exponent ν lay in the range 0.63--0.64, with the most probable value ν = 0.638, as previously estimated for the spin-half model.[6] Again the correction exponent was estimated to be $\Delta_1 \simeq 0.6$, as expected from Eq. (7). Finally, for the second moment also, if the assignment γ = 1.25, ν = 0.638 were made, the apparent amplitude of the leading correction term decreased rapidly as S decreased to one half.

The principal remaining quantity which has been studied in detail is the specific heat. Camp et al.[12] examined the spin-S Ising specific heat with a mind to universality and hyperscaling. Here the twelfth order series available for the internal energy (from identification of the internal energy with the nearest-neighbor pair correlations) are very short indeed. In addition, specific heat series are extremely difficult to analyze. This is because it behaves (at best) as

$$C_{H=0} \approx A(t) + B(t)\,(t^{\alpha} - 1)/\alpha \tag{18}$$

with $|\alpha| \ll 1$, and with A(t) and B(t) analytic amplitude functions. Thus, the singularity in $C_{H=0}$ literally hides in the background. Its extraction is to say the least an untrustworthy proposition. Nevertheless, Camp et al.[12] were able to say with some confidence that α lay in the range 0.10--0.13 with perhaps the best estimate $\alpha \approx 0.11$ for both S = 1/2 and S = ∞. Even so, using longer series available from spin-half free energy expansions,[6] some evidence was found for the traditional estimate $\alpha \simeq 0.125$.[6]

So, until recently the case seemed quite clear: universality was found to hold quite nicely for the Ising model. Corrections to scaling have been found which also were universal. However, the importance of such corrections diminished or disappeared altogether at spin half. This was consistent with the conclusion that the universal exponents were closer to those estimated by traditional methods for the spin-half model than to those found more recently by the same methods for spin infinity. Furthermore, the best exponent estimates seemed in violation of hyperscaling (which itself seems a nearly[30] inescapable conclusion of RGT[2,22]). In particular the hyperscaling relation

$$d\nu = 2 - \alpha \tag{19}$$

seemed to be violated for the three dimensional Ising model for which the best estimates were $3\nu = 1.91$ and $2 - \alpha = 1.88$--1.89, leading to a discrepancy of about 0.02 in the hyperscaling relation. However, as noted by Camp et al. the best estimates for γ and ν could be shifted easily by $\approx \pm 0.005$ without affected apparent convergence of confluent-singularity analysis. Furthermore, the value of α seemed particularly poorly determined. Thus, while breakdown of hyperscaling seemed a real concern, the evidence for it was not forcing.

More recently some of the detailed conclusions of the corrections-to-scaling analysis have been questioned as longer series have become available. McKenzie[27] extended the n-fit analysis of Camp and VanDyke[12] for the spin-half Ising susceptibility on the fcc lattice to fifteenth order. In particular she fit the series to

$$R_n = R_\infty \left(1 + \frac{\gamma - 1}{n} + \frac{a}{n^{1+\Delta_1}} + \frac{b}{n^2}\right) \tag{20}$$

with <u>both</u> γ <u>and</u> Δ_1 specified as input. A comparison was made between the estimates obtained using $\gamma = 1.250$, $\Delta_1 = 0.50$ and those obtained using $\gamma = 1.241$, $\Delta_1 = 0.496$ as estimated by LeGuillou and Zinn-Justin[23] from RGT calculations. It was found that through fifteenth order, the latter set of exponents yielded better apparent convergence for estimates of R_∞, a and b than did the former. Even so, the estimates of R_∞ differed only by 0.01% in the two cases.

In addition, Camp and VanDyke's conclusion that corrections to scaling are apparently very small for $S = 1/2$[12] was reconfirmed, although the size of the corrections were found to be of opposite sign and two to three times as large as estimated by Camp and VanDyke (χ_{11}/χ_0 in Eq. (9) was estimated by McKenzie[27] to be $\simeq -0.05$ as compared with Camp and VanDyke's estimate $\chi_{11}/\chi_0 \simeq 0.02$.[12])

The recent work by Nickel,[20] who has obtained twenty-first order series for the spin-S Ising susceptibility on the bcc lattice raises even more doubt about the conclusions of confluent singularity analysis. Nickel estimates that $\gamma(S) \simeq 1.235$--1.239 for all S in strong contrast to previous estimates. Furthermore, he finds evidence for $\gamma(1/2) \simeq 1.235$--1.239 on the fcc lattice also, based on analysis of the fifteenth-order susceptibility series. Nickel questions neither the universality of γ, nor that of Δ_1, nor the importance of corrections to scaling. Rather, he points to the difficulty of resolving the correct asymptotic behavior from short series, and cites evidence that twelfth-order series are short indeed for such purposes.[20]

Although further analysis of the bcc series is called for, it appears difficult to envisage resolving differences smaller than 0.01 or 0.02 in critical exponents based on analysis of currently available (or projected) series. If this pessimistic conclusion holds up other avenues of investigating such interesting questions as hyperscaling must be pursued (such as those opened up by Nickel and Sharpe[14] and by Baker and Kincaid[14] based on direct analysis of the effective field theoretic coupling constant). On further reflection, if we accept this limit of accuracy the problem of hyperscaling disappears into the noise (at least to the extent it has arisen from series estimates of critical exponents).[31]

HEISENBERG MODELS

The problem of universality for the spin-S Heisenberg model was reviewed and discussed in the context of corrections to scaling by Camp and VanDyke.[28] In this case, series are shorter than those available for the Ising model, and apparently much noiser. Nonetheless, some evidence for universality in γ was found. The best previous estimate for $\gamma(S = 1/2)$ was that of Baker et al.,[32] namely $\gamma(S = 1/2) \approx 1.43$. Camp and VanDyke[28] found no evidence for confluent corrections to scaling in $\chi(S = 1/2)$ but noted that the uncertainty in the above estimate was rather large. They quoted $\gamma(S = 1/2) \approx 1.41$--$1.51$ with perhaps a "best guess" $\gamma(S = 1/2) = 1.43$ in accord with Baker et al.[32] For spin infinity, the susceptibility series are longer and apparently better behaved. Previous estimates of $\gamma(S = \infty)$ were 1.375 ± 0.02 by Ritchie and Fisher[33] and 1.405 ± 0.02

by Ferer et al.[34] In contrast, Camp and VanDyke[28] found evidence for a confluent correction term with correction exponent $\Delta_1 \approx 0.54 \pm 0.10$. When the confluent corrections were allowed for the best estimate for $\gamma(S = \infty)$ was found to be $1.42^{+0.02}_{-0.01}$.

Therefore, the best estimates for spin-half and spin-infinity are at least consistent with universality; and a best universal value $\gamma \approx 1.42$--1.43 might be estimated.

For the correlation functions, universality cannot be investigated because series of adequate length exist only at $S = \infty$.[28] However, one can investigate the series to see if they are consistent with the confluent correction picture painted for the susceptibility. The Heisenberg model has a relevant field not present for Ising models; namely, the anisotropy fields, g, mentioned above. This means that another scaling exponent ϕ enters which describes the crossover from Heisenberg (n = 3) to XY (n = 2) or Ising (n = 1) symmetry. The scaling function becomes

$$\mu_n \approx t^{-(\gamma + n\nu)} M_n \left(H/t^{\Delta}, g/t^{\phi}, ut^{\Delta_1}, \ldots\right) \quad . \tag{21}$$

Thus, one can form crossover functions

$$C_n = \lim_{g \to 0} \mu_n^{-1} \frac{\partial \mu_n}{\partial g} \tag{22}$$

which diverge at $t^{-\phi}$. Camp and VanDyke[28] then looked for corrections to scaling in both the moments and crossover functions. For both cases, the series were consistent with the presence of such corrections. The values of Δ_1 estimated were quite scattered but consistent with the estimate 0.54 ± 0.10 found from the susceptibility. When such corrections were allowed for the values for ν and ϕ were both somewhat higher than previous estimates. The correlation exponent ν was estimated to be $\nu = 0.725 \pm 0.02$; and ϕ was found to be 1.30 ± 0.03. In contrast previous series estimates were $\nu \approx 0.705$[33] or $\nu \approx 0.717$[34] and $\phi \approx 1.25$.[35] It should be noted that RGT estimates also tend to be lower than the Camp-VanDyke estimates.[22]

To summarize, the spin-infinity Heisenberg series are at least consistent with the presence of confluent corrections. Allowance for such corrections leads to estimates for γ consistent with best estimates for $\gamma(S = 1/2)$. However, in all cases the series are both too short and too noisy to draw reliable conclusions.

+ACKNOWLEDGEMENT

This work was supported in part by the United States Department of Energy. The author thanks J. P. VanDyke, M. E. Fisher and J. Rehr for useful discussions.

References

1. L. D. Landau, JETP 7 , 19 (1937).
2. K. G. Wilson, Phys. Rev. D2, 1438 (1970); see also K. G. Wilson and J. Kogut, Phys. Repts. 12C, 76 (1974).
3. L. Onsager, Phys. Rev. 65, 117 (1944).
4. L. P. Kadanoff, Physics 2, 263 (1966).
5. C. N. Yang and T. D. Lee, Phys. Rev. 93, 1131 (1954).
6. C. Domb in Phase Transitions and Critical Phenomena Vol. III, C. Domb and M. S. Green, eds. (Academic Press, London, 1974).
7. M. Wortis, in Phase Transitions and Critical Phenomena Vol. III, C. Domb and M. S. Green, eds. (Academic Press, London, 1974).
8. C. Domb, Adv. Phys. 9, 149-361 (1960).
9. G. S. Rushbrooke, G. A. Baker, Jr., and P. J. Wood in Phase Transitions and Critical Phenomena Vol. III, C. Domb and M. S. Green, eds. (Academic Press, London, 1974).
10. C. Domb and M. F. Sykes, Phys. Rev. 128, 168 (1962).
11. D. M. Saul, M. Wortis and D. Jasnow, Phys. Rev. B 11, 2571 (1975).
12. W. J. Camp and J. P. Van Dyke, Phys. Rev. B 11, 2579 (1979). W. J. Camp, D. M. Saul, J. P. Van Dyke and M. Wortis, Phys. Rev. B 14, 3990 (1976).
13. W. J. Camp and J. P. Van Dyke, AIP Conf. Proc. 24, 322 (1975).
14. B. G. Nickel and B. Sharpe, J. Phys A 12, 1819 (1979), G. A. Baker, Jr. and J. M. Kincaid, Phys. Rev. Lett. 42, 1431 (1979); and J. Stat. Phys. 24, 471 (1981).
15. H. E. Stanley in Phase Transitions and Critical Phenomena Vol. III, C. Domb and M. S. Green, eds. (Academic Press, London, 1974).
16. D. Jasnow and M. Wortis, Phys. Rev. 176, 739 (1968).
17. P. Pfeuty, D. Jasnow and M. E. Fisher, Phys Rev B 10, 2088 (1974).
18. See the lectures by D. S. Gaunt and S. MacKenzie, this volume.
19. Tenth order series for the correlations and moments of the anisotropic Heisenberg model have been obtained by W. J. Camp and J. P. Van Dyke, 1974, unpublished.
20. See B. G. Nickel's lecture, this volume.
21. M. E. Fisher and R. J. Burford, Phys. Rev. 156, 583 (1967).
22. See the articles in Phase Transitions and Critical Phenomena Vol. VI, C. Domb and M. S. Green, eds. (Academic Press, London, 1976).
23. G. A. Baker, Jr., B. G. Nickel, M. S. Green and D. I. Meiron Phys. Rev. Lett. 36, 1351 (1976); J. C. LeGuillou and J. Zinn-Justin, Phys. Rev. Lett. 39, 95 (1977).
24. M. Wortis, Newport Beach Conference on Phase Transitions, 1970 (unpublished).
25. F. J. Wegner, Proceedings of the Conference on the Renormalization Group in Critical Phenomena and Quantum Field Theory, Chestnut Hill, 1973 (Temple University, Philadelphia, 1974).
26. G. A. Baker, Jr. and D. L. Hunter, Phys. Rev. B7, 3377 (1973).

27. S. McKenzie, J. Phys. A12, L185 (1979).
28. W. J. Camp and J. P. VanDyke, J. Phys. A9, 731 (1976).
29. Even traditional Euler transform techniques fail to improve matters on the bcc and sc lattices--see Camp and VanDyke, Ref. 12.
30. M. E. Fisher in Nobel 24 (1973) Collective Properties of Physical Systems has suggested several ways in which hyperscaling may be violated in a generalized scaling theory.
31. However, it should be noted that the accuracy with which discrepancies in hyperscaling have been found is comparable to that in the determination of η. [The point is that sometimes the intuition an investigator develops about a problem is more trustworthy than his numerical estimates would seem to justify.]
32. G. A. Baker, Jr., M. E. Gilbert, J. Eve and G. S. Rushbrooke, Phys. Rev. 164, 800 (1967).
33. D. S. Ritchie and M. E. Fisher, Phys. Rev. B5, 2668 (1972).
34. M. Ferer, M. A. Moore and M. Wortis, Phys. Rev. B4, 3954 (1971).
35. P. Pfeuty, D. Jasnow and M. E. Fisher, Phys. Rev. B10, 2088, (1974).

BICRITICALITY AND PARTIAL DIFFERENTIAL APPROXIMANTS

Michael E. Fisher and Jing-Huei Chen

Baker Laboratory
Cornell University
Ithaca, New York 14853, U.S.A.

ABSTRACT

The question of understanding multicriticality is explained with emphasis on bicriticality and the anticipated scaling behavior. The problem is posed of checking theoretical ideas and obtaining reliable estimates for model properties on the basis of series expansions in two variables of restricted length. Inhomogeneous partial differential approximants are defined and shown to be capable of meeting the challenge. Their principal properties are explained and techniques for their use are illustrated by application to bicritical behavior where both a crucial amplitude ratio and the full scaling function can be estimated with good precision.

1. CRITICALITY AND MULTICRITICALITY

Critical behavior in, say, a fluid or ferromagnet is characterized by two "relevant" variables: thus for a ferromagnet both variation of the magnetic fields, H, and the temperature, T, should be studied. Nevertheless, in many cases much of the important physics associated with an ordinary critical point can be extracted simply from the variation of various properties with the reduced temperature deviation $t = (T-T_c)/T_c$. In the case of multicriticality, however, the interplay of two (or more) variables constitutes an essential feature of the problem, both theoretically and experimentally. This is illustrated schematically in Figure 1 for the case of bicriticality in anisotropic antiferromagnets such as $GdAlO_3$ and MnF_2. Excellent portrayals of the experimental data will be found, in particular, in the work of Rohrer and coworkers [1-5]. (See also Refs. [6-12] for earlier experiments and data on other systems and different aspects of the phenomenon.) One finds that at high temperatures and general magnetic fields ($H \equiv H_{\parallel}$ parallel to the "easy" magnetic axis of the crystal), the magnetic spins are always disordered, thus defining a paramagnetic phase. On cooling the system in low fields across the critical line $T_c^{\parallel}(H)$ (see Figure 1) it undergoes a standard ordering transition, into an antiferromagnetic phase of $\parallel$ order with the spins predominantly aligned parallel or antiparallel to the easy axis. However, on cooling in a high field and crossing the distinct critical line, $T_c^{\perp}(H)$, the system passes into a so-called "spin-flopped" phase in which the spins are still ordered, predominantly, in antiferromagnetic fashion except that the spatial direction of the order is now perpendicular to the easy axis. The two distinct critical lines meet at the bicritical point (T_b, H_b) . Below T_b the two phases, antiferromagnetic and flopped (or $\parallel$ and $\perp$), are separated by a first order transition line, the "spin flop line," $H_{\varphi}(T)$, which is shown bold in the figure.

Now the two basic bicritical variables may be taken as

$$t = (T-T_b)/T_b \tag{1.1}$$

$$g = (H^2-H_b^2) , \tag{1.2}$$

where H^2 is used in place of H for theoretical reasons [13]. What is to be expected for the thermodynamic behavior in the vicinity of the bicritical point? The answer to this is provided by scaling theory through the appropriate scaling hypothesis [13, 14] which is confirmed theoretically by renormalization group, $\varepsilon = 4-d$ dimensionality expansions [15, 16].

First, it is essential to recognize the existence of special scaling axes in the (t, g) plane which may be defined by specifying

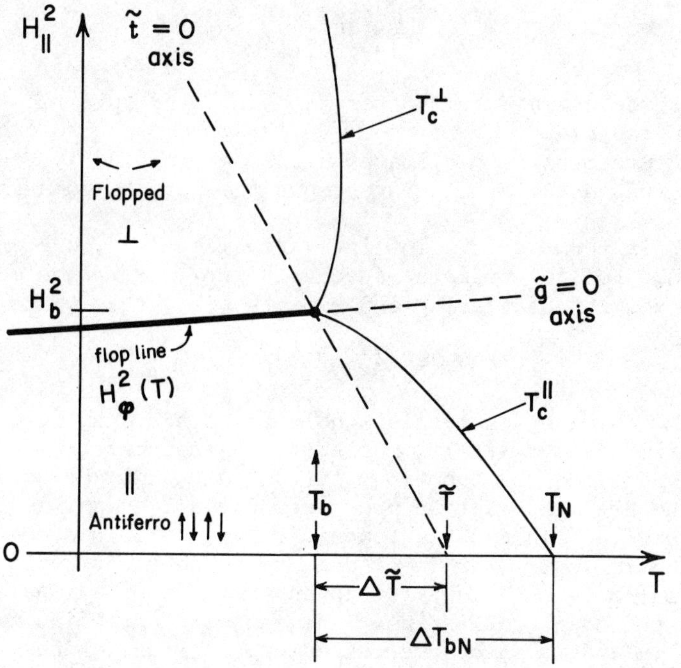

FIG. 1. Schematic phase diagram of an anisotropic antiferromagnet in a field, $H_{\parallel}$, parallel to the easy axis showing two distinct critical lines meeting at a bicritical point (T_b, H_b). Below T_b a first order spin flop transition line, $H_\varphi(T)$, separates a uniaxial, "parallel" ordered antiferromagnetic phase from a "perpendicular" ordered spin flopped phase at high fields. The dashed lines labelled $\tilde{t} = 0$ and $\tilde{g} = 0$ denote the bicritical scaling axes.

the <u>scaling variables</u>

$$\tilde{g} = g - pt \quad \text{and} \quad \tilde{t} = t + qg \quad . \tag{1.3}$$

These axes are shown in Figure 1 as straight dashed lines passing through the bicritical point. The slope of the $\tilde{g}=0$ axis is given simply by

$$p = T_b(\partial H_\varphi^2/\partial T)_b \quad , \tag{1.4}$$

and so can be determined from observations of the spin flop line. The value of the slope q, of the $\tilde{t}=0$ axis, raises more subtle questions but theoretical estimates can be given [13] for the point $\tilde{T}$ at which this scaling axis intercepts the zero field axis, $H=0$.

Second, in terms of the scaling variables (1.3), one expects a scaling form. For the ordering susceptibility, χ, which may be observed in neutron scattering experiments [11], this reads

$$\chi(T,H) \approx A\tilde{t}^{-\gamma} Z(C\tilde{g}/\tilde{t}^{\phi}) \quad , \tag{1.5}$$

as $t,g \to 0$. Similar expressions hold for other quantities, such as the specific heat, with γ replaced, in that case, by α and with a different scaling function, $Z(z)$, although the scaled form, $\tilde{g}/\tilde{t}^{\phi}$, of the argument must be preserved for all quantities in the bicritical region. In studying specific systems or models, experimentally, or theoretically, it is evident that one first wants to locate the bicritical point (T_b, H_b); then one needs to find the slopes, p and q, of the scaling axes; next one wants to determine the critical exponent, γ, and the crossover exponent, ϕ; and, finally, one should obtain the scaling function, $Z(z)$, and the two critical amplitudes (or "metrical factors") A and C. To fix A, the normalization condition

$$Z(0) = 1 \quad , \tag{1.6}$$

proves convenient. Various choices of normalization are possible to fix C (which need not be positive): one will be mentioned below.

A third feature of importance in our expectations for bicritical behavior is that the scaling function, $Z(z)$, must itself display singularities at two values of its argument, one positive and one negative, say at $z = \dot{z}_+$ and $z = -\dot{z}_-$. The reason for this is simply that $\chi(T, H)$ must diverge not only <u>at</u> the bicritical point [which is looked after by the prefactor $\tilde{t}^{-\gamma}$ in (1.5)] but also <u>on</u> both the critical lines $T_c^{\parallel}(H)$ and $T_c^{\perp}(H)$. Thus for $\dot{z}_+ > z > -\dot{z}_-$ we expect the scaling function to have a form like

$$Z(z) = P_+/(\dot{z}_+ - z)^{\gamma_+} + P_-/(\dot{z}_- + z)^{\gamma_-} + P_0(z) \quad , \tag{1.7}$$

where P_+ and P_- are constant amplitudes while $P_0(z)$ represents a less singular "background" term. The susceptibility exponents γ_+ and γ_- must be those appropriate to critical point transitions into the $\parallel$ and $\perp$ phases, respectively. In the case of $GdAlO_3$ both these phases have uniaxial anisotropy and so are Ising-like i.e. characterized by an (n=1)-component order parameter [1-3]: then we have $\gamma_+ = \gamma_I = \gamma_-$. For MnF_2, however, the spin-flopped phase has planar or XY-like character, with an (n=2)-component vector order parameter and then we have $\gamma_+ = \gamma_I$ and $\gamma_- = \gamma_{XY}$. Theoretically, this second case is somewhat more interesting since the bicritical point, which should then have Heisenberg-like or (n=3)-component character, is less symmetrical since it describes crossover from Ising-like to XY-like critical behavior. This situation will be the object of the specific theoretical studies we will report. In the former case, relevant to $GdAlO_3$, the bicritical point itself should exhibit XY-like character.

A little reflection shows that the scaling form (1.5) with $Z(z)$ singular at $\dot{z}_+$ and $-\dot{z}_-$ implies that the critical loci, $T_c^{\parallel}$ and $T_c^{\perp}$, must be described asymptotically, as $t \to 0$, by

$$\tilde{g}_c(T) = g_c(T) - pt \approx \pm(\dot{z}_{\pm}/C)\tilde{t}^{\phi} = \pm w_{\pm}\tilde{t}^{\phi} . \tag{1.8}$$

If $\phi \neq 1$ this implies that the two critical loci approach and meet at the bicritical point in a singular manner: specifically if, as predicted theoretically by ε-expansion calculations [13-16], one has $\phi > 1$ the critical loci must meet <u>tangentially</u> to one another and to the spin-flop line. Although such a singular feature in a simple phase diagram is, at first sight, rather surprising, the prediction is, in fact, fully confirmed by careful and precise experimentation [1-5]. The degree of tangency is found to be described by $\phi \simeq 1.17$ for $GdAlO_3$ and $\phi \simeq 1.27$ for MnF_2. Within the experimental and theoretical uncertainties of ±0.02 and ±0.03, respectively, the agreement between theory and experiment for the two cases is very encouraging.

A final significant feature of the theoretical expectations is that the (normalized) bicritical scaling function, $Z(z)$, should be <u>universal</u> i.e. independent of the details of the particular experimental system or theoretical model being considered, other than the question of the bicritical symmetry embodied (Heisenberg-like for MnF_2 or XY-like for $GdAlO_3$). This expectation means, specifically, that the amplitude ratio

$$Q = w_-/w_+ = \dot{z}_-/\dot{z}_+ , \tag{1.9}$$

which describes the relative rate of approach of the two critical loci to the bicritical point, should also be universal! In the XY-like case, describing Ising-to-Ising crossover, theory predicts simply $Q_{XY} \equiv 1$ [13, 14] and this is quite well confirmed by experiment [1, 2]. The value in the Heisenberg case is nontrivial and, indeed, has been somewhat problematical. It thus represents an important parameter we wish to estimate as part of a full determination of the scaling function. A check of the universality predictions should, of course, also be part of any systematic study.

As a last comment on the scaling formulation, notice that a natural normalization of the scaling function argument is provided by $\dot{z}_+ = 1$. This fixes the amplitude C in (1.5) which, in application to anti-ferromagnets as in Figure 1, then turns out to be negative.

2. THE ANALYSIS OF TWO-VARIABLE SERIES EXPANSIONS

Having established the importance of singular functions of two variables to the study of multicritical phenomena, let us consider how one might hope to estimate the multicritical, or <u>multisingular</u> behavior of some function, $f(x,y)$, of two (real) variables x and y if it has a power series expansion

$$f(x,y) \equiv \sum_{i,i'}^{\infty} f_{i,i'}\, x^{i} y^{i'} , \tag{2.1}$$

but, as will commonly be the case, one knows the expansion coefficients, $f_{i,i'}$, only for the labels (i,i') contained in some finite label set $\underset{\sim}{I}$ with $I = |\underset{\sim}{I}|$ distinct elements. In the practical cases we will discuss later, where f represents the total susceptibility of a classical spin model, one has, for example, $I = 36$, corresponding to the $\frac{1}{2}k(k+1)$ coefficients describing the expansion to $(k=8)$th order in x and y jointly.

Thus we are interested in methods of series analysis and extrapolation that would allow us to estimate the parameters in the asymptotic expression

$$f(x,y) \approx A\,|\Delta\tilde{x}|^{-\gamma}\, Z(C\Delta\tilde{y}/|\Delta\tilde{x}|^{\phi}) + B, \tag{2.2}$$

which, generalizing (1.5) slightly, is conjectured to be valid as (x,y) approaches "from below" some multisingular point (x_c, y_c). Here the deviations from multisingularity are defined by

$$\Delta x = x - x_c \quad \text{and} \quad \Delta y = y - y_c , \tag{2.3}$$

while the scaling combinations are

$$\Delta\tilde{x} = \Delta x - \Delta y/e_2 \quad \text{and} \quad \Delta\tilde{y} = \Delta y - e_1 \Delta x, \tag{2.4}$$

where e_1 and e_2 represent the slopes in the (x,y) plane of the first and second scaling axes. (Compare with the dashed lines in Figure 1.) Evidently, from the knowledge of the I coefficients we would like to estimate the six bicritical parameters x_c, y_c, e_1, e_2, γ, ϕ, the three amplitudes, A, B, and C, and the multisingular scaling function, Z(z), together with the location of any singularities, $z = \dot{z}_+$ etc., it may have and their nature.

One response to this challenge is simply to reduce the two-variable series (2.1) to a power series in a single variable by, for example, setting $y = kx$, etc. and then using established methods of single-variable analysis (see e.g. Refs. [17-22]). If there is some known symmetry and if the method is used judiciously, by examining various cross derivatives, $(\partial^{j+k}f/\partial x^j \partial y^k)$, it can provide useful information (see e.g. Ref. [23]). In other cases, however, it proves quite unsatisfactory and, clearly, by such means it is difficult to see how to estimate the scaling function singularities effectively.

Thus one turns to methods which more directly utilize the full information contained in the array, $\underset{\sim}{I}$, of known coefficients. One approach to the extrapolation of functions of two variables closely follows the well known method of constructing direct Padé approximants [19]. To explain this and the further developments, a notation for a finite polynomial of two variables is useful: to this end, if $\underset{\sim}{L} = \{(l,l')\}$ denotes a fixed set or array of coefficient labels we write

$$\sum_{(l,l')\in\underset{\sim}{L}} p_{l,l'}\, x^l y^{l'} \equiv P_{\underset{\sim}{L}}(x,y) \quad , \tag{2.5}$$

and so on for $Q_{\underset{\sim}{M}}(x,y)$, etc. .

Then the so-called Canterbury or Chisholm approximants [24, 25, etc.] are defined by approximating the required function as

$$f(x,y) \simeq P_{\underset{\sim}{L}}(x,y)/Q_{\underset{\sim}{M}}(x,y) \quad , \tag{2.6}$$

where the coefficients $p_{l,l'}$ and $q_{m,m'}$ are chosen so that the expansion of the ratio, $P_{\underset{\sim}{L}}/Q_{\underset{\sim}{M}}$ in powers of x and y matches the series (2.1) for f, as far as possible. However, it is not hard to see that such an approach is not helpful if behavior of the multisingular form (2.2) is anticipated or suspected [26]. Nor does one gain anything significant by the device of looking at the series for the logarithmic derivative, $D\log f = (\partial \ln f/\partial x)$ [26]. Thus some other idea must be sought!

To gain inspiration, consider the scaling form (2.2) in the simple case where the scaling axes are parallel to the original

Cartesian axes, i.e. $e_1 = 0$ and $e_2 = \infty$. On absorbing the amplitudes A and C into the definition of $Z(z)$ and taking (2.2) as a strict equality we have

$$f(x,y) = |\Delta x|^{-\gamma} Z(\Delta y/|\Delta x|^{\phi}) + B \quad . \tag{2.7}$$

Now let us attempt to derive a "scaling differential equation" in which the scaling function, Z, does not appear explicitly. Taking partial derivatives yields

$$f_x \equiv \frac{\partial f}{\partial x} = [-\gamma|\Delta x|^{-\gamma-1} Z - \phi\, |\Delta x|^{-\gamma-\phi-1}\Delta y\, Z'] \,\mathrm{sgn}(\Delta x) \;, \tag{2.8}$$

$$f_y \equiv \frac{\partial f}{\partial y} = |\Delta x|^{-\gamma-\phi} Z' \;, \tag{2.9}$$

where $Z' \equiv (dZ/dz)$. Now between these three equations we can eliminate Z and Z'. On rewriting the result we obtain the partial differential equation

$$U_0 + P_0 f = Q_1(x-x_c)\frac{\partial f}{\partial x} + R_2(y-y_c)\frac{\partial f}{\partial y} \;, \tag{2.10}$$

in which the constant parameters must satisfy the simple relations

$$P_0/Q_1 = -\gamma, \quad R_2/Q_1 = \phi, \quad U_0/P_0 = -B \;. \tag{2.11}$$

Since the equation (2.10) is linear one of the coefficients may be chosen arbitrarily: thus in general we may specify

$$P_0 = 1 \;. \tag{2.12}$$

From this small exercise we conclude that the scaling form (2.7) satisfies a simple linear inhomogeneous partial differential equation in which the multisingular point corresponds to the simultaneous vanishing of the coefficients, say Q and R, of the derivative terms $(\partial f/\partial x)$ and $(\partial f/\partial y)$. (It is worth remarking that the Callan and Symanzik approach to renormalized field theory [27, 28], and its developments used in constructing $\varepsilon = 4\text{-}d$ expansions for critical point behavior [29], yields a linear partial differential equation which is then seen to have solutions of scaling form!)

To handle the case where the scaling form holds only asymptotically close to the multisingular point, our observation suggests that we generalize the equation (2.10) by allowing the coefficients U_0, P_0, $Q_1\Delta x$, and $R_2\Delta y$ to become polynomials in x and y. The coefficients of these polynomials are then to be chosen, following the example of the Padé and Canterbury approximants, so that the power series solution of the partial differential equation matches the known expansion coefficients of $f(x, y)$ as far as possible.

This serves to construct inhomogeneous partial differential approximants (or PDA's for short). We will define these approximants more fully and study their properties and uses.

3. INHOMOGENEOUS PARTIAL DIFFERENTIAL APPROXIMANTS

Let us specify inhomogeneous partial differential approximants more precisely [22]: We propose the approximation of f as

$$f(x,y) \simeq F(x,y) \equiv F_{\underset{\sim}{J}/\underset{\sim}{L};\underset{\sim}{M},\underset{\sim}{N}}(x,y) \equiv [\underset{\sim}{J}/\underset{\sim}{L};\underset{\sim}{M},\underset{\sim}{N}]_f , \qquad (3.1)$$

where $F(x,y)$ is the solution, satisfying appropriate boundary conditions, of the generating equation

$$U_{\underset{\sim}{J}}(x,y) + P_{\underset{\sim}{L}}(x,y)F = Q_{\underset{\sim}{M}}(x,y)\frac{\partial F}{\partial x} + R_{\underset{\sim}{N}}(x,y)\frac{\partial F}{\partial y} , \qquad (3.2)$$

in which the coefficients $u_{j,j'}$, $p_{l,l'}$, $q_{m,m'}$, and $r_{n,n'}$ of the (finite) polynomials $U_{\underset{\sim}{J}}$, $P_{\underset{\sim}{L}}$, $Q_{\underset{\sim}{M}}$, and $R_{\underset{\sim}{N}}$ with specified label sets

$$\underset{\sim}{J} \equiv \{(j,j')\}, \quad \underset{\sim}{L} \equiv \{(l,l')\}, \quad \underset{\sim}{M} \equiv \{(m,m')\} \text{ and } \underset{\sim}{N} \equiv \{(n,n')\},$$

are chosen so that (3.2) holds as an identity with $F(x,y)$ replaced by the expansion (2.1) for $f(x,y)$ on a maximal specified matching label set, $\underset{\sim}{K} \equiv \{(k,k')\}$.

Since the generating equation, (3.2), is linear the polynomials U, P, Q and R may be normalized by imposition of an extra condition. The normalization relation

$$P_{\underset{\sim}{L}}(0,0) = p_{0,0} = 1 , \qquad (3.3)$$

is usually, but not always, appropriate. The precise nature of the "appropriate boundary conditions" on the solution $F(x,y)$ will vary somewhat with the nature of the equation for small x and y: the essential point, however, is that they must be chosen to ensure that those expansion coefficients of the approximant $F(x,y)$ which are not specified unambiguously by recurrence relations generated by (3.2), are set equal to the values dictated by matching to the expansion of $f(x,y)$.

The order of an approximant is specified by the "shapes" of the label sets $\underset{\sim}{J}$, $\underset{\sim}{L}$, $\underset{\sim}{M}$ and $\underset{\sim}{N}$ regarded as coefficient arrays. If one takes $\underset{\sim}{J} \equiv \emptyset$, the empty set, one obtains homogeneous partial differential approximants which may be denoted simply

$$F_{\underset{\sim}{L};\underset{\sim}{M},\underset{\sim}{N}}(x,y) \equiv [\underset{\sim}{L};\underset{\sim}{M},\underset{\sim}{N}]_f .$$

(These actually represent the approximants originally proposed [26, 30, 31].) Strictly, the matching array, $\underset{\sim}{K}$, should also be specified however, it is usually most natural to choose this to be as compact as possible and so, to avoid overburdening the notation, $\underset{\sim}{K}$ will normally be omitted in refering to a particular approximant. Likewise, when the arrays are triangular (e.g. $\underset{\sim}{J} = \{(j,j\;) : j+j' \le j_0\}$) or close-to-triangular, or close to other prespecified shapes, such as rectangular (e.g. $\underset{\sim}{J} = \{(j,j') : j \le j_1,\ j' \le j_2\}$), it often proves convenient to refer to approximants merely by the "sizes" or number of elements in the arrays $J = |\underset{\sim}{J}|$, $L = |\underset{\sim}{L}|$, $M = |\underset{\sim}{M}|$, and $N = |\underset{\sim}{N}|$ [where one has $J = \frac{1}{2}(j_0+1)(j_0+2)$ for triangular arrays and $J = (j_1+1)(j_2+1)$ for rectangular arrays, etc.].

As in the case of ordinary, single-variable Padé approximants [18, 19], the calculation of the coefficients $u_{j,j'}$, $p_{l,l'}$, etc. requires only the solution of a set of linear algebraic equations. It is thus a procedure that is easily programmed for a computer. However, one should be warned that double precision (or more, in the case of long series) is usually essential if sufficiently accurate and reliable results are to be obtained. An example may be helpful. Consider the low order approximant $[\emptyset/1\,;3,3]_f \equiv [1;3,3$ with arrays

$$\underset{\sim}{J} = \emptyset, \qquad \underset{\sim}{L} = \begin{array}{|l}\hline 1\end{array}, \qquad \underset{\sim}{M} = \underset{\sim}{N} = \begin{array}{|ll}\hline * & * \\ * & \end{array},$$

$$\underset{\sim}{K} = \begin{array}{|lll}\hline * & * & * \\ * & * & \\ * & & \end{array},$$

where the "1" in the equation for $\underset{\sim}{L}$ indicates imposition of the normalization condition (3.3): thus there are a total of six undetermined coefficients. The equations for these read, in matrix form,

$$\begin{bmatrix} f_{10} & \cdot & \cdot & f_{01} & \cdot & \cdot \\ 2f_{20} & f_{10} & \cdot & f_{11} & f_{01} & \cdot \\ f_{11} & \cdot & f_{10} & 2f_{02} & \cdot & f_{01} \\ 3f_{30} & 2f_{20} & \cdot & f_{21} & f_{11} & \cdot \\ 2f_{21} & f_{11} & 2f_{20} & 2f_{12} & 2f_{02} & f_{11} \\ f_{12} & \cdot & f_{11} & 3f_{03} & \cdot & 2f_{02} \end{bmatrix} \begin{bmatrix} q_{00} \\ q_{10} \\ q_{01} \\ r_{00} \\ r_{10} \\ r_{01} \end{bmatrix} = \begin{bmatrix} f_{00} \\ f_{10} \\ f_{01} \\ f_{20} \\ f_{11} \\ f_{02} \end{bmatrix} \tag{3.4}$$

where the dots denote zero entries. Provided the determinant of the matrix does not vanish, these equations may be solved to find the required polynomials $Q(x,y)$ and $R(x,y)$.

Of course, it may happen (as in constructing Padé approximants) that the determinant of the generating matrix, which evidently reflects directly the series coefficients $f_{i,i'}$, does indeed vanish. Then, either (a) the defining equations are underdetermined, so that there exists a continuum of approximants of the chosen order, or (b) the equations are overdetermined (or inconsistent), in which case no approximant of the chosen order exists. In the former case, one may frequently specify one or more of the extra coefficients in U, P, Q or R at one's convenience. (See a further remark below about circumstances in which the determinant vanishes.)

Although the arrays chosen for this example are compact (without 'holes') and triangular, this is certainly not necessary. Thus we might also have examined approximants with, say,

$$\underset{\sim}{M} = \begin{array}{|ll} \cdot & * \\ * & \end{array} \quad \text{and} \quad \underset{\sim}{N} = \begin{array}{|lll} \cdot & * & * \\ * & * & \\ * & & \end{array} ,$$

in which the dot denotes an absent (or zero) coefficient in $Q_{\underset{\sim}{M}}$ and $R_{\underset{\sim}{N}}$. Other special types of noncompact arrays will be mentioned below.

The generalization of (3.1) and (3.2) to three or more variables is obvious. It is also worthwhile here to comment on the reduction to a single variable, x (in which case, all terms involving R and $\underset{\sim}{N}$ are to be dropped). Then (i) the approximants $[\underset{\sim}{J}/\underset{\sim}{L};\emptyset]_f$ reduce to

$$f(x) \simeq F_{\underset{\sim}{J}/\underset{\sim}{L}}(x) = [\underset{\sim}{J}/\underset{\sim}{L}]_f = -U_{\underset{\sim}{J}}(x)/P_{\underset{\sim}{L}}(x), \qquad (3.5)$$

which (provided $\underset{\sim}{J}$ and $\underset{\sim}{L}$ denote the first consecutive J and L powers respectively, including x^0), are simply the ordinary, direct Padé approximants $[J-1/L-1]_f$ [17-22]. Likewise, (ii) on rearranging the generating equation (3.2), which is now an ordinary differential equation, the approximant $[\emptyset/\underset{\sim}{L};\underset{\sim}{M}]$, yields

$$\text{Dlog } f(x,y) \equiv \frac{d}{dx}\ln f(x) \simeq \frac{d}{dx}\ln F(x) = P_{\underset{\sim}{L}}(x)/Q_{\underset{\sim}{M}}(x) . \qquad (3.6)$$

This corresponds to the standard Dlog-Padé approximant which has proved so useful in the theory of critical phenomena in studying the behavior of functions with strong singularities [17-20]. Finally, (iii) in the general one-variable case, the approximants $[\underset{\sim}{J}/\underset{\sim}{L};\underset{\sim}{M}]$ are just the recently studied inhomogeneous ordinary differential approximants [22] (or "integral approximants" [21]), which are particularly useful for analyzing functions with weak singularities superimposed on significant "backgrounds", such as critical point specific heats for ferromagnets and fluids, as illustrated graphically in Ref. [22] . These differential approximants are, in turn, special cases of the general higher-order single-variable approximants proposed by Gaunt, Gammel, Guttmann and Joyce [32-34] which can be effective in studying functions with distinct but confluent singularities such as $f(x) \approx A_1|\Delta x|^{-\gamma_1} + A_2|\Delta x|^{-\gamma_2} + \cdots$ as $\Delta x \to 0$.

4. THE MULTISINGULAR BEHAVIOR OF PARTIAL DIFFERENTIAL APPROXIMANTS

Of most interest for application to bicriticality and other sorts of multicriticality is the fact that partial differential approximants can represent multisingular scaling behavior of the sort expressed in (2.2) and thereby provide an effective means of estimating the unknown multisingular parameters of central interest. Let us first summarize the essential formulae and then indicate their derivation.

Suppose that, in the vicinity of some anticipated or suspected multicritical point, (x_c, y_c), the polynomials $Q(x, y)$ and $R(x, y)$ of a given approximant have a common zero, (x_0, y_0), so that

$$Q(x_0, y_0) = R(x_0, y_0) = 0 . \tag{4.1}$$

[Here and below, as above, we omit explicit reference to the orders $\underset{\sim}{J}$, $\underset{\sim}{K}$, $\underset{\sim}{L}$, $\underset{\sim}{M}$ and $\underset{\sim}{N}$ unless they are of some special significance.] Then it transpires that x_0 and y_0 provide estimates for the true multisingular coordinates, x_c and y_c. Further, if we write

$$P_0 = P(x_0, y_0) \quad \text{and} \quad U_0 = U(x_0, y_0) , \tag{4.2}$$

the "background term", B in (2.2) is estimated simply by

$$B_0 = -U_0/P_0 . \tag{4.3}$$

If one uses a homogeneous approximant (with $\underset{\sim}{J} = \emptyset$) one, of course, obtains $B_0 \equiv 0$: this amounts to neglecting any background contributions to the leading multisingularity. If the singularity is strong (say, $\gamma \underset{\sim}{>} 1$) this may not be a bad approximation. In most cases, however, such a contribution should be allowed for and it will normally be essential if γ is small or negative.

Next make a Taylor expansion of Q and R about (x_0, y_0): this may be written

$$Q(x,y) = Q_1\Delta x + Q_2\Delta y + O(\Delta x^2, \Delta x\Delta y, \Delta y^2) , \tag{4.4}$$

$$R(x,y) = R_1\Delta x + R_2\Delta y + O(\Delta x^2, \Delta x\Delta y, \Delta y^2) , \tag{4.5}$$

where we now take

$$\Delta x = x - x_0 \quad \text{and} \quad \Delta y = y - y_0 , \tag{4.6}$$

while the numerical coefficients Q_1, Q_2, R_1 and R_2 are easily evaluated once x_0 and y_0 are found. Then the two slopes, e_1 and e_2, of the multisingular scaling axes [see (2.4)] are approximated by the appropriate roots, e_+ and e_- of the quadratic equation

$$Q_2\tilde{e}^2 + (Q_1-R_2)\tilde{e} - R_1 = 0 . \tag{4.7}$$

In terms of these roots one may define two <u>flow eigenvalues</u> (see below)

$$\lambda_1 = Q_1 - R_1/e_- \quad \text{and} \quad \lambda_2 = R_2 - e_+Q_2 . \tag{4.8}$$

The crossover and leading exponent in (2.2) are finally given by

$$\phi = \frac{\lambda_2}{\lambda_1} = \frac{R_2-e_+Q_2}{Q_1-R_1/e_-} \quad \text{and} \quad \gamma = \frac{-P_0}{\lambda_1} = \frac{-P_0}{Q_1-R_1/e_-} . \tag{4.9}$$

Note that a "misidentification" of e_+ and e_- with e_1 and e_2 merely amounts to the opposite choice of scaling axis convention which yields $\phi \Rightarrow \bar{\phi} = 1/\phi$ and $\gamma \Rightarrow \bar{\gamma} = \gamma/\phi$.

To establish these results and to see how to solve the generating partial differential equation (3.2) efficaciously, let us introduce a time-like flow variable, τ. (In an analogy with renormalization group techniques, τ corresponds to the spatial rescaling variable $l = \ln b$ [e.g. 29, 35, 36].) Then we define <u>flow trajectories</u> (or characteristics) in the (x,y) plane, $[x = x(\tau), y = y(\tau)]$, by the coupled first order, ordinary differential equations

$$\frac{dx}{d\tau} = Q(x, y) , \quad \frac{dy}{d\tau} = R(x, y) . \tag{4.10}$$

The generating equation (3.2) for the approximant $F(x, y) \equiv F(\tau)$ itself then reduces to the ordinary differential equation

$$\frac{dF}{d\tau} = P(x, y)F + U(x, y) , \tag{4.11}$$

which is easily integrated once the trajectories are known. In the case of a homogeneous approximant with $U \equiv 0$ the result may be written more explicitly as

$$\ln F(x, y) = \ln F(x_a, y_a) + \int_0^{\tau} P[x(\tau'), y(\tau')]d\tau' \qquad (4.12)$$

where (x_a, y_a) denotes the starting point of the trajectory, at which F is presumed known (or otherwise calculable — see below) and $x \equiv x(\tau)$ and $y \equiv y(\tau)$ specify the end points of the trajectory. As explained further in the discussion of the applications below, one normally envisages integrating inwards towards the 'unknown' multisingular region from the origin or axes where $F(x,y)$ is 'known'. However one may equally well integrate outwards from a point (x, y) where F is unknown until one reaches some boundary on which F is known and can be 'matched'. In short, the sign of τ is purely a matter of convenience.

To study a multisingular region — note that there may be more than one — using the trajectory equations (4.10), let us substitute the Taylor expansions (4.4) and (4.5) and neglect the quadratic terms. As a first step we seek special trajectories which are asymptotically linear through (x_0, y_0). Thus, putting $\Delta y(\tau) = \tilde{e}\Delta x(\tau)$ we obtain

$$\frac{(dx/d\tau)}{(dy/d\tau)} = \frac{Q_1 + \tilde{e}Q_2}{R_1 + \tilde{e}R_2} = \frac{\Delta x}{\Delta y} = \frac{1}{\tilde{e}} . \qquad (4.13)$$

This yields the quadratic equation (4.7) whose two roots give the scaling axis slopes. This in turn enables us to define the scaling combinations

$$\Delta\tilde{x} = \Delta x - \Delta y/e_- \quad \text{and} \quad \Delta\tilde{y} = \Delta y - e_+\Delta x , \qquad (4.14)$$

for which variables the trajectory equations (to linear order) yield simply

$$\frac{d}{d\tau}\Delta\tilde{x} = \lambda_1\Delta\tilde{x} \quad \text{and} \quad \frac{d}{d\tau}\Delta\tilde{y} = \lambda_2\Delta\tilde{y} , \qquad (4.15)$$

where λ_1 and λ_2 are given by (4.8). Solving these gives

$$\Delta\tilde{x} \approx ae^{\lambda_1\tau} \quad \text{and} \quad \Delta\tilde{y} \approx be^{\lambda_2\tau} , \qquad (4.16)$$

where a and b are the constants of integration. Eliminating τ between these equations yields the family of trajectories in the (x, y) plane described by

$$\Delta\tilde{y} \approx c\,|\Delta\tilde{x}|^{\phi} , \qquad (4.17)$$

where c can take any value while ϕ, the crossover exponent is given by (4.9). The reader will find it instructive to sketch the trajectories for various values of e_+, e_-, and ϕ.

Finally the differential equation (4.11) for $F(x,y)$ on the trajectory reduces to

$$dF/d\tau \quad = \quad P_0 F + U_0 \quad , \tag{4.18}$$

with, for a trajectory (4.17) with given c, the solution

$$F \;=\; A_0(c)e^{P_0\tau} \;-\; (U_0/P_0) \, , \tag{4.19}$$

where $A_0(c)$ is the constant of integration. This result can be written

$$F(x, y) \approx |\Delta\tilde{x}|^{-\gamma} Z(\Delta\tilde{y}/|\Delta\tilde{x}|^{\phi}) \;+\; B_0 \, , \tag{4.20}$$

where we have used (4.16) and (4.17) to eliminate τ and c, and invoked the relations (4.3) and (4.9) for B_0 and γ. This is just the expected scaling form in which, evidently, the scaling function, $Z(z)$, is simply proportional to the 'constant' or, more realistically, function of integration, $A_0(c)$, with $c \equiv z$.

It may be remarked that although we may restrict attention to real values of x_0, y_0, U_0, P_0, Q_1, Q_2, R_1 and R_2 it cannot be guaranteed in advance that e_+ and e_-, and thence λ_1 and λ_2 will always be real. The analysis still goes through if they are complex but the direct, physical interpretation of the scaling is lost.

Before using the above results to study bicriticality we will examine some other general analytical properties of partial differential approximants which will give some insight into their power and flexibility. However the reader more interested in the techniques of application may prefer to proceed directly to Section 6.

5. OTHER PROPERTIES OF PARTIAL DIFFERENTIAL APPROXIMANTS

An ordinary Padé approximant [17-20] which approximates a function $f(x)$ by a ratio of two polynomials, $[L/M]_f = P_L(x)/Q_M(x)$, can obviously provide an <u>exact</u> representation of the function if it is rational and if M, the order of the denominator, is sufficiently large. Likewise a Dlog-Padé approximant of sufficiently high order will be exact for any function which is a finite product of factors of the form $[C_k(x)]^{\Gamma_k}$ where the $C_k(x)$ are polynomials and the Γ_k are arbitrary exponents. It is of interest to ask what classes of functions of two variables can similarly be represented exactly as partial differential approximants. A result in this direction is:

<u>Theorem 1</u> Any two-variable function of the form

$$f(x,y) = [\tilde{C}(x,y) + \tilde{D}(x,y)]^{\zeta} , \quad (5.1)$$

where $\tilde{C}$ and $\tilde{D}$ are functions of the type

$$\tilde{C}(x,y) = \exp[C_0(x,y)/C_{-1}(x,y)] \prod_{k=1}^{K} [C_k(x,y)]^{\Gamma_k} , \quad (5.2)$$

in which the $C_k(x,y)$ [and the corresponding $D_l(x,y)$] are polynomials, can be represented exactly by a homogeneous $(\underset{\sim}{J} \equiv \emptyset)$ partial differential approximant.

To prove this result it suffices to exhibit the polynomials P, Q, and R which specify a generating equation, (3.2) [with $U \equiv 0$], for which (5.1) is a solution. To this end, let us for brevity put $1/\zeta = \bar{\zeta}$ and denote partial derivatives by subscripts x and y. Then we have

$$f^{\bar{\zeta}} = \tilde{C} + \tilde{D} , \quad (5.3)$$

$$\bar{\zeta} f^{\bar{\zeta}-1} f_x = \tilde{C}_x + \tilde{D}_x , \quad \bar{\zeta} f^{\bar{\zeta}-1} f_y = \tilde{C}_y + \tilde{D}_y , \quad (5.4)$$

where, from (5.2), we have

$$\tilde{C}_x(x,y) = \tilde{C}\left[\frac{C_{0,x}}{C_{-1}} - \frac{C_0 C_{-1,x}}{C_{-1}^2} + \sum_{k=1}^{K} \Gamma_j \frac{C_{j,x}}{C_j}\right] ,$$

$$= \tilde{C}(x,y)\, A_1(x,y)/A(x,y) , \quad (5.5)$$

where $A(x,y) = (C_{-1})^2 \prod_{k=1}^{K} C_k$ is clearly a polynomial. Evidently $A_1(x,y)$ is also a polynomial as are $A_2(x,y)$, defined similarly for $\tilde{C}_y(x,y)$, and $B_1(x,y)$, $B_2(x,y)$, $B(x,y)$ defined in like fashion for $\tilde{D}_x(x,y)$ and $\tilde{D}_y(x,y)$. On making the analogous substitutions for $\tilde{C}_x$, $\tilde{C}_y$, $\tilde{D}_x$ and $\tilde{D}_y$ in (5.4) one may eliminate the function D by using (5.3). Multiplication by $f^{1-\bar{\zeta}}$, then yields

$$ABf_x = \zeta f^{1-\bar{\zeta}} \tilde{C}(BA_1 - AB_1) + \zeta AB_1 f , \quad (5.6)$$

$$ABf_y = \zeta f^{1-\bar{\zeta}} \tilde{C}(BA_2 - AB_2) + \zeta AB_2 f . \quad (5.7)$$

Finally, we may eliminate the factor $f^{1-\bar{\zeta}}\tilde{C}$ to obtain a linear equation in f, f_x and f_y which is just the generating equation for an approximant with coefficients which are found to be

$$P(x,y) = \zeta(A_1B_2 - A_2B_1), \quad (5.8)$$

$$Q(x,y) = AB_2 - A_2B , \quad R(x,y) = A_1B - AB_1 . \quad (5.9)$$

By construction, all these are polynomials as required.

It is apparent from this proof that the approximant polynomials required for the exact representation may be of rather high order. However, in the case of the typical "test functions" one is inclined to think of, the required orders may actually be quite low. The example

$$f_1(x,y) = [(1 - 2x + y)^{-\mu} + e^{x-2y}]^{\frac{1}{2}}, \tag{5.10}$$

is quite instructive in this respect. It can be represented exactly by a [1;3,3] approximant of the sort considered in (3.4). The normalization $P(0,0) \equiv p_{00} = -3\mu/2$ proves more convenient than (3.3) and yields

$$P(x,y) = -3\mu/2, \qquad Q(x,y) = \mu-2 + 4x - 2y \quad ,$$

$$\text{and} \qquad R(x,y) = 2\mu - 1 + 2x - y \quad . \tag{5.11}$$

The trajectory equations are easily solved explicitly for this approximant and one finds the family

$$x(\tau) = \tfrac{2}{3}(1+y_a)(1-e^{3\tau}) + \mu\tau \; , \tag{5.12}$$

$$y(\tau) = y_a + \tfrac{1}{3}(1+y_a)(1-e^{3\tau}) + 2\mu\tau \quad , \tag{5.13}$$

which have been parametrized by their intersection with the y axis. It is easy to check that the trajectory with $y_a = -1$ describes simply the linear locus $1-2x+y = 0$ on which $f_1(x,y)$ evidently has a singularity. A little reflection shows that, in general, <u>a locus of singularities should be described by a particular trajectory.</u> This observation is important in applications.

The example (5.10) is also of anecdotical interest because its square, namely $f_2 = (f_1)^2$, was used [37] to test the Canterbury approximants (2.6) which, even in Dlog form, certainly cannot represent $f_2(x,y)$ or $f_1(x,y)$ exactly.

If a function, such as $f_1(x,y)$, can be exactly represented by a given approximant, say [1;3,3] as here, what happens if one attempts to calculate for $f_1(x,y)$ an approximant of higher order than needed? The answer is that when the arrays $\underset{\sim}{L}$, $\underset{\sim}{M}$ and $\underset{\sim}{N}$ are too large the determinants of the associated defining equations vanish identically! Thus a vanishing determinant can merely be a signal that a lower order approximant already provides an exact representation! If the function is known in closed form such a guess can, of course, be checked explicitly: more generally however, one would have to devise a means of showing that the expansion coefficients satisfied the recursion relations implied by the particular generating equation to all orders: and, certainly, in practise vanishing determinants for particular approximants often occur in circumstances where there is no exact representation! Nevertheless in studying test functions this feature should be kept in mind.

Some other results for homogeneous approximants in a slightly different vein are worth recording. Suppose we know one or more solutions of the generating equations

$$P_{(h)}f = Qf_x + Rf_y \quad , \tag{5.14}$$

for $h = 0, 1, 2, \cdots, H$. Are there other functions, $\bar{f}$, constructed from these solutions for which one can give a new generating equation,

$$\bar{P}\,\bar{f} = \bar{Q}\,\bar{f}_x + \bar{R}\,\bar{f}_y \quad , \tag{5.15}$$

which will provide exact representation? Some answers are provided by:

Theorem 2 (a) If $f_{(i)}$ for $i = 0, 1, 2, \cdots$ are particular solutions of (5.14) for $h = 0$, then

$$\bar{f}(x,y) = \sum_{i=1} a_i\,[f_{(i)}(x,y)]^{\zeta} \tag{5.16}$$

is exactly representable by (5.15) with $\bar{P} = \zeta P_{(0)}$, $\bar{Q} = Q$ and $\bar{R} = R$; (b) If $f_{(0)}$ is a solution of (5.14) for $h = 0$ and $C(x,y)$ is a polynomial, then

$$\bar{f}(x,y) = [C(x,y)]^{\zeta} f_{(0)}(x,y) \tag{5.17}$$

is exactly representable by (5.15) with

$$\bar{P} = CP_{(0)} + \zeta C_x Q + \zeta C_y R \quad ,$$

$$\bar{Q} = CQ \;, \qquad \text{and} \qquad \bar{R} = CR \;; \tag{5.18}$$

(c) If $f_{(h)}$ is a solution of (5.14) for specified h, then

$$\bar{f}(x,y) = a\prod_{h=0}^{H}[f_{(h)}(x,y)]^{\zeta_h} \quad , \tag{5.19}$$

is exactly representable by (5.15) with $\bar{P} = \sum_{h=0}^{H} P_{(h)}$, $\bar{Q} = Q$, and $\bar{R} = R$.

The proofs of the various parts of this theorem proceed straightforwardly by calculating $(Q\bar{f}_x + R\bar{f}_y)$ and using (5.14) to reexpress the result in the form (5.15).

Inhomogeneous partial differential approximants allow more complicated functions than (5.1) to be represented exactly. As an example, we quote the function

$$f_3(x,y) = (1-x-\tfrac{1}{2}y)^{-\frac{3}{2}}(1-2x-\tfrac{1}{5}y)^{-2} + e^{-x-2y}\cos xy \quad , \qquad (5.20)$$

which has been utilized in a recent test of homogeneous approximants [38]. It is not hard to show that an approximant with polynomials U, P, Q and R satisfying the equations

$$yQ + xR = 0 \quad , \qquad xP + (x-2y)Q = 0 \; , \qquad (5.21)$$

$$\left[\frac{3(2x-y)}{4(1-x-\frac{1}{2}y)} + \frac{x-(y/10)}{1-2x-\frac{1}{5}y} + x-2y\right] Q = xU \; , \qquad (5.22)$$

will provide an exact representation of $f_3(x,y)$.

Another result [22] which bears on the generalization of the asymptotic scaling form (2.2) to allow for nonlinear scaling fields [35, 36] is:

Theorem 3 If $A(x,y)$, $B(x,y)$, $C(x,y)$ and $D(x,y)$ are polynomials and $Z(z)$ is a general function, then the two-variable function

$$f(x,y) = A(x,y)\,|\Delta\tilde{x}|^{-\gamma} Z(C\Delta\tilde{y}/|D\Delta\tilde{x}|^{\phi}) + B(x,y) \; , \qquad (5.23)$$

in which $\Delta\tilde{x}$ and $\Delta\tilde{y}$ are monomials [see (2.4)] is exactly representable by an inhomogeneous partial differential approximant; if $B \equiv 0$ then a homogeneous approximant suffices.

This theorem is proved along the lines used in Section 2 and for Theorem 1. We will omit the details but quote the results

$$Q(x,y) = A\,\Delta\tilde{y}\,[\phi CD/e_2 - (\phi CD_y - C_yD)\Delta\tilde{x}] + ACD\Delta\tilde{x} \quad , \qquad (5.24)$$

$$R(x,y) = A\,\Delta\tilde{y}\,[\phi CD + (\phi CD_x - C_xD)\Delta\tilde{x}] + ACDe_1\Delta\tilde{x} \quad . \qquad (5.25)$$

Similar, but even longer explicit polynomial expressions specify $U(x,y)$ and $P(x,y)$ [22].

It will be evident to the reader at this point (if not before) that, in solving the generating equation for an approximant, one important role played by the boundary conditions is, ultimately, that of determining the nature of the scaling function, $Z(z)$, at a multisingular point. This must be so because Z itself does not appear directly in the expressions for the approximant coefficients even though the multisingular location, axes, exponents, and background do enter directly into U, P, Q and R. Conversely, while knowledge of U, P, Q, and R near a multisingular point provides values (or estimates) for x_c, y_c, e_1, e_2, γ and ϕ the scaling function requires further information.

In many applications the meaning of the two variables x and y in the functions of interest will have no similarity so that there may be little sense in considering combinations of the variables. Even then, it is worth pointing out that an <u>upper triangular expansion</u> of the form

$$\begin{aligned} f(x,y) = f_{00} &+ f_{10}x + f_{20}x^2 + f_{30}x^3 + \dots \\ &+ f_{11}xy + f_{21}x^2y + f_{31}x^3y + \dots \\ &\qquad + f_{22}x^2y^2 + f_{32}x^3y^2 + \dots \\ &\qquad\qquad + f_{33}x^3y^3 + \dots, \end{aligned} \tag{5.26}$$

in which all coefficients below the diagonal vanish ($f_{ii'} \equiv 0$ for $i < i'$) can be converted into full rectangular form (with no imposed vanishing of coefficients) by the transformation

$$\bar{x} = x, \qquad \bar{y} = xy \ . \tag{5.27}$$

Such functions often crop up and this transformation and its obvious extensions may prove a significant convenience in practise.

On the other hand in many applications the two variables will enter on essentially equal footing so that various linear combinations have sensible meaning. (In the applications discussed below x and y measure the strength of similar but potentially competing or cooperating interactions in a model ferromagnet.) An obvious question to ask, in such cases, is whether there is any advantage to be gained by some rotation of the axes or more general linear transformation

$$\begin{aligned} x &\Rightarrow \bar{x} = v_{11}x + v_{12}y \ , \\ y &\Rightarrow \bar{y} = v_{21}x + v_{22}y \ , \end{aligned} \tag{5.28}$$

before partial differential approximants are computed? The following invariance result [39] shows that in certain cases there is nothing to gain!

<u>Theorem 4</u> The generating partial differential equation for a homogeneous approximant $[\underset{\sim}{L};\underset{\sim}{M},\underset{\sim}{N}]_f$ to a function f(x,y) with triangular arrays of order

$$L = \tfrac{1}{2}(l_0 + 1)(l_0 + 2) \quad \text{and} \quad M = N = \tfrac{1}{2}(m_0 + 1)(m_0 + 2),$$

matched on a triangular array $\underset{\sim}{K}$ of order satisfying the condition

$$K \equiv \tfrac{1}{2}(k_0 + 1)(k_0 + 2) = \tfrac{1}{2}l_0(l_0 + 3) + m_0(m_0 + 3) + 2, \tag{5.29}$$

is, after transformation by (5.28), identical to the corresponding generating equation for the approximant $[\underset{\sim}{L}; \underset{\sim}{M}, \underset{\sim}{M}]_{\bar{f}}$ for the function $\bar{f}(\bar{x}, \bar{y})$ obtained by transforming the expansion for $f(x, y)$ by (5.28).

The restriction to triangular arrays in such cases is quite natural since, if x and y enter a problem on an equal footing, the coefficients, $f_{i,i'}$, will typically be computable at a given stage up to some <u>overall order</u> determined by $i + i' \leq i_0$ which, indeed specifies a triangular array. Likewise the transformation (5.28), in general, preserves only triangular arrays. However the <u>full balance matching condition</u>, (5.29), is rather more restrictive. Since l_0, $m_0 \equiv n_0$ and k_0 must be integers the condition represents a Diophantine equation. The general solutions may be constructed [39] but are surprisingly sparse! The first few are $[L; M, N] = [1; 3, 3]$, $[10; 3, 3]$, $[10; 6, 6]$, $[55; 6, 6]$, $[36; 10, 10]$, $[171; 10, 10]$, $[91; 15, 15]$, etc. . Since the approximants of these orders are invariant to a linear transformation of the variables one may suspect that they will be more accurate, in general, than approximants with other arrays. A similar argument is invoked for direct Padé approximants to functions of a single variable to suggest that the diagonal approximants, $[L/L]_f$, which are invariant under the Euler transformation, $\bar{x} = v_0 x/(1+v_1 x)$, should be more reliable than off-diagonal approximants [18, 19]. In practice, at least for the classes of two-variable functions studied so far, approximants with near-triangular arrays and M close to N do seem to give better results than other choices. However, a flexible numerical approach, which allows for computation of a wide range of approximants, is to be prefered as a general rule since one can then hope to gauge accuracy and reliability by observing the concurrence, or lack thereof, between nonequivalent approximants.

There are probably also invariance properties of partial differential approximants under Euler transformations but these have not yet been elucidated. (Compare with Ref. [21].)

6. BICRITICALITY IN THE ISING-HEISENBERG-XY MODEL

To illustrate the application of the partial differential approximant technique, to explain some of its extensions, to expose some of its practical limitations, and to see what it can teach us, more specifically, about bicritical behavior, we will discuss one of the simplest statistical mechanical models exhibiting a bicritical point, namely, the classical anisotropic Heisenberg model in three dimensions. To specify this model, consider a regular space lattice — results for the sc, bcc, and fcc lattices will be displayed — with sites, i, at each of which a classical (or $S = \infty$) spin variable $\vec{s}_i \equiv (s_i^x, s_i^x, s_i^z)$ is located. If $[i,j]$ denotes a nearest neighbor pair of sites the Hamiltonian for the model may be written

$$\mathcal{H} = -\sum_{[i,j]} [J_\perp (s_i^x s_j^x + s_i^y s_j^y) + J_\parallel s_i^z s_j^z] \quad ,$$

$$= -\sum_{[i,j]} \bar{J}\{ \vec{s}_i \cdot \vec{s}_j + g[\tfrac{1}{3}(s_i^x s_j^x + s_i^y s_j^y) - \tfrac{2}{3} s_i^z s_j^z]\}, \qquad (6.1)$$

where the second form of writing the Hamiltonian is dictated by symmetry considerations [23, 42]. The two alternative forms for anisotropic couplings are related by

$$\bar{J} = \tfrac{2}{3}J_\perp + \tfrac{1}{3}J_\parallel \quad , \qquad g = (J_\perp - J_\parallel)/\bar{J} \, , \qquad (6.2)$$

or, equivalently, by

$$J_\perp = \bar{J}(1 + \tfrac{1}{3}g), \qquad J_\parallel = \bar{J}(1 - \tfrac{2}{3}g) \; . \qquad (6.3)$$

All interactions will be supposed ferromagnetic so that $\bar{J}$, $J_\parallel$, $J_\perp \geq 0$.

The properties of the model in more than one dimension cannot, of course, be calculated exactly; however systematic expansions in inverse powers of the temperature, T, may be constructed [20]. The natural variable for this, which determines the order of the expansion, is

$$w = \bar{J}/k_B T \quad , \qquad (6.4)$$

(where as always, k_B denotes Boltzmann's constant). The anisotropy field, g, then appears as a secondary variable and the expansions have the upper triangular form illustrated in (5.26). In fact the expansions of the various thermodynamic properties are known to $O(w^8)$ for the fcc lattice and to $O(w^9)$ for the bcc and sc lattices [23]. However, from the first form of the Hamiltonian, one sees that the variables

$$x = J_\parallel / k_B T \qquad \text{and} \qquad y = J_\perp / k_B T \qquad (6.5)$$

are also quite natural; furthermore, they enter the problem on an essentially identical footing. Some initial trials of the partial differential (or PDA) method were made using the variables w and g and met with reasonable success [26]. However further study showed, as might have been guessed *a priori*, that the analysis is better undertaken in terms of the physically similar variables x and y.

It is appropriate to discuss the general features of the Hamiltonian (6.1) and to review briefly what can be learned by single-variable methods of analysis, before plunging into the study of two-variable PDA's! Thus consider, first, the pure Ising-like limit defined by $J_\perp \equiv 0$ (or $y = 0$). In that case only the z-components of the spins are coupled so that at low temperatures ordering takes place with the spins parallel ($\parallel$) to the z-axis. Correspondingly, at high temperatures one should see critical behavior through the divergence of the parallel (reduced) susceptibility at zero field

$$\chi_\parallel(T,g) = k_B T(\partial M_z/\partial H_z) \quad , \tag{6.6}$$

where $M_z = \langle s_i^z \rangle$ is the parallel or z-component of the magnetization while H_z is the corresponding component of the total magnetic field $\vec{H}$. [Note that the ordering field enters into the Hamiltonian via an extra term $-\vec{H} \cdot \vec{s}_i$ for each spin in (6.1).] The divergence (for $J_\perp = 0$) should be Ising-like as

$$\chi_\parallel \sim (x_I - x)^{-\gamma_I} \quad \text{as} \quad x \rightarrow x_I- \quad , \tag{6.7}$$

and this is confirmed by single-variable series analysis [23] with the results shown in Table I. (It should be noted that since the series available are comparatively short, it is not worthwhile to make special allowance for the effects of possible confluent singularities [35, 36, 40].)

Pure XY-like (or $n = 2$) coupling corresponds to $J_\parallel = 0$ (or $x = 0$). The spins will then order in the plane perpendicular ($\perp$) to the z axis and the appropriate susceptibility to examine above the transition is evidently

$$\chi_\perp(T,g) = k_B T(\partial M_x/\partial H_x) \equiv k_B T(\partial M_y/\partial H_y) \quad . \tag{6.8}$$

This will diverge with an exponent, γ_{XY}, having the estimated value shown in Table I. Note, however, that the parallel susceptibility $\chi_\parallel$, will not diverge in this case (although it will display an energy-like singularity).

TABLE I Critical parameters estimated for the classical spin models with pure couplings of various symmetry by single-variable series extrapolation techniques [23, 31].

Symmetry	Susceptibility Exponents	Crossover Exponents	Critical Points (fcc)
$n = 1$	$\gamma_I \simeq 1.23$		$x_I \simeq 0.2850$
$n = 2$	$\gamma_{XY} \simeq 1.31$	$\phi = 1.18 \pm 0.02$	$y_{XY} \simeq 0.2993$
$n = 3$	$\gamma_H \simeq 1.38$	$\phi = 1.25 \pm 0.02$	$w_H \simeq 0.3147$

Finally there is the pure, fully isotropic Heisenberg limit, $g = 0$ or $J_\parallel = J_\perp$ and $x = y = w$. In this case $\chi_\parallel = \chi_\perp$ and both susceptibilities will diverge with a still different exponent γ_H (see Table I). Now, any nonzero value of anisotropy, g, will destroy the global spin rotation symmetry which characterizes the pure Heisenberg limit. Because of this and the results of the corresponding renormalization group analysis [41, 42] one firmly believes that asymptotic Heisenberg behavior will be destroyed as soon as g departs from zero (or $x \neq y$): for $g < 0$ (i.e. $J_\parallel > J_\perp$ or $x > y$) we expect Ising-like critical behavior, with a divergence only of $\chi_\parallel(T,g)$, on a critical locus $T_c^\parallel(g)$, while for $g > 0$ (i.e. $J_\perp > J_\parallel$ or $x < y$) the critical behavior should be XY-like with only $\chi_\perp(T,g)$ divergent on a locus $T_c^\perp(g)$. In short, we have concluded that the point $w = w_H$, $g = 0$ (or $x = y = w_H$) is a bicritical point of Heisenberg (or $n = 3$) character: compare with Figure 1 and the discussion in Section 1 and see also Figures 5 and 6 below. Granted this identification of the bicritical point, the corresponding crossover exponent, ϕ, can be estimated reliably using single-variable techniques [23] by studying the expansions for the derivatives $(\partial^m \chi_\perp / \partial g^m)$, etc., on the Heisenberg axis, $x = y = w$, where they should diverge as

$$(w_H - w)^{-\gamma_H - m\phi}$$

This method yields the estimates shown in Table I. It is worth remarking that these values for ϕ are well confirmed by $\varepsilon = 4-d$ dimensionality expansions [41-43, 35, 36].

The symmetry analysis of the Hamiltonian at the bicritical point [24, 42] which is embodied in the second line of (6.1) also shows that the asymptotic scaling axes should be specified by $w = w_H$ and $g = 0$ (or $x = y$). In terms of the variables x and y this means the bicritical scaling axis slopes are known exactly as $e_1 = 1$ $(g = 0)$ and $e_2 = -1/2$ $(w = w_H)$. This feature will serve to provide a test of the PDA technique: see Figure 4 below.

We are now ready to embark on the two-variable PDA analysis in order to learn more about the phase diagram in general and the bicritical region in particular. However, one should pause a moment to ask "Which thermodynamic function ought to be the first object of study?" Now one learns from extensive experience with single-variable series analysis [17-22] that functions with strong divergences, like ordering susceptibilities, can be analysed more easily and extrapolated more reliably than functions with weak singularities, like specific heats. Thus the susceptibilities $\chi_\parallel(T,g)$ and $\chi_\perp(T,g)$ are obvious candidates: but, while both of these will diverge at the bicritical point, the latter remains finite for $g < 0$ $(x > y)$ even on the Ising-like critical line, $T_c^\parallel(g)$, while the former, likewise, remains finite for $g > 0$ $(x < y)$ even on the XY-like critical line, $T_c^\perp(g)$. Both functions, therefore, behave in a strongly nonuniform manner in the immediate bicritical region. On the other hand, the total susceptibility

$$\chi(T,g) = \frac{1}{3}\left[\left(\frac{\partial M_x}{\partial H_x}\right) + \left(\frac{\partial M_y}{\partial H_y}\right) + \left(\frac{\partial M_z}{\partial H_z}\right)\right] k_B T ,$$

$$= \frac{2}{3}\chi_{\perp}(T,g) + \frac{1}{3}\chi_{\parallel}(T,g) , \qquad (6.9)$$

although a less natural quantity from a purely physical viewpoint, diverges strongly not only at the bicritical point, but also on both critical lines. One might therefore guess that χ is the best function to study in the first step of locating the bicritical point, and estimating its exponents and axis slopes. Indeed the practical calculations strongly endorse this conclusion. On the other hand, once the nature of the bicritical point has been well established one can use this information, as we will explain, to go back and estimate reliably the individual functions $\chi_{\parallel}(T, g)$ and $\chi_{\perp}(T, g)$ (as well as more delicately singular functions like the specific heats and nonordering susceptibilities which are not even divergent at the bicritical point!) [44, 45].

7 PREDICTIONS FROM UNBIASSED APPROXIMANTS

As explained, certain features of the bicritical point in the Ising-Heisenberg-XY model are known exactly a priori, namely, $x_c = y_c$, $e_1 = 1$, $e_2 = -1/2$, while others are known to good accuracy, namely the values of x_c, γ and ϕ. To test the PDA technique, however, let us first look at unbiassed approximants in which this information is not utilized in any way [31].

Figure 2 represents a scatter diagram in the (x,y) plane showing estimates of the bicritical point for the fcc lattice obtained by locating the common zeros of $Q_{\underset{\sim}{M}}(x,y)$ and $R_{\underset{\sim}{N}}(x,y)$ [see (4.1)] for a range of homogeneous PDA's with near-triangular arrays constructed from the expansion of $\chi(x,y)$ using from $K = |\underset{\sim}{K}| = 21$ to 36 coefficients [i.e. through orders w^6, w^7, and w^8 in (6.9)]. The exact scaling axes (accepting the estimate of w_H in Table I) are drawn in the figure. It is surprizing that the estimates tend to cluster strongly along a line parallel to (although slightly lower than) the second scaling axis of slope $e_2 = -1/2$. The precise reason for this is not presently understood but it is a general feature of the technique that the various estimates usually do lie close to definite curvilinear loci in the parameter planes rather than being scattered about in a "dart board pattern". The precision of these unbiassed estimates of x_c and y_c can be gauged by the ±0.1% and ±1% "boxes" shown in the figure. Considering the comparatively short length of the available series, the accuracy is really quite satisfying. One does, however, observe (in support of the speculations based on Theorem 4 of Section 5) that the better estimates tend to derive from approximants with label sets $\underset{\sim}{M}$ and $\underset{\sim}{N}$ which match fairly closely.

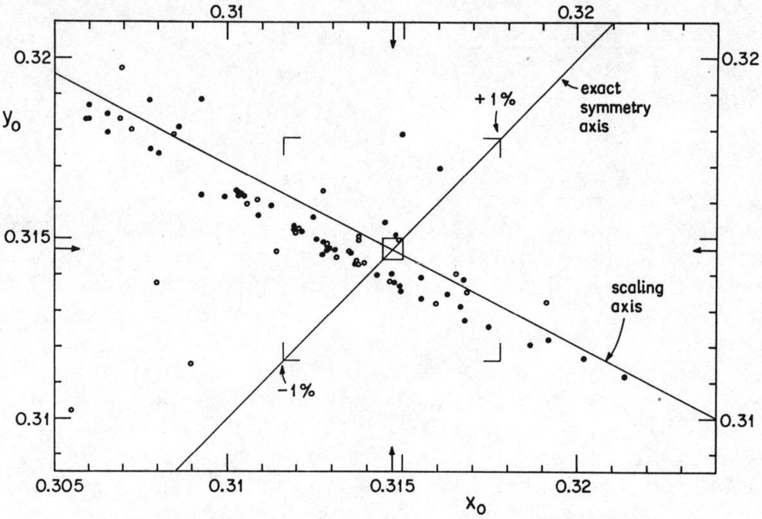

FIG. 2. Estimates $x_0 = J_{\parallel}/k_BT_c$, $y_0 = J_{\perp}/k_BT_c$ of the multicritical point for the Ising-Heisenberg-XY model on the fcc lattice. Open circles denote lower order PDA estimates; closed circles those of higher order.

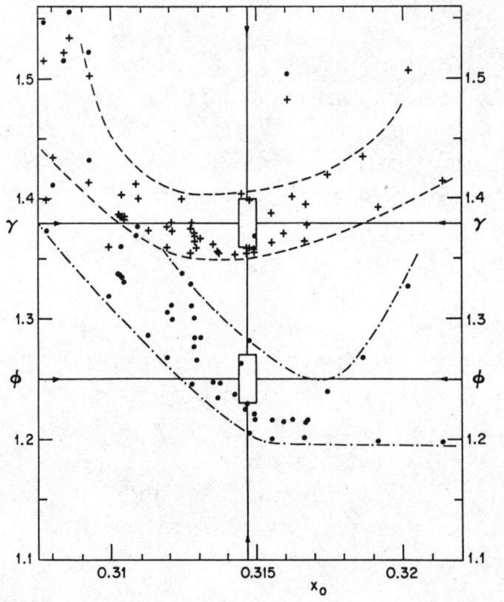

FIG. 3. Unbiassed estimates of the exponents γ and φ vs. x_0 for PDA's with K ≥ 28. The dashed and dot-dash curves serve only to indicate trends.

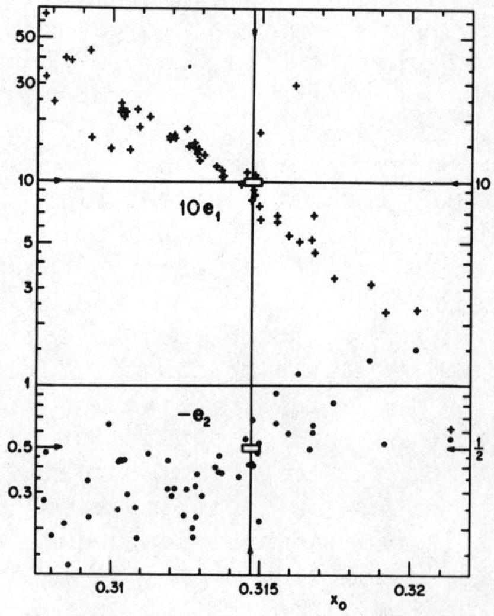

FIG. 4. Unbiassed estimates of the axis slopes, e_1 and e_2, vs. x_0 as in Figure 3. Note that $10e_1$ and $-e_2$ are plotted.

In Figure 3 the unbiassed estimates of the exponents γ and ϕ, corresponding to the chosen set of approximants, are plotted versus the estimate x_0. Because of the correlation mentioned in Figure 2, this is almost equivalent to a plot versus g. The "boxes" in the figure denote the "accepted" values embodied in Table I. Again a pronounced curvilinear correlation is observed (which resembles in scale and form the correlation normally seen between exponent and critical point estimates in standard Dlog-Padé analysis [17-22]). Nevertheless, the central exponent estimates, with $x_0' \simeq y_0$ (or $g \simeq 0$), are remarkably good.

A similar plot for the estimates of the axis slopes, e_1 and e_2, is presented in Figure 4. The agreement with the exact results, indicated by the "boxes", is again quite encouraging for those approximants yielding $x_0 \simeq y_0$, but a much stronger overall variation with x_0 is observed than for the exponents. Of course, in the case of the analysis of single-variable functions there is no analog at all to these slope parameters!

Figure 5 displays the full ferromagnetic quadrant $(x,y \geq 0)$ of the phase diagram, together with an enlarged view, in the inset, of the bicritical region. The data exhibited derive from one of the better homogeneous approximants with $L = 13$ and $M = N = 12$. Note, first, the loci of zeros of $Q_{\underset{\sim}{M}}(x,y)$ and $R_{\underset{\sim}{N}}(x,y)$ (dotted curves) whose intersection (in this case at $x_0 = 0.3148$ and $y_0 = 0.3151$) locates the approximate bicritical point. The corresponding exponent estimates are $\gamma = 1.398$ and $\phi = 1.282$. The approximate scaling axes, with estimated slopes $e_1 = 1.07$ and $e_2 = -0.405$, are represented by the broken straight lines. The remaining dashed and solid curves, bearing arrowheads, correspond to selected trajectories calculated by integrating the flow equations (4.10) in towards the bicritical point.

The two trajectories plotted as solid curves originate on the x and y axes at zeros

$$Q_{\underset{\sim}{M}}(x_I^0, 0) = 0 \qquad \text{and} \qquad R_{\underset{\sim}{N}}(0, y_{XY}^0) = 0 \quad , \qquad (7.1)$$

which, in fact, one can show* should represent estimates of the

---

*To demonstrate this for the x axis it is only necessary to assume (which in this case one can do with confidence by considering the physics of the transformation $y \Rightarrow -y$) that $f(x,0)$ diverges at $x = x_I$ with an exponent γ_I and that the derivative $(\partial f/\partial y)_{y=0}$ also diverges at $x = x_I$ but with an exponent, say, ψ which satisfies $\psi < \gamma + 1$. (In fact, for the case in hand $\psi = 0$ and $(\partial f/\partial y)$ actually remains finite.) Then one sees that one should expect $Q(x, 0) \approx Q_{I,1}(x-x_I^0)$ with $x_I^0 \simeq x_I$ and, furthermore, $-P(x_I^0, 0)/Q_{I,1} \simeq \gamma_I$. Parallel considerations apply, of course, to the pure XY singularity on the y axis.

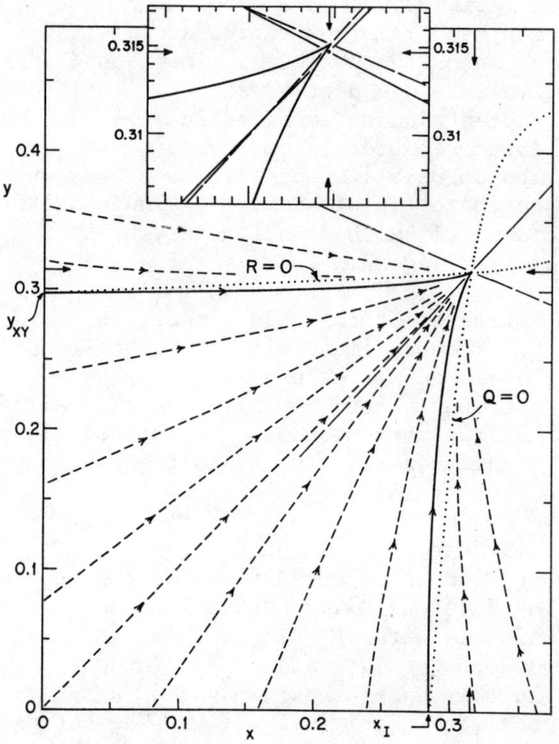

FIG. 5. Trajectories in the (x,y) plane for a [13;12,12] approximant for the fcc lattice. The inset shows the (approximate) critical loci in detail in the bicritical region: the solid lines here and the dot represent the best estimates of the scaling axes and bicritical point.

FIG. 6. Trajectories for a [17;14,14] approximant showing a different flow pattern than in Figure 5, achieved by special choice of the forms of $Q_M(x,y)$ and $R_N(x,y)$ which ensures reduction to ordinary Dlog Padé approximants on the axes.

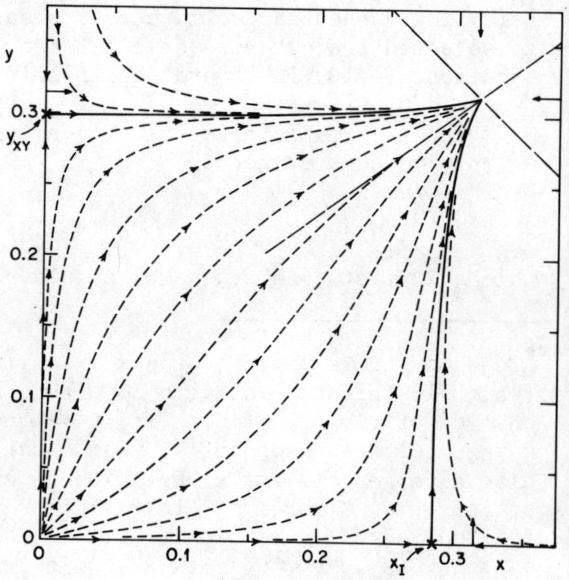

pure Ising-like and pure XY-like critical points, x_I and y_{XY}, listed in Table I. These estimates also prove to be quite good but for optimal accuracy one may prefer to replace them by the "accepted" values from Table I. In any event, it is clear from the flow equation (4.12) that since the total susceptibility $\chi(x,y)$ must diverge at $(x_I, 0)$ and $(0, y_{XY})$ it must also diverge all along the two solid trajectories. These two trajectories thus represent estimates for the two critical loci, $T_c^{\parallel}(g)$ and $T_c^{\perp}(g)$. As evident from the inset in Figure 5, the two critical loci (like all the trajectories) meet at the multicritical point with a characteristic cusp of precise shape determined by the crossover exponent ϕ. This then represents the practical realization of the singular behavior of the critical lines at a bicritical point originally anticipated (in Section 1) on the basis of phenomenological scaling arguments and embodied in the formula (1.8). It should be noted that the nature and shape of this cusp is almost impossible to elucidate reliably using single variable techniques [23], even in a case like this where the location of the bicritical point is well-determined (see further below).

8 ALTERNATIVE FLOW PATTERNS AND DEFECTIVE APPROXIMANTS

The general form of the trajectories illustrated in Figure 5 is essentially unique in the multicritical region since it is dictated by the scaling properties. The global aspects of the flows, however, can be quite different as we now explain. The first point to notice is that certain restrictions on the shape of the polynomials $Q(x,y)$ and $R(x,y)$, which together determine all the trajectories, can have a profound effect on the global character of the flows. Thus suppose the array $\underset{\sim}{N}$ for $R(x,y)$ is assigned so that

$$R_{\underset{\sim}{N}}(x,y) = y\bar{R}_{\underset{\sim}{N}}(x,y) , \tag{8.1}$$

where $\bar{R}$ is again just a polynomial. It is evident from the flow equations (4.10), that this implies $(dy/d\tau) = 0$ for $y = 0$ which, in turn, means that there must be a trajectory which runs directly along the x axis. On the x axis $(y = 0)$ the generating equation (3.2) then reduces simply to

$$\frac{d}{dx} \ln F(x) = Q_{\underset{\sim}{M}}(x, 0)/P_{\underset{\sim}{L}}(x,0) , \tag{8.2}$$

where, for simplicity, we have restricted attention to homogeneous approximants with $U \equiv 0$. But this equation just defines a standard Dlog Padé approximant [17-22] for the function $f(x,0)$! A PDA with (8.1) imposed might thus be called a "Dlog PDA".

Likewise if $Q_{\underset{\sim}{M}}$ is restricted by

$$Q_{\underset{\sim}{M}}(x, y) = x\bar{Q}_{\underset{\sim}{M}}(x, y) , \tag{8.3}$$

there is a trajectory which coincides with the y axis $(x = 0)$ and the homogeneous PDA's on the y axis reduce to standard Dlog Padé approximants for $f(0,y)$.

If both the conditions (8.1) and (8.3) are imposed to define "double Dlog PDA's", matters are slightly more complicated because in (8.2), for example, $Q_M(x,0)$ must be replaced by $x\bar{Q}_M(x,0)$ which defines only a restricted class of Dlog Padé approximants. To avoid this limitation one may replace the normalization condition (3.3) by the requirement $P(0,0) = p_{00} = 0$, which amounts only to excluding the label (0,0) from the set $\underset{\sim}{L}$, and then normalizing by, say, the condition $q_{1,0} = 1$. (It is also then appropriate to exclude the label (0,0) from the matching array $\underset{\sim}{K}$.)

In the present bicritical example where the function, χ, under study has a natural decomposition into a sum of two functions, each dominant on one of the axes [see (6.9)], an alternative procedure is possible. Thus to allow for the prefactor, x, appearing in the Dlog expression (8.2), and the similar prefactor, y, in the corresponding formula for the y axis, we may consider the modified total susceptibility

$$\tilde{\chi}(x,y) = \frac{1}{3} x \chi_{\parallel}(x,y) + \frac{2}{3} y \chi_{\perp}(x,y) \quad . \tag{8.4}$$

Then, when both (8.1) and (8.3) are used, the homogeneous PDA's will reduce to ordinary Dlog Padé approximants on both axes! Note, furthermore, that on the Heisenberg symmetry axis, $x = y = w$, one obtains simply $\tilde{\chi} = w\chi(w,w)$, so that behavior in the asymptotic bicritical region is quite unchanged.

An example of the trajectories arising when this procedure is adopted for the Ising-Heisenberg-XY model on the fcc lattice is shown in Figure 6. This new flow topology (which evidently represents the pure Ising and pure XY critical points as 'unstable' fixed points) offers, in principle, an advantage in evaluating $f(x,y)$ and the scaling function, $Z(z)$, since all but the limiting (or critical) trajectories flow out of (or into) the origin where appropriate boundary conditions may be directly imposed. By contrast, the flow topology shown in Figure 5 clearly requires boundary values of $f(x,y)$ along, say, both the x and y axes. In practise, however, as we illustrate later, this is not really a serious drawback. And, conversely, one discovers that the convergence of the double Dlog approximants for $\tilde{\chi}$ is not as good in the multicritical region as found the ordinary, unrestricted PDA's.

Another annoying feature observed more frequently with the double Dlog PDA's in the bicritical application [31], is the occurrence of defective approximants. The problem of defects or tears is often seen in ordinary, single-variable Padé approximant analysis: a spurious zero of the denominator appears [18,19], implying a singularity in the corresponding approximant, $F_{L/M}(x) \equiv [L/M]_f$, in a quite inappropriate region where the function $f(x)$ is expected or, sometimes, even proven rigorously to be quite smooth and analytic! Figure 7 illustrates the analogous phenomenon in the case of partial differential approximants (in this case for a near-triangular [15; 11, 11] approximant with the 'super triangular' labels (2, 2) included in $\underset{\sim}{M}$ and $\underset{\sim}{N}$, and (6,1) and (4,3) in $\underset{\sim}{K}$, and with six extra bicritical conditions imposed — see below). Note that, in addition to the expected loci of zeros of Q and R corresponding closely to those illustrated in Figure 5, there are two completely new and superfluous loci of zeros. Surprisingly, the estimates of multicritical point properties themselves are often not much disturbed by such defects, as can be seen from the form of the critical lines shown in the figure (as solid curves). Much the same is found in the case of ordinary Dlog Padé approximants with defects. However the extra, unwanted multisingular points implied by the further multiple zeros, clearly cast doubt on the reliability of this particular approximant. More generally the phenomenon of defects shows that examination of the global features of an approximant may be a wise precaution even if the focus of interest is confined to some particular multicritical region.

A still more complicated example of a defective approximant shown in Figure 8, which illustrates a [21; 10, 10] double Dlog approximant for $\tilde{\chi}$ with triangular arrays for P, $\bar{Q}$, and $\bar{R}$ [and supertriangular labels (1, 7), (3, 5), (5, 3) and (7, 1) included in $\underset{\sim}{K}$]. Note the form of the zero loci and the corresponding complexity of the trajectories shown in Figure 9, which reveal a completely spurious new, 'unstable' multicritical point in the vicinity of $(x,y) = (0.12, 0.03)$. The form of the trajectories offers no reasonable way of integrating towards the multicritical point from a known region near the origin or on the axes. Nevertheless it is interesting that the polynomial $P_{\underset{\sim}{L}}(x, y)$, which is usually free of zeros in the physical region, has, in this case, a locus of zeros (shown as a dotted curve in Figure 8) which almost coincides with parts of the spurious zero loci of $Q(x,y)$ and $R(x,y)$. This might be interpreted as an attempt by the approximant to cancel the harm done by the misplaced zeros since, on integrating the trajectory equation for $F(x,y)$ there will be a corrective effect. A similar phenomenon is, indeed, seen in ordinary Padé approximants where a defective zero in the denominator is often almost identically cancelled by a corresponding zero in the numerator [19]. This can be understood since an <u>exact</u> cancellation would merely reduce an $[L/M]$ Padé approximant to an equivalent one, $[L-1/M-1]$, of lower order. It would be interesting to gain a similar insight into the mechanism by which defects are (almost) corrected in partial differential approximants; but the task promises to be more difficult and not very rewarding as regards practical applications!

As a last remark on the topic of the nature of the PDA flows

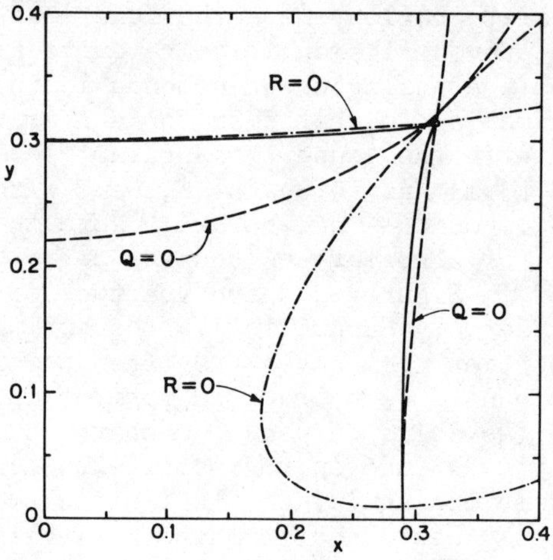

FIG. 7. The locus of zeros of the polynomials Q(x,y) and R(x,y), and the critical loci, for a defective [15;11,11] approximant for χ on the fcc lattice (with six constraints imposed).

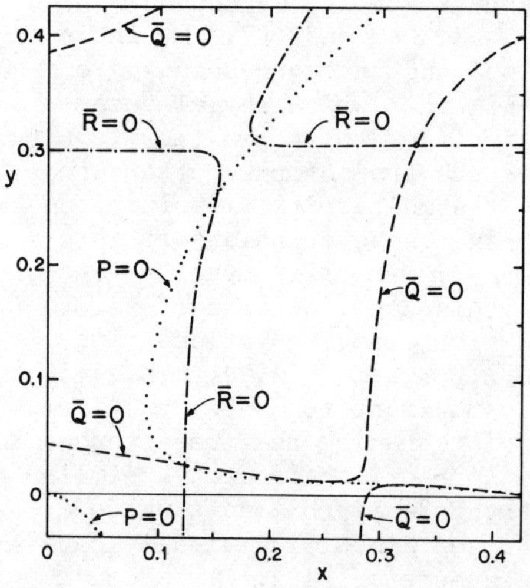

FIG. 8. The loci of zeros of P(x,y), $\bar{Q}(x,y)$, and $\bar{R}(x,y)$ for a defective double Dlog PDA of order [21;10,10] for $\tilde{\chi}$ on the fcc lattice.

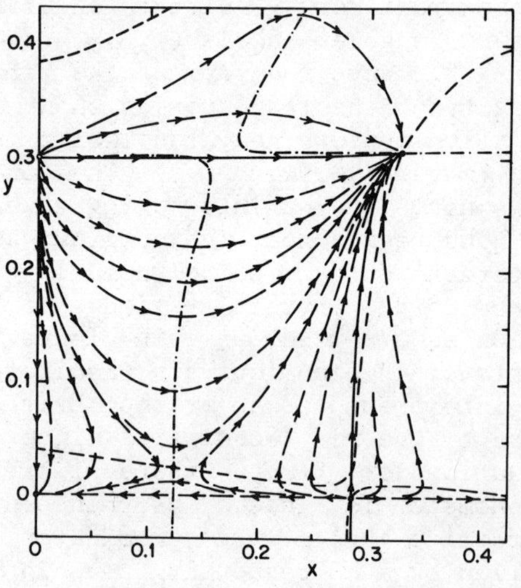

FIG. 9. The trajectories for the defective [21;10,10] approximant with the zero loci shown in Figure 8, demonstrating how the bicritical point is isolated from the origin.

we point out that it is possible to ensure that a homogeneous PDA reduces to an ordinary Dlog Padé approximant on an arbitrary linear locus, say, $y = cx + b$. Basically, as is easily checked, it is only necessary to impose the relation

$$R_{\underset{\sim}{N}}(x, cx+b) = c\,Q_{\underset{\sim}{M}}(x, cx+b) \quad , \tag{8.5}$$

on the generating polynomials. This requires that the shapes of $\underset{\sim}{M}$ and $\underset{\sim}{N}$ be appropriately matched so that, by comparing powers of x, this equation can be interpreted merely as a (finite) set of extra linear equations for the coefficients $q_{m,m'}$ and $r_{n,n'}$. Matching on the generating partial differential equation provides the remaining equations needed as in the standard case. This procedure is just a special example of the general technique of imposing desired constraints on approximants which we take up in the next section.

To end this section we comment on a problem, alluded to above, which has arisen in some applications, namely, that for many selected approximants the determinants of the defining equations vanish. As mentioned, this can be an indication that a low order approximant actually provides an exact representation of the function: in many cases, however, this is clearly not the case and the problem arises, rather, from a large number of vanishing coefficients, $f_{i,i'}$, in the array defining the function of interest. One approach which may be worth exploring in such cases is to add a known stabilizer function to the function under study. Such a stabilizer function should be "harmless", in the sense that it is an entire function or, at least, has no singularities in the interesting regions of the (x,y) plane. It should also be relatively small so that its properties do not dominate those of the function being studied. Finally it must not be too simple in analytic structure since, as the theorems in Section 5 above illustrate, the determinants may still vanish! This approach which is, perhaps, something of a last resort is still to be explored in practical applications. In other cases, however, experience shows that simply increasing the size of the inhomogeneous polynomial, $U_{\underset{\sim}{J}}(x,y)$, in the generating equation may be an effective way of finding good approximants. Certainly, as we illustrate further below, the partial differential approximant technique has reserves of flexibility which are likely to reward exploration and ingenuity!

9 CONSTRAINED APPROXIMANTS

Up to this point we have discussed only unbiassed approximants in which no knowledge about the multisingular point under study was used in constructing approximants. However, to obtain optimal accuracy in the estimation of quantities of interest, it is desirable to try to incorporate into the approximants as much information, exact or reliably estimated, as is available. The only proviso is that in building-in the appropriate knowledge the calculation of the new, constrained approximants should not be unduly complicated. In practise this condition will be met if, but only if, the constrained approximants can still be calculated by solving only linear algebraic equations. The general strategy, then, is to allow for extra coefficients in the polynomials U, P, Q, and R over and above those needed to match the known expansion coefficients of $f(x,y)$, and to determine these additional coefficients by the imposition of further linear constraints on the polynomials. The following cases arise naturally and are important for applications:

(a) Imposition of a multisingular point. If two extra coefficients are added to one or more of the polynomials U, P, Q and R, a multisingular point at (x_c,y_c) may be ensured by imposing the two constraints

$$Q_{\underset{\sim}{M}}(x_c,y_c) = \sum_{(m,m')\varepsilon \underset{\sim}{M}} q_{m,m'} x_c^m y_c^{m'} = 0 , \tag{9.1}$$

and

$$R_{\underset{\sim}{N}}(x_c,y_c) = \sum_{(n,n')\varepsilon \underset{\sim}{N}} r_{n,n'} x_c^n y_c^{n'} = 0 ; \tag{9.2}$$

which clearly, each represent a single linear equation for the coefficients. In trials on the total susceptibility $\chi(T,g)$ for the Ising-Heisenberg-XY model imposition of $x_c = y_c$ with the best single-variable critical point estimate (see Table I) yields improved estimates for γ, in the range 1.36 to 1.39, and for ϕ, in the range 1.22 to 1.28. The results agree well with the direct single-variable analysis for the Heisenberg limit $(x = y = w)$ but the improvement is not especially striking.

Incidentally, reference to Figure 5 and the accompanying discussion shows that it may also be worthwhile to impose the single condition $Q_{\underset{\sim}{M}} = 0$ at the pure Ising critical point $x = x_I$ and $y = 0$ (see Table I) since this should ensure that the locus of zeros of $Q_{\underset{\sim}{M}}(x,y)$ crosses the x axis at the optimal position. Likewise it may be helpful to impose the condition $R_{\underset{\sim}{N}} = 0$ at $x = 0$, $y = y_{XY}$ on the y axis.

(b) Multisingular axes. When information is available regarding the slopes, e_1 and e_2, of the multisingular axes this may also be incorporated provided a reasonable estimate, say (x_0, y_0), is available for the location of the multisingular point itself. Thus, by allowing for one extra coefficient the product, e_1e_2, of the two slopes may be imposed via the constraint

$$(\partial R_{\underset{\sim}{N}}/\partial x)_0 + e_1e_2(\partial Q_{\underset{\sim}{M}}/\partial y)_0 = 0 , \tag{9.3}$$

as may be seen from the quadratic equation (4.7). The subscripts zero denote evaluation of the polynomials at (x_0, y_0). Likewise the sum, $(e_1 + e_2)$, may be imposed via

$$(\partial Q_{\underset{\sim}{M}}/\partial x)_0 + (e_1 + e_2)(\partial Q_{\underset{\sim}{M}}/\partial y)_0 - (\partial R_{\underset{\sim}{N}}/\partial y)_0 = 0 . \quad (9.4)$$

Imposition of the two constraints (9.3) and (9.4) will clearly fix the values of both e_1 and e_2. On the other hand, if only one slope, say e_0, is known, e.g. through some special symmetry of the problem, it may be imposed by requiring

$$e_0^2(\partial Q_{\underset{\sim}{M}}/\partial y)_0 + e_0[(\partial Q_{\underset{\sim}{M}}/\partial x)_0 - (\partial R_{\underset{\sim}{N}}/\partial y)_0] = (\partial R_{\underset{\sim}{N}}/\partial x)_0 , \quad (9.5)$$

as follows directly from (4.7).

(c) Multisingular exponents. Suppose the values of $(x_c, y_c) \equiv (x_0, y_0)$ and e_2 (but not necessarily e_1) are known exactly or to good accuracy: the value of the exponent γ can be fixed through the constraint

$$P_{\underset{\sim}{L}}(x_0, y_0) + \gamma(\partial Q_{\underset{\sim}{M}}/\partial x)_0 = \gamma(\partial R_{\underset{\sim}{N}}/\partial x)/e_2 , \quad (9.6)$$

which again clearly represents a linear equation in the coefficients $p_{l,l'}$, $q_{m,m'}$, and $r_{n,n'}$.

If both axis slopes e_1 and e_2 are available [as well as (x_0, y_0)] the crossover exponent may be imposed by requiring

$$\phi(\partial Q_{\underset{\sim}{M}}/\partial x)_0 + e_1(\partial Q_{\underset{\sim}{M}}/\partial y)_0 = \phi(\partial R_{\underset{\sim}{N}}/\partial x)_0/e_2 + (\partial R_{\underset{\sim}{N}}/\partial y)_0 . \quad (9.7)$$

In the practical studies of the Ising-Heisenberg-XY model [46] we have used the combinations (a)+(b), imposing x_c, y_c, e_1 and e_2, which yields improved estimates for γ and ϕ, and (a)+(b)+(c), fixing also γ and ϕ. (See Figure 10 below.)

(d) Imposition of the multisingular trace. If a good estimate (x_0, y_0) of the multicritical point is at hand but the slopes of the scaling axes are not known, the combination of exponents

$$\mu = (1 + \phi)/\gamma = -(\lambda_1 + \lambda_2)/P_{\underset{\sim}{L}}(x_0, y_0) , \quad (9.8)$$

which might be called the multisingular trace, can be imposed through the constraint

$$(\partial Q_{\underset{\sim}{M}}/\partial x)_0 + (\partial R_{\underset{\sim}{N}}/\partial y)_0 + \mu P_{\underset{\sim}{L}}(x_0, y_0) = 0 . \quad (9.9)$$

Trials of this condition demonstrate that it does yield some improvement in the estimates of the scaling axis slopes; this could be useful in some applications where the exponents may be known through separate theoretical considerations (as, for example, is the case for tricritical exponents in more than three dimensions: see e.g. Refs. [16, 35, 36]).

(e) Specification of the background. Finally we remark, as is obvious from (4.3), that the value of the background contribution, B_0, at a given multisingular point can be fixed via the relation

$$B_0 P_{\underset{\sim}{L}}(x_0,y_0) + U_{\underset{\sim}{J}}(x_0,y_0) = 0 \quad . \tag{9.10}$$

The above possibilities are certainly not exhaustive: the reader may easily devise others. However it is worth noting that certain desirable combinations of information cannot, apparently, be imposed without sacrificing the requirement that the equations determining the approximants be linear. Thus it would certainly be valuable to be able to impose the value of one or both exponents, γ and ϕ, without at the same time fixing the location of the multisingular point itself. [One might consider imposing only the constraints (9.6) and (9.7) using a "rough" approximation to (x_c,y_c), without also requiring (9.1) and (9.2); however, since exponent estimates usually vary quite rapidly with estimates of the multisingular location, it seems unlikely that such an approach would be very useful. Nevertheless it probably deserves some trials.] Again one might wish to impose only, say, the x coordinate of a multisingular point: but since y_c appears nonlinearly as an unknown in the constraint equation the corresponding approximants cannot be calculated efficiently. In practice, however, these difficulties may be circumvented in more pedestrian way simply by running trials with a range of choices of estimate (x_0,y_0) etc. and observing the various corelations as in Figures 2, 3, and 4.

10 THE UNIVERSAL BICRITICAL AMPLITUDE RATIO

As explained in Section 1, a quantity of prime interest for testing the general theories of bicriticality [13-16] is the amplitude ratio

$$Q = \dot{z}_-/\dot{z}_+ = w_-/w_+ \quad , \tag{10.1}$$

which determines how the two critical lines, of XY-like and Ising-like character, respectively, come together and meet at the bicritical point. In terms of the scaling fields, this is determined asymptotically by

$$\tilde{g}_c(T) \approx \pm w_\pm |\tilde{t}|^\phi \quad \text{or} \quad \Delta\tilde{y}_c \propto \pm \dot{z}_\pm |\Delta\tilde{x}|^\phi \quad , \tag{10.2}$$

[see (1.8), (1.9) and (2.2) - (2.4)]. Renormalization group and general scaling ideas suggest that Q should assume a universal value: as mentioned, for Ising-to-Ising crossover or XY-like bicriticality, the prediction is simply $Q_{XY} = 1$ and this agrees well with the experimental avidence for $GdAlO_3$ [1,2].

In the case of n-component bicriticality with crossover from n_+-like criticality to n_--like criticality (with $n = n_+ + n_-$) the simplest extension of mean field theory, which just allows for the symmetrically correct way of introducing the anisotropy field [as illustrated in the second line of (6.1)], yields $Q_n = (n_-/n_+) + O(\varepsilon)$ where, as usual $\varepsilon = 4-d$ and d is the spatial dimensionality [13]. However a fuller renormalization group analysis [47] gives

$$Q_n = (n_-/n_+)^{\phi(\varepsilon)} + O(\varepsilon^3) \quad , \tag{10.3}$$

where $\phi(\varepsilon)$ is the appropriate n-component crossover exponent known to second order in ε [42, 35, 36]. In our case of Heisenberg bicriticality we have $n_+ = 1$, $n_- = 2$ and $n = 3$: then (10.3) indicates $Q_H = 2.3$ to 2.4^+ with, surely, $Q_H > 2.2$. (Bruce [47] suggested $Q_H \simeq 2.33$, which we will see is surprisingly close to our final conclusion!). However, on the basis of single-variable analysis [23, 48] (utilizing the value for ϕ_H in Table I) the significantly higher estimate, $Q_H \simeq 2.51$ had been advanced [13]: but the uncertainties which should realistically have been attached to this value were obscure because of the difficulties, already alluded to, of making such an estimate by single-variable series analysis techniques. In view of this, the agreement with the ε expansion calculations seemed reasonable. A cause for consternation, however, was that most painstaking and precise experiments by Rohrer and King [4, 5] on MnF_2, which should be described as Heisenberg-like in the bicritical region, suggested the much lower value $Q = 1.56 \pm 0.35$. Apparently the largest possible value for a reasonable fit to the data was $Q \simeq 1.91$, which falls even below the limiting, mean field value $Q_H = 2$!

In order to remove these discrepancies a careful study was undertaken using partial differential approximants [46]. Both the constraint combinations (a)+(b), fixing $x_c = y_c = w_H$ and $e_1 = 1$, $e_2 = -1/2$, and (a)+(b)+(c) in which γ and ϕ were assigned the values 1.380±0.005 and 1.250±0.005 were explored. A variety of high order homogeneous approximants were constructed and the critical lines were calculated by integrating the trajectory equations from the x and y axes using, respectively, the accepted values of x_I and y_{XY} from Table I (and analogously for the sc and bcc lattices), or the alternative pure critical point estimates, x_I^0 and y_{XY}^0 [obtained from (7.1) above]. The differences found by using x_I^0 in place of x_I were only in the fourth decimal place of Q, which is much smaller than other uncertainties. In particular changing the assignment of $x_c = y_c$ from 0.3147 to 0.3146 yields in sets of [21; 11, 11] approximants the estimates $Q \simeq 2.557, 2.563, 2.560, 2.509,$ or $2.555, 2.587, 2.254, 2.285, 2.526,$ etc. Changes in the assignments of the exponents γ and ϕ of ±0.005 likewise lead to changes in the estimates for Q of ±0.05.

An overview of the results can be obtained from Figure 10 which shows results for all three cubic lattices, for a variety of high order approximants, and for imposition of both the sets of constraints (a)+(b) (open symbols) and (a)+(b)+(c) (solid symbols).

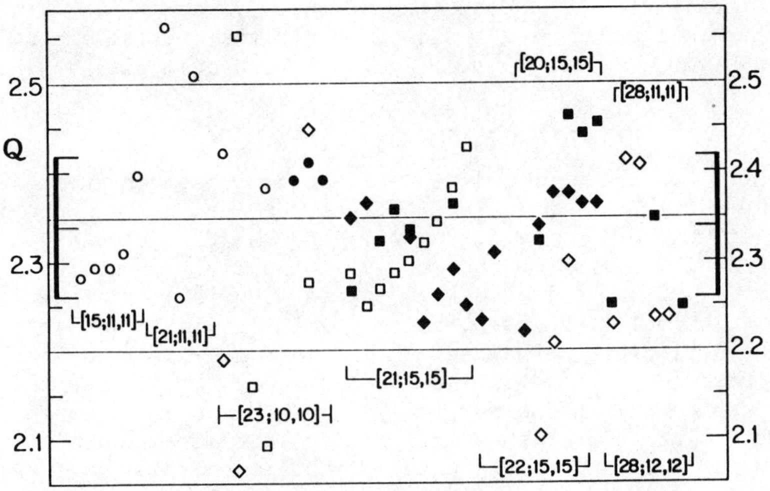

FIG. 10. Estimates of the bicritical amplitude ratio Q for the Ising-Heisenberg-XY model on fcc(o), bcc(◇), and sc(□) lattices from homogeneous approximants [L;M,N] with four or six constraints imposed (open or full symbols, respectively).

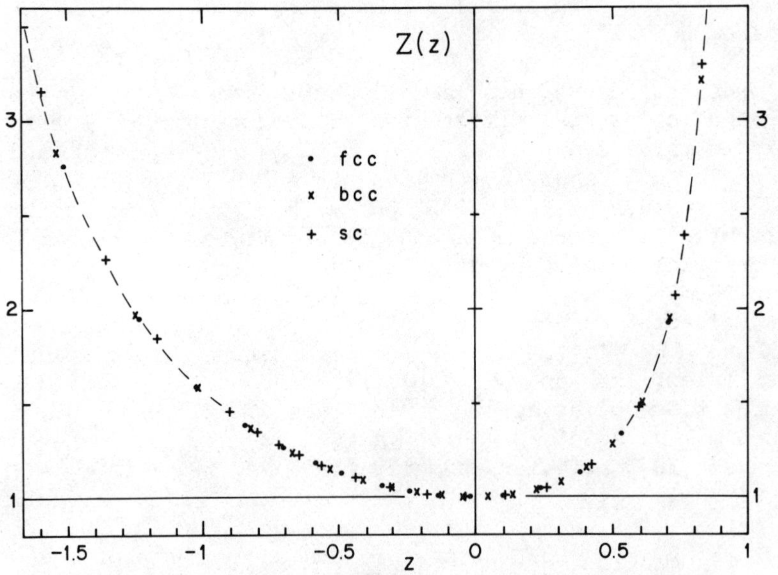

FIG. 11. Estimates for the bicritical scaling function for Ising-Heisenberg-XY crossover demonstrating universality as evidenced by the concordance between the results for different lattices. Note the Ising-like singularity occurs at $z = 1$ by the normalization convention.

From this evidence and the sensitivity to the various other factors mentioned above the final estimate adopted [46] is

$$Q_H = 2.34 \pm 0.08 \quad , \tag{10.4}$$

which is in excellent agreement with the renormalization group ε expansion result (10.3) and also reveals directly the shortcomings of the single-variable series analyses [23, 48, 13].

It might be commented that in integrating the trajectory equations (4.10) into the bicritical region it is essential to take precautions in the computer routines employed if accuracy is not to be sacrificed: this is merely a reflection of the mathematically singular nature of a multicritical point. Incidentally, in determining the precise location of a multisingular point estimate, it may also prove efficient to use the trajectory integral routines rather than adapting standard zero-finding programs to the case of two polynomials.

Since the theoretical result (10.4), which we feel is rather reliable, still disagrees violently with the value derived from the experiments, what is one to conclude about the physics of the situation? In the first place, it turns out that fitting a cusped form such as (10.2) with <u>unknown scaling axes</u> and <u>unknown multicritical point</u>, to statistically noisy data can be a rather sensitive matter. In particular, numerical studies [46] in which random noise of various amplitudes was added to accurate results obtained from PDA analysis, demonstrated that standard least squares fitting routines may entail very significant <u>bias</u> in a parameter such as Q; i.e. the value of Q providing the "best fit" at a given noise level deviates (even "on average") from the true value! However the experimental data analysis is able to avoid this problem because the slope of the spin flop line is comparatively small (and quite well determined). The issue thus remains unresolved: but we suspect that small systematic errors in assigning the precise temperature $T_c^{\perp}(H)$ to the inevitably somewhat rounded data and a comparably small shift in $H_c^{\parallel}(T)$ are needed. At any rate, such small reinterpretations of the data do suffice to yield full consistency with the theory!

11. CALCULATION OF SCALING FUNCTIONS

Once one has obtained reliable estimates for the location (x_c, y_c), for the axis slopes, e_1 and e_2, for the exponents, γ and ϕ, and for the background term B_0 of a multisingular point, a complete asymptotic description still demands estimates for the full scaling function, $Z(z)$ [see (1.5), (2.2), or (4.20)]. As seen, the behavior of the function of interest in the multicritical region, say the total susceptibility $\chi(T,g)$, can be found by integrating the trajectory equations (4.10) into the multisingular region and using the solutions to integrate the companion equation

(4.11) [see also (4.12)] for the function itself. Appropriate boundary conditions outside the multicritical region are essential, but the form these should take depends on the nature of the trajectories away from multicriticality which, as demonstrated in Section 8, is certainly not a universal feature. In our detailed studies of the bicritical scaling functions [46, 49] we have utilized approximants (both homogeneous and inhomogeneous) with trajectories of the general form illustrated in Figure 5 above. (It would still be worth exploring the application of approximants with trajectories of the type shown in Figure 6: conversely, defective approximants with anomalous trajectories, such as illustrated in Figure 9, must be discarded.) When the trajectories resemble those in Figure 5 it is natural to impose boundary conditions by specifying the values of the function — to be concrete, the total susceptibility, χ — on the x and y axes i.e. in the pure Ising-like and pure XY-like limits. The first step, then, is to obtain good estimates for χ on the axes. This can clearly be done by single-variable methods using either ratio techniques [17, 18], Dlog Padé methods [17-20], or preferably, especially when weak singularities are involved (see below), the newer inhomogeneous differential approximant approach [21, 22, 44].

Given an initial value $\chi(\tau=0) = \chi_a$ corresponding to initial values, say, $x(\tau=0) = x_a$, $y(\tau=0) = 0$, one must then integrate (using appropriate care with the computer routines, as mentioned) into the multicritical region. At the same time the effective scaling variable

$$z(\tau) = C\Delta\tilde{y}(\tau)/|\Delta\tilde{x}(t)|^{\phi} \quad , \tag{11.1}$$

defined through (2.4), and the effective scaling function

$$Z[z(\tau)] = |\Delta\tilde{x}(\tau)|^{\gamma}\chi(\tau)/A \quad , \tag{11.2}$$

are to be calculated. Of course, the values of x_c, y_c, e_1 and e_2 appropriate to the particular approximant chosen must be used in computing $\Delta\tilde{x}$ and $\Delta\tilde{y}$: in practice it is advantageous to specify these in advance by using a constrained approximant (Section 9). Likewise the particular estimates of ϕ and γ entailed in, or specified for the approximant must be used.

The amplitudes C and A appearing in (11.1) and (11.2) are to be fixed via the agreed normalization conditions, on our case,

$$\dot{z}_+ = 1 \qquad \text{and} \qquad Z(0) = 1 \quad . \tag{11.3}$$

As explained, neither C nor A are universal parameters: rather they will depend on the lattice structure being studied. In practice, however, both C and A will also vary slightly from approximant to approximant for a given lattice, as is implicit in the scatter seen in Figure 10 for the universal amplitude ratio Q. Thus C should be determined for each particular approximant by integrating along the singular trajectory originating at $(x_I, 0)$.

When the trajectory integrations have been carried sufficiently far, so that $[x(\tau),y(\tau)]$ is close to the multisingular point, the values of $z(\tau)$ and $Z[z(\tau)]$ (say, with an assigned value of A) will be observed to approach constant values (say z_∞ and Z_∞). At that stage the integration may be stopped since the pair (z_∞,Z_∞) specify one point on the graph of $Z(z)$. By repeating the process with different initial conditions further points are obtained as desired.

To enforce the normalization condition $Z(0) = 1$ and fix A, one may simply scan through the initial integration points until a value of $z_\infty = z(\tau\rightarrow\infty)$ is found which is sufficiently close to zero. Alternatively, since the appropriate scaling axis $(\Delta\tilde{y} = 0)$ is known precisely, one may choose a point on the scaling axis close to the multicritical point and integrate outwards until the boundaries are reached and impose the "known" value of χ at the appropriate point. If one employs approximants in which the symmetric, Heisenberg axis, $x=y$, is imposed as a trajectory (see Section 8) this procedure is particularly straightforward.

The effectiveness of this approach to calculating the scaling function may be gauged from Figure 11, which displays results for the three cubic lattices and for a range of high order approximants. A striking consistency is observed across the different lattices and over the various approximants even when the shapes of the trajectories leading to similar values of z differ strongly in the individual cases. This represents, of course, both a validation of the partial differential approximant approach and a confirmation of the full universality hypothesis [16, 17, 35, 36].

To study the scaling function in further detail a more sensitive representation is appropriate. As was explained in Sections 1 and 6, it is anticipated that the scaling function for the total susceptibility will itself display divergent singularities of Ising-like character at $\dot{z}_+$ and of XY-like character at $\dot{z}_-$ [see (1.7)]. In fact, granted trajectories like those in Figure 5 and proper approximants for the pure Ising and XY limits [see (6.7) and (6.8)] the calculated scaling function must necessarily display such behavior. Now because of the symmetries of the Hamiltonian (6.1) about the Heisenberg limit $g = 0$, one can show that $(\partial\chi/\partial g)$ vanishes at $g = 0$. Consequently the scaling function obeys

$$(dZ/dz)_{z=0} = 0 \quad , \tag{11.4}$$

that is, it has zero slope at the origin. It is thus appropriate to embody both the expected singularities and this fact in introducing a reduced scaling function, $P(z)$, via the definition

$$Z(z) = P(z)\left\{\frac{1}{3}\left[(1-z)^{-\gamma_I} - \gamma_I z\right] + \frac{2}{3}\left[\left(1+\frac{z}{Q}\right)^{-\gamma_{XY}} + \gamma_{XY}\frac{z}{Q}\right]\right\}. \qquad (11.5)$$

Here we have also embodied the normalization condition for $\dot{z}_+$ [see (11.3)] so that Q again denotes the universal critical locus amplitude ratio discussed in the previous section. (However, recall that the predicted value of Q will vary somewhat from approximant to approximant.) The conditions (11.3) and (11.4) translate into

$$P(0) = 1 \qquad \text{and} \qquad (dP/dz)_{z=0} = 0 \ . \qquad (11.6)$$

Naturally, plots of $P(z)$ for different approximants will show greater variability than visible in Figure 11 for $Z(z)$. However for the fcc lattice three different [23;10,10] homogeneous approximants (with imposed values of x_c, y_c, e_1 and e_2, etc.) yield values for $P(z)$ differing by at most 0.3% [49]. Likewise two different [15;11,11] approximants for $P(z)$ coincide to within better than 0.1%; however a third one, predicting a significantly larger value of Q, is found to exhibit 1 to 4% deviations. The overall agreement between the two different sets is at the 1% level except near the ends of the range (i.e. $z \gtrsim 0.8$ or $z \lesssim -1.8$) where it rises to 3 or 4%. Results for the sc lattice are similar but somewhat less regular. Surprisingly, the bcc results found appeared to be considerably less well converged with limiting values, $P(1)$, ranging from 0.95 up to 1.07, in contrast to the fcc estimates which suggest $P(1) \simeq 0.94$ to 0.97.

It is certainly possible that the more erratic behavior of the bcc series reflects a slightly inaccurate assignment of the bicritical point value $x_c = y_c$; however, it could also mean that the global behavior of $\chi(T,g)$ for the bcc lattice is somewhat different and, coincidentally, somewhat harder to represent accurately by homogeneous approximants. To check that possibility a range of inhomogeneous approximants were calculated for each lattice [49]. Some results are shown in Figure 12 for approximants with polynomials $U_J(x,y)$ having $J = 3$, 4, and 6 coefficients. Although there is a good deal of variability — and the bcc estimates still behave more erratically than the fcc and sc results — there is a clear overall consensus regarding the behavior of the reduced scaling function. The agreement between the different lattices, even in this sensitive representation, is rather encouraging. A reasonable overall or universal representation of the results described, examined, and displayed [49] is provided by the form

$$P(z) \simeq 1 - \frac{az^3}{1 - bz} \ , \qquad (11.7)$$

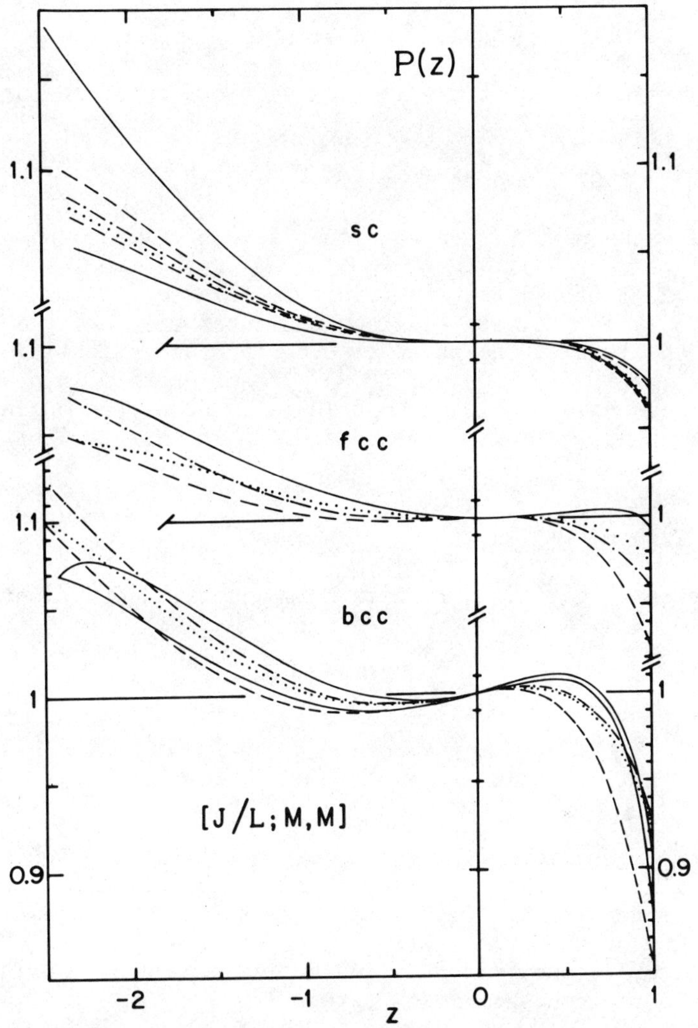

FIG. 12. Estimates for the reduced scaling function, P(z), for Ising-Heisenberg-XY crossover (see text) derived from various high order inhomogeneous partial differential approximants, [J/L;M,M], with J = 3, 4, and 6 for the sc, bcc, and fcc lattices.

where both the alternative sets of parameter values

$$a = 0.012795, \qquad b = 0.57350 \quad , \tag{11.8}$$

$$a = 0.011456, \qquad b = 0.618131 \quad , \tag{11.9}$$

prove satisfactory. At the limits, where we estimate the universal values

$$P(1) \simeq 0.97 \qquad \text{and} \qquad P(-Q) \simeq 1.07 \quad , \tag{11.10}$$

we believe the results are good to within a few percent.

The form (11.5) and the results (11.7) to (11.10) apply, of course, only to the total susceptibility as originally defined in (6.9). Physically, the separate ordering susceptibilities $\chi_{\parallel}(T,g)$ and $\chi_{\perp}(T,g)$ [see (6.6) and (6.8)] are of greater interest. The same methods have been used to calculate the corresponding scaling functions [49]. Simpler reduced scaling representations such as [23]

$$\chi_{\parallel}(T,g) \approx A_{\parallel}|\Delta\tilde{x}|^{-\gamma} P_{\parallel}(z)(1-z)^{-\gamma_I} \quad , \tag{11.11}$$

are then quite satisfactory with normalization conditions (11.3), so that

$$P_{\parallel}(0) = 1 \quad . \tag{11.12}$$

Then we find [49]

$$P_{\parallel}(1) \simeq 0.965 \quad , \tag{11.13}$$

which compares favorably with the estimate $P_{\parallel}(1) \simeq 0.9548$ obtained by renormalization group $\varepsilon = 4-d$ expansion calculations correct to first order in ε [50]. Note that $P(z)$ should exhibit a singularity of the type $(z+\dot{z}_-)^{1-\alpha_{XY}}$ as $z \to -\dot{z}_- = -Q$: this will be correctly accounted for by imposition of the appropriate behavior of $\chi_{\parallel}$ in the pure XY limit $(x=0)$.

In this general connection the idea of <u>imposed-trajectory approximants</u> should be mentioned. As we have seen, the correct trajectories in the multicritical region are fixed for <u>all</u> properties of a given model, by the values of x_c, y_c, e_1, e_2 and the exponent ϕ. Furthermore, the correct singular trajectories, corresponding e.g. to $\dot{z}_+$ and $\dot{z}_-$, are also fixed since they must describe the critical loci. But it is evident from (4.1) and (4.4) to (4.10), that all these features are determined, for a given approximant $[\underset{\sim}{J}/\underset{\sim}{L};\underset{\sim}{M},\underset{\sim}{N}]$, solely by the polynomials $Q_{\underset{\sim}{M}}(x,y)$ and $R_{\underset{\sim}{N}}(x,y)$: the polynomials $P_{\underset{\sim}{L}}(x,y)$ and $U_{\underset{\sim}{J}}(x,y)$ play no role. This observation suggests that a pair of polynomials, $Q_{\underset{\sim}{M}}$ and $R_{\underset{\sim}{N}}$ (found, say, by studying a strongly divergent function like the total susceptibility), which are "good" in the sense that they yield optimal values of the multicritical parameters, accurate critical loci, and otherwise smooth and regular trajectories, should be "saved" and used in further applications to more weakly singular properties of the model. Specifically, in studying, say, the parallel susceptibility, $\chi_{\parallel}(T,g)$, or a nonordering susceptibility [44, 45, 49, 50], the chosen "good" values of the coefficients $q_{m,m'}$ and $r_{n,n'}$ are imposed and the equations defining the approximants are solved only for the coefficients $p_{l,l'}$ and $u_{j,j'}$ of $P_{\underset{\sim}{L}}$ and $U_{\underset{\sim}{J}}$. Of course, to take full advantage of the length of the available series, the arrays $\underset{\sim}{L}$ and $\underset{\sim}{J}$ must be chosen sufficiently large. In some cases it has also been found [49] that the array $\underset{\sim}{J}$ of coefficients $u_{j,j'}$ should not be taken too small, particularly when weak or nondivergent singularities are involved. Approximants for different properties calculated this way will be precisely <u>consistent</u> in their multicritical and critical parameters: this feature could be especially advantageous in studies where different properties of a model are to be combined to estimate various derived functions of interest.

Trials of this imposed-trajectory approach shows that it works quite well in most cases but does not necessarily lead to improved estimates of the multicritical scaling functions and amplitudes for the particular property under study (as judged in the bicritical models by comparison with single-variable estimates on the Heisenberg locus x=y).

In conclusion, we have demonstrated that partial differential approximants provide effective and incisive tools for studying bicriticality in all its aspects. One may be optimistic that these approximants will be equally successful in application to other types of multicritical behavior, particularly, as seen near tricritical points and Lifshitz points. However, it must be recognized that these examples will provide challenges differing in various ways from those encountered in our attack on bicriticality: this reflects the fact that the characters of functions of two variables and their natural singularities, are much more diverse than experience with functions of only a single variable might lead one to believe! Perhaps the new challenges will require novel extensions of the technique or completely fresh ideas. To stimulate you, our reader, let us finish with a question by asking: "How can <u>confluent multicritical singularities</u>, or "corrections-to-scaling" of the sort suggested by renormalization group theory [41, 51, 35, 36] be incorporated effectively into two-variable series analysis?"

ACKNOWLEDGMENTS

We are indebted to Dr. Helen Au-Yang for her contributions, help, and advice in nearly all aspects f the work reported here. Gratitude is due to the National Science Foundation for their support of the work, principally through the Mathematics and Computer Science Division but also, in part, through the Materials Science Center at Cornell University. One of us (MEF) is grateful to the Scientific Committee of NATO for making possible the meeting in Corsica at which the material was originally presented.

REFERENCES

[1] H. Rohrer, Phys. Rev. Lett. 34, 1638 (1975).

[2] H. Rohrer and Ch. Gerber, Phys. Rev. Lett. 38, 909 (1977).

[3] H. Rohrer and Ch. Gerber, J. Appl. Phys. 49, 1341 (1977).

[4] A. R. King and H. Rohrer, AIP Conf. Proc. 29, 420 (1976).

[5] A. R. King and H. Rohrer, Phys. Rev. B 19, 5864 (1979).

[6] Y. Shapira, S. Foner, and A. Misetich, Phys. Rev. Lett. 23, 92 (1969).

[7] Y. Shapira and S. Foner, Phys. Rev. B 1, 3083 (1970).

[8] J. H. Shelleng and S. A. Friedberg, Phys. Rev. 185, 728 (1969).

[9] N. F. Oliveira, Jr., A. Paduan Filho and S. R. Salinas, Phys. Lett. 55A, 293 (1975).

[10] Y. Shapira and C. C. Becerra, Phys. Lett. 57A, 483 (1976).

[11] J. A. J. Basten, E. Frikkee, and W. J. M. de Jonge, Phys. Rev. Lett. 42, 897 (1979).

[12] Y. Shapira, N. F. Oliveira, Jr., and T. S. Chang, Phys. Rev. Lett. 42, 1292 (1979).

[13] M. E. Fisher, Phys. Rev. Lett. 34,1634 (1975).

[14] M. E. Fisher, and D. R. Nelson, Phys. Rev. Lett. 32, 1350 (1974).

[15] D. R. Nelson, J. M. Kosterlitz, and M. E. Fisher, Phys. Rev. Lett. 33, 813 (1974).

[16] M. E. Fisher, AIP Conf. Proc. 24, 273 (1975).

[17] M. E. Fisher, Rept. Prog. Phys. 30, 615 (1967).

[18] M. E. Fisher, Rocky Mtn. J. Math. 4, 181 (1974).

[19] G. A. Baker, Jr. "Essentials of Padé Approximants," (Academic Press, New York, 1975).

[20] C. Domb and M. S. Green, Eds. "Phase Transitions and Critical Phenomena" vol. 3 (Academic Press, London, 1974).

[21] D. L. Hunter and G. A. Baker Jr. Phys. Rev. B 19, 3808 (1979).

[22] M. E. Fisher and H. Au-Yang, J. Phys. A: Math. Gen., 12, 1677 (1979).

[23] P. Pfeuty, D. Jasnow and M. E. Fisher, Phys. Rev. B 10, 2088 (1974).

[24] J. S. R. Chisholm, Math. Comput. 27, 841 (1973).

[25] J. S. R. Chisholm and J. McEwan, Proc. Roy. Soc. A 336, 421 (1974).

[26] M. E. Fisher, pp. 3-32 in "Statistical Mechanics and Statistical Methods in Theory and Application," Ed. U. Landman (Plenum Press, New York, 1977).

[27] C. G. Callan, Jr., Phys. Rev. D 2, 1541 (1970).

[28] K. Symanzik, Commun. Math. Phys. 118, 227 (1970).

[29] E. Brézin, J. C. Le Guillou, and J. Zinn-Justin, Phys. Rev. D 8, 434, 2418 (1973).

[30] M. E. Fisher, Physica 86-88, 590 (1977).

[31] M. E. Fisher and R. M. Kerr, Phys. Rev. Lett. 39, 667 (1977).

[32] J. L. Gammel, pp. 3-10 in "Padé Approximants and their Applications," Ed. P. R. Graves-Morris (Academic Press, New York, 1973).

[33] A. J. Guttmann and G. S. Joyce, J. Phys. A: Math. Gen. 5, L81 (1972).

[34] J. J. Rehr, G. S. Joyce, and A. J. Guttmann, J. Phys. A: Math Gen. 13, 1587 (1980).

[35] M. E. Fisher, Rev. Mod. Phys. 46, 597 (1974).

[36] C. Domb and M. S. Green, Eds. "Phase Transitions and Critical Phenomena," vol. 6 (Academic Press, London, 1976).

[37] D. Roberts, H. P. Griffiths and D. W. Wood, J. Phys. A: Math. Gen. 8, 421 (1974).

[38] J. F. Stilck and S. R. Salinas: Univ. of São Paulo, preprint (Jan. 1981).

[39] H. Au-Yang and M. E. Fisher [unpublished].

[40] B. G. Nickel in these Proceedings.

[41] K. G. Wilson and M. E. Fisher, Phys. Rev. Lett. 28, 240 (1972).

[42] M. E. Fisher and P. Pfeuty, Phys. Rev. B 5, 1889 (1972).

[43] K. G. Wilson, Phys. Rev. Lett. 28, 548 (1972).

[44] J.-H. Chen and M. E. Fisher, J. Phys. A: Math. Gen. [in press].

[45] P. R. Gerber and M. E. Fisher, Phys. Rev. B 13, 5042 (1976).

[46] M. E. Fisher, J.-H. Chen and H. Au-Yang, J. Phys. C: Solid St. Phys. 13, L459 (1980).

[47] A. D. Bruce, J. Phys. C: Solid St. Phys. 8, 2992 (1975).

[48] S. Singh and D. Jasnow, Phys. Rev. B 12, 493 (1975).

[49] J.-H. Chen and M. E. Fisher [to be published].

[50] E. Domany, D. R. Nelson and M. E. Fisher, Phys. Rev. B 15, 3493 (1977).

[51] F. J. Wegner, Phys. Rev. B 5, 4529 (1972).

HIGH TEMPERATURE SERIES ANALYSIS FOR THE THREE-DIMENSIONAL ISING MODEL: A REVIEW OF SOME RECENT WORK

David S. Gaunt

Physics Department, King's College, University of London
Strand, London WC2R 2LS, England

ABSTRACT

Traditional high temperature series estimates of critical exponents for the three-dimensional Ising model differ from renormalization group theory estimates but by only a small amount. However, the quoted uncertainties in each method are too small to explain the difference in a convincing manner. Reanalysis of the high temperature series has been undertaken by several researchers, whose work we review here, in an attempt to resolve this discrepancy.

1. INTRODUCTION

A problem which has attracted a great deal of interest recently is the apparent discrepancy in three dimensions ($d = 3$) between values of critical exponents obtained, on the one hand, from exact series expansions and, on the other hand, using renormalization group (RG) theory. My aim in these Lecture Notes is to review recent series expansion work on this problem for the simple (spin-$\frac{1}{2}$, nearest neighbour) Ising model, since it is for this model that series estimates of critical exponents are the most precise. The exponent which has been studied most extensively and was considered to be the best established is the exponent γ describing the dominant zero-field behaviour of the reduced isothermal susceptibility χ_0 just above the critical temperature T_c,

$$- \frac{1}{kT} \left(\frac{kT}{m}\right)^2 \frac{\partial^2 F}{\partial H^2} \bigg|_{H=0} \equiv \chi_0 \sim C\, t^{-\gamma} \quad . \tag{1.1}$$

Here t is the reduced temperature $(T-T_c)/T_c$ and C is a critical amplitude. A typical series estimate of the exponent γ is $\gamma = 1.250 \pm 0.003$, which value is taken from the well-known review article by Fisher[1]. In view of the exact result $\gamma = 1\frac{3}{4}$ in two dimensions and $\gamma = 1$ for mean-field theory, it was natural to conjecture that $\gamma = 1\frac{1}{4}$ for the three-dimensional Ising model.

More recently RG ideas have led to new methods of calculation based upon the study of the ϕ^4 continuous field theory. The first precise RG estimates of critical exponents were obtained by Baker et al.[2,3] who calculated perturbation expansions for the coefficients of the Callan-Symanzik equation in fixed dimension. The expansions which are in powers of a dimensionless coupling constant g, say, are through seventh order for $\beta(g)$ and one order less for the other coefficients. They were summed using the Padé-Borel method yielding $\gamma = 1.241 \pm 0.004$. The more precise estimate of $\gamma = 1.2402 \pm 0.0009$ was made by Le Guillou and Zinn-Justin[4] by making further plausible, although unproven, analyticity assumptions.

And now you perceive the problem: the series estimates of γ are centred around 1.250, the RG estimates between 1.240 and 1.241, and although this discrepancy is only about $\frac{3}{4}$% the uncertainties in each estimate (although only confidence limits) are too small to explain away the difference in a convincing manner. The resolution of this discrepancy is clearly an important theoretical problem. Increasingly it is becoming a matter of some experimental importance since it appears[5,6] that in some systems it is now possible to measure critical exponents to within this sort of precision.†

A similar difficulty arises with regard to some of the other high temperature exponents Δ, α and ν. The 'gap' exponent Δ describes the asymptotic divergence as $T \to T_c+$ of the $2n^{th}$ field derivative of the free energy F evaluated in zero magnetic field,

$$-\frac{1}{kT}\left(\frac{kT}{m}\right)^{2n} \frac{\partial^{2n}F}{\partial H^{2n}}\bigg|_{H=0} \equiv f_0^{(2n)} \sim C_{2n}\, t^{-[\gamma+2(n-1)\Delta]} . \tag{1.2}$$

On setting n = 1 one recovers (1.1) with the identification $f_0^{(2)} \equiv \chi_0$ and $C_2 \equiv C$. Symmetry with respect to field reversal ensure that all odd derivatives are identically zero. The exponents α and ν describe the asymptotic divergence of the zero-field specific heat C_H and correlation length ξ as $T \to T_c+$,

$$C_H/k \sim (A/\alpha)t^{-\alpha} , \quad \xi/a \sim Et^{-\nu} , \tag{1.3}$$

---

†See also the Lecture Notes by J. V. Sengers, D. Beysens and G. Ahlers.

where a is the lattice spacing. The most useful definition of ν hinges on the second moment μ_2 of the spin-spin correlation function with asymptotic behaviour

$$\mu_2 \sim M_2 t^{-(\gamma+2\nu)} , \tag{1.4}$$

which, as can easily be checked, is consistent with the first definition. RG estimates of Δ, α and ν are given in Table 1, as are the best series estimates[1,7-9] - based, in general, upon rather fewer terms than were used for γ. The analogous (exact) values for d = 2 lattices and for mean-field theory are also given for comparison. The quoted confidence limits are definitely too small to explain the discrepancy in ν, while α is marginal and for Δ the series and RG estimates are in very good agreement.

A different but not unrelated problem to be discussed† at this School is the apparent failure of hyperscaling for the d = 3 Ising model. While relations such as

$$2\Delta - \gamma - d\nu = 0 \tag{1.5}$$

are satisfied exactly in RG theory and appear to hold for the d = 2 Ising model, for d = 3 one has from Table 1

$$2\Delta - \gamma - d\nu = -0.038 \begin{smallmatrix} +\ 0.012 \\ -\ 0.015 \end{smallmatrix} , \tag{1.6}$$

i.e. a small but clear failure of hyperscaling[10]. This view has been challenged in recent work by Nickel and Sharpe[11]. Although not

Table 1 High temperature critical exponents for the simple Ising model. (The uncertainty in parentheses is the Author's estimate.)

				d=3(RG)	
Exponent	Mean Field	d=2	d=3(series)	Refs 2,3	Ref 4
γ	1	$1\frac{3}{4}$	1.250 ± 0.003	1.241 ± 0.004	1.2402 ± 0.0009
Δ	$1\frac{1}{2}$	$1\frac{7}{8}$	1.563 ± 0.003	1.566 ± 0.006	1.565 ± 0.002
α	0(disc.)	0(log)	$0.125\pm(0.015)$	0.110 ± 0.008	0.110 ± 0.002
ν	$\frac{1}{2}$	1	$0.638^{+0.002}_{-0.001}$	0.630 ± 0.002	0.6300 ± 0.0008

† See the Lecture Notes by B.G. Nickel and G.A. Baker.

conclusive, these authors found, by analysing existing high temperature series in terms of a certain "renormalized" coupling constant variable, that hyperscaling could quite plausibly be satisfied.

Let us begin our reassessment of the extant but recent high temperature series work by giving in Table 2 the number of available coefficients, N, for each of the four standard lattices, namely, the face-centred cubic (FCC), body-centred cubic (BCC), simple cubic (SC) and diamond (D) lattices. Most of this work has focussed on the exponent γ which, as previously stated, was thought to be the best established. For convenience, therefore, we give in Table 3 the coefficients a_n of the susceptibility expansion

$$\chi_0 = \sum_{n=0}^{\infty} a_n v^n \ . \qquad (1.7)$$

The expansion variable is the usual high temperature counting variable $v = \tanh(J/kT)$. Several methods† have been used for the derivation of the coefficients a_n, the most successful of which are:

(i) the susceptibility counting theorem[17],

(ii) the star graph method[18],

(iii) the linked-cluster expansion method[19].

The extension of these expansions to their current length has a long and interesting history, and we conclude this section with a brief summary.

Table 2 Number, N, of high temperature series coefficients. (References in parentheses.)

	FCC	BCC	SC	D
χ_0	15 (12)	15 (13)	19 (14)	22 (15)
$f_0^{(2n)}$ $(n \leq 6)$	8 (7)	12 (7)	12 (7)	16 (7)
C_H	14 (8)	16 (8)	18 (8)	18 (16)
μ_2	12 (9)	12 (9)	12 (9)	...

† See the Lecture Notes by S. McKenzie.

Table 3 High temperature susceptibility coefficients a_n

n	D	SC	BCC	FCC
0	1	1	1	1
1	4	6	8	12
2	12	30	56	132
3	36	150	392	1 404
4	108	726	2 648	14 652
5	324	3 510	17 864	151 116
6	948	16 710	118 760	1 546 332
7	2 772	79 494	789 032	15 734 460
8	8 076	375 174	5 201 048	159 425 580
9	23 508	1 769 686	34 268 104	1 609 987 708
10	67 980	8 306 862	224 679 864	16 215 457 188
11	196 548	38 975 286	1 472 595 144	162 961 837 500
12	566 820	182 265 822	9 619 740 648	1 634 743 178 420
13	1 633 956	852 063 558	62 823 141 192	16 373 484 437 340
14	4 697 412	3 973 784 886	409 297 617 672	163 778 159 931 180
15	13 501 492	18 527 532 310	2 665 987 056 200	1 636 328 839 130 860
16	38 742 652	86 228 667 894		
17	111 146 820	401 225 368 086		
18	318 390 684	1 864 308 847 838		
19	911 904 996	8 660 961 643 254		
20	2 608 952 940			
21	7 463 042 916			
22	21 328 259 716			

The diamond lattice was a relative latecomer to the ranks of the four standard lattices. Block[20] noticed that, unlike the three cubic lattices, the coefficients of the low temperature series were all positive. To facilitate the analysis of the low temperature series requires an accurate estimate of the critical point and this is usually provided by analysis of the high temperature susceptibility series. Consequently, Essam and Sykes[21], followed by Gaunt and Sykes[15], calculated $\chi_0(v)$ through v^{16} and v^{22}, respectively, using the susceptibility counting theorem.

The first workers to calculate series of significant length for the three cubic lattices were Domb and Sykes[22] using elementary methods. Using series through v^9 for the SC and BCC lattices, and through v^8 for the FCC lattice, they made the initial suggestion that

$$\gamma = 1.244 \text{ (BCC)}, \quad \gamma = 1.250 \text{ (SC, FCC)} \quad . \tag{1.8}$$

The series for the SC lattice were extended by two further terms following Sykes' discovery of the counting theorem[17] and reanalysis of the series for all three cubic lattices led Domb and Sykes[23] to conjecture as early as 1961 that $\gamma = 1\frac{1}{4}$ for all three-dimensional lattices. <u>This conjecture constituted an important step towards the formulation of the universality hypothesis</u>.

The next major advance was the application of the linked- cluster expansion method by Moore et al.[9] and the consequent extension of all three cubic lattice series to order v^{12}. A small error in the coefficient of v^{12} for the FCC lattice was detected by Sykes et al.[13] who used the susceptibility counting theorem to extend the series through orders v^{17} (SC), v^{15} (BCC) and v^{12} (FCC). The series for the BCC lattice are still the longest available†. For the FCC lattice the series were subsequently extended to orders v^{13} and then v^{15} by Rapaport[24] and by McKenzie[12], respectively. Their work represents a significant application of the star-graph method. For the SC lattice, Sykes and McKenzie corrected a small error in the coefficient of v^{17} and calculated two further terms using both the susceptibility counting theorem and the star graph method. The new coefficients are given in the analysis paper of Gaunt and Sykes[14].

In summary, we have seen that the counting theorem has proved very successful for the derivation of susceptibility expansions for the simple (spin $\frac{1}{2}$) Ising model. It has yielded the longest series yet derived† for all but the FCC lattice. Of all the series, that for the SC lattice is the most reliably known, in the sense that for this lattice <u>all</u> of the coefficients have also been derived, quite independently, by the star graph method. In this type of work, where

† B. G. Nickel (see this Volume) has recently extended the series for the BCC lattice to order v^{21} using the linked-cluster expansion.

great importance attaches to accuracy, independent deviations of this kind are extremely valuable.

The existence of a discrepancy between traditional series estimates and RG estimates of critical exponents has inspired a number of recent attempts at reanalysing the high temperature susceptibility expansions. This work is reviewed in the next two sections: Padé-based methods in §2 and ratio-based methods in §3. We assume familiarity with the basic techniques of series analysis as reviewed by Hunter and Baker[25,26] and by Gaunt and Guttmann[27]. We conclude in §4 with a brief summary.

2. PADE-BASED ANALYSIS

2.1 Dlog Padé Approximants

According to RG theory, χ_0 is expected for d = 3 to have correction-to-scaling terms confluent with the dominant singularity,

$$\chi_0 \sim C\, t^{-\gamma} \{1 + A\, t^{\Delta_1} + \ldots\} \tag{2.1}$$

where the correction-to-scaling exponent[3] $\Delta_1 = 0.496 \pm 0.004$. In general, as is well-known[25-27], dlog Padé approximants are not well-suited for dealing with singularities of this form. However, since the dominant term is strongly singular ($\gamma \simeq 1.25$) and the correction-to-scaling amplitude A is expected[28,29] to be, if not exactly zero, rather small, then the dlog Padé table might nevertheless converge quite well. The poles and residues (in parentheses) of the main- and near-diagonal $[n + j/n]$ approximants ($j = 0, \pm 1$) to the $(d/dv)\ell n\, \chi_0$ series are given in Tables 4 to 7 for the D, SC, BCC and FCC lattices, respectively. Defects[25,27] on the positive and negative real axes are denoted by † and ‡ respectively, and defects in the complex plane by *. The apparent convergence caused by defective approximants should be entirely discounted[25,27].

For comparison, the previously accepted best estimates of the critical point (biased by the assumption $\gamma = 1.25$) are as follows[13,15]:

$$\begin{aligned} v_c = \ & 0.35381 \pm 0.00003\ (\mathrm{D}) \ , \quad 0.21813 \pm 0.00001\ (\mathrm{SC})\ , \\ & 0.15612 \pm 0.00003\ (\mathrm{BCC})\ , \quad 0.10174 \pm 0.00001\ (\mathrm{FCC}). \end{aligned} \tag{2.2}$$

These estimates are very well supported by the results in Tables 4-7.

Regarding the residues, the apparent convergence is particularly good for the D lattice giving an estimate of γ very close to 1.250.

Table 4 Dlog Padé analysis of $\chi_0(v)$, D lattice

n	[n-1/n]	[n/n]	[n+1/n]
4	0.352 297 (-1.219 04)	0.353 412 (-1.240 94)	0.353 777 (-1.249 44)
5	0.354 020 (-1.256 50)	0.353 793 (-1.249 89)	0.353 774 (-1.249 38)†
6	0.353 876 (-1.252 38)	0.353 866 (-1.252 05)‡	0.353 828 (-1.250 80)
7	0.353 878 (-1.252 43)†	0.353 801 (-1.249 86)	0.353 806 (-1.250 06)
8	0.353 807 (-1.250 09)	0.353 804 (-1.249 97)	0.353 805 (-1.250 01)‡
9	0.353 805 (-1.250 02)‡	0.353 803 (-1.249 95)†‡	0.353 814 (-1.250 40)
10	0.353 823 (-1.250 92)	0.353 819 (-1.250 65)	0.353 817 (-1.250 57)
11	0.353 815 (-1.250 42)		

Table 5 Dlog Padé analysis of $\chi_0(v)$, SC lattice

n	[n-1/n]	[n/n]	[n+1/n]
1	0.250 000 (-1.500 00)	0.190 476 (-0.870 75)	0.238 636 (-1.712 30)
2	0.215 614 (-1.196 31)	0.216 107 (-1.204 82)	0.217 874 (-1.245 39)
3	0.215 441 (-1.194 65)†	0.218 968 (-1.280 87)	0.218 385 (-1.259 38)
4	0.218 053 (-1.246 78)	0.218 151 (-1.250 53)	0.218 193 (-1.252 32)
5	0.218 257 (-1.255 82)	0.218 165 (-1.251 07)‡	0.218 128 (-1.249 22)
6	0.218 145 (-1.250 18)*	0.218 135 (-1.249 65)	0.218 141 (-1.250 00)
7	0.218 138 (-1.249 79)	0.218 138 (-1.249 84)	0.218 136 (-1.249 62)
8	0.218 138 (-1.249 80)*	0.218 142 (-1.250 05)†	0.218 123 (-1.248 31)
9	0.218 135 (-1.249 59)	0.218 091 (-1.241 83)‡	

Table 6 Dlog Padé analysis of $\chi_0(v)$, BCC lattice

n	[n-1/n]	[n/n]	[n+1/n]
1	0.166 667 (-1.333 33)	0.139 535 (-0.934 56)	0.170 635 (-1.709 08)
2	0.153 300 (-1.180 25)	0.155 931 (-1.242 46)	0.156 300 (-1.254 52)
3	0.156 374 (-1.257 66)	0.156 182 (-1.250 02)	0.156 122 (-1.247 23)
4	0.156 132 (-1.247 78)	0.156 016 (-1.239 23)	0.156 102 (-1.246 15)‡
5	0.156 108 (-1.246 58)	0.156 110 (-1.246 70)	0.156 112 (-1.246 84)
6	0.156 118 (-1.247 51)	0.156 111 (-1.246 80)	0.156 112 (-1.246 86)†
7	0.156 110 (-1.246 69)	0.156 109 (-1.246 62)	

Table 7 Dlog Padé analysis of $\chi_0(v)$, FCC lattice

n	[n-1/n]	[n/n]	[n+1/n]
1	0.100 000 (-1.200 00)	0.101 010 (-1.224 36)	0.101 852 (-1.255 23)
2	0.111 111	0.101 877 (-1.256 47)	0.101 857 (-1.255 43)‡
3	0.101 776 (-1.250 31)	0.101 711 (-1.245 16)	0.101 754 (-1.249 04)
4	0.101 742 (-1.247 96)	0.101 741 (-1.247 87)	0.101 739 (-1.247 67)
5	0.101 744 (-1.248 07)†	0.101 737 (-1.247 37)	0.101 741 (-1.247 81)†
6	0.101 717 (-1.240 410)	0.101 748 (-1.247 90)†	0.101 726 (-1.245 13)
7	0.101 761 (-1.245 614)†	0.101 724 (-1.244 42)	

For the SC lattice, convergence appears to be almost as good and a small range between 1.248 and 1.250 is indicated. For the BCC and FCC lattices, still smaller values are preferred around 1.246 to 1.248 and 1.240 to 1.248, respectively. (Note that the apparent convergence for the FCC lattice is far inferior to that observed for the D lattice.) We shall see this trend many times again; namely, the apparent value of γ is about 1.250 for the D lattice and decreases steadily with increasing lattice coordination number, q. We do not attribute the apparent variation of γ with coordination number to a breakdown of universality, but to convergence difficulties associated with too short a series. See §4 for further discussion.

2.2 Padé Analysis of Non-standard Lattices

High temperature series expansions have also been derived for a number of non-standard (regular, but non-Bravais) d = 3 lattices. These are given in Table 8. Recently, it has been claimed[33] that the series for the tetrahedron lattice is better behaved than the ones for the four standard lattices. To check this claim we have performed a dlog Padé analysis of the $\chi_0(v)$ series. In our opinion, the results, which are given in Table 9, are less well converged than they were for any of the standard lattices. This supports the tradition view[34] that the series for all the lattices in Table 8 are less regularly behaved than for the four standard lattices but that in all cases the results are quite consistent with $\gamma = 1.250$ as required by the universality hypothesis.

Table 8 Number, N, of high temperature susceptibility coefficients for non-standard d = 3 lattices. (References in parentheses.)

Lattice	q	N
Hydrogen peroxide	3	27 (30)
Hyper-triangular	6	13 (30)
Crystobalite B-site spinel Tetrahedron	6	16 (31)
Octahedral	8	9 (32)

Table 9 Dlog Padé analysis of $\chi_0(v)$, Tetrahedron lattice

n	[n-1/n]	[n/n]	[n+1/n]
3	0.233 156 (-1.225 37)	0.226 989 (-1.079 81)	0.245 193 (-2.007 11)
4	0.229 696 (-1.151 10)	0.236 961 (-1.434 75)	0.188 866 (-0.117 61)
5	0.233 476 (-1.270 08)	0.232 451 (-1.225 24)	0.233 299 (-1.266 91)
6	0.232 911 (-1.246 55)	0.233 102 (-1.256 29)	0.233 029 (-1.251 92)
7	0.233 061 (-1.253 94)	0.232 899 (-1.241 84)	0.232 993 (-1.249 58)
8	0.233 008 (-1.250 74)		

Since our experience with the tetrahedron lattice differs from that of Oitmaa and Ho-Ting-Hun[33], we have repeated their analysis. Thus, the series for $[\chi_0(v)]^{1/\gamma}$ was calculated for values of γ in the range $1.23 < \gamma < 1.26$, Padé approximants were formed and the location of the pole corresponding to v_c plotted against the assumed γ for three of the highest-order approximants as shown in Figure 1. By arguing that the best value of γ is that for which all Padé's give the same estimate of the location, they estimated[33]

$$\gamma \simeq 1.250 \pm 0.001, \quad v_c = 0.23300 \pm 0.00001 . \tag{2.3}$$

However, on closer inspection, we find that only those approximants denoted by $+$ in Figure 1 are free of defects on the positive real axis in the interval $(0, v_c)$. These non-defective approximants appear to define a single smooth curve upon which entries from all other Padé approximants, based upon sixteen terms and having no defects, also fall. This curve describes the way in which the values of v_c and γ are correlated but is of no help in ascertaining the best values. However, Figure 1 suggests to us, and it is confirmed by other high-order approximants (not shown), that the number of defective approximants is least for values of γ in the range 1.250 to 1.252

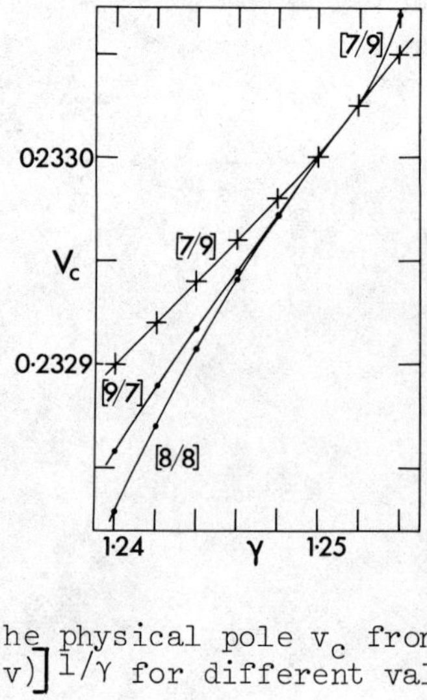

Fig.1 Location of the physical pole v_c from high-order Padé approximants to $[\chi_0(v)]^{1/\gamma}$ for different values of γ.

and this feature does provide a useful criterion for estimating optimum values of γ and v_c. In this way, we find for the tetrahedron lattice

$$\gamma = 1.251 \pm 0.001, \quad v_c = 0.233\,015 \pm 0.000\,015 \tag{2.4}$$

as the best estimates given the number of terms currently available.

2.3 Baker-Hunter Analysis

In this Section we outline the work of Bessis et al[35] who have reanalysed the $\chi_0(v)$ series using the Baker-Hunter method[26]. This method was designed to take account of confluent singularities of the type

$$\chi_0(v) = C_0(1 - \frac{v}{v_c})^{-\gamma_0} + C_1(1 - \frac{v}{v_c})^{-\gamma_1} + C_2(1 - \frac{v}{v_c})^{-\gamma_2} + \ldots ,$$

$$(\gamma_0 = \gamma) . \tag{2.5}$$

For d = 2 the dominant confluence is analytic (i.e. $\gamma_1 = \gamma - 1$) and the Baker-Hunter method seems to work quite well[35]. For d = 3, Baker and Hunter themselves[26] found no evidence for confluent singularities but the series available to them had between 3 and 7 fewer coefficients. The longest series are for the D lattice for which the results of Bessis et al[35] are presented in Table 10 assuming $v_c^{-1} = 2.82641$. While only 12 terms are needed to reach a reasonably stable and precise value of γ it is necessary to go to 16 terms to stabilize the leading subdominant exponent. Using all 22 terms one sees that

$$\gamma_0^{(22)} = 1.2506, \quad \gamma_1^{(22)} = 0.424 \tag{2.6}$$

and

$$C_0^{(22)} = 1.045, \quad C_1^{(22)} = 0.048 . \tag{2.7}$$

Notice that the amplitude of the leading subdominant singularity is very small (less than 5% of the dominant amplitude) compared to the situation in two dimensions (around 45% for the square lattice[36]).

With the exception of the FCC lattice which exhibits considerably less stability, all the standard d = 3 lattices give similar results. (The lack of stability for the FCC lattice seems to mirror the poor convergence of the dlog Padé approximants that was noted in §2.1). The estimates of γ_0 and γ_1 in (2.6), together with the analogous estimates for the other three standard lattices, are given in Table 11. The uncertainties indicate the changes induced by changing, within the usual confidence limits, the assumed estimate of v_c. In

<u>Table 10</u> Baker-Hunter analysis of $\chi_0(v)$, D lattice using $v_c^{-1} = 2.82641$

No. of terms	γ_0	C_0	γ_1	C_1	γ_2	C_2
12	1.2439	1.066	-1.3 - 0.9i	-0.03 - 0.003i	-1.3 + 0.9i	-0.03 + 0.003i
14	1.2436	1.067	-1.4 - 0.9i	-0.03 - 0.002i	-1.4 + 0.9i	-0.03 - 0.002i
16	1.2509	1.043	0.457	0.048	-1.1	-0.09
18	1.2503	1.046	0.385	0.048	-1.0	-0.09
20	1.2509	1.043	0.468	0.048	-1.1	-0.09
22	1.2506	1.045	0.424	0.048	-1.1	-0.09

Table 11 Baker-Hunter analysis. Highest-order estimates of $\gamma_0(=\gamma)$ and γ_1.

Lattice	γ_0	γ_1
D	$1.2506^{-0.014}_{+0.0015}$	$0.42^{-0.11}_{+0.10}$
SC	$1.2493^{-0.0008}_{+0.0012}$	$0.36^{-0.30}_{+0.30}$
BCC	$1.2478^{-0.0040}_{+0.0042}$	$0.385^{-0.15}_{+0.19}$
FCC	$1.2458^{-0.0015}_{+0.0005}$	$0.135^{(\text{Complex})}_{+0.10}$

this connection, it should be remembered that most estimates of v_c, like those in (2.2), are biased, being based upon some assumption about the value of γ such as $\gamma = 1.250$.

The values of $\gamma(\equiv\gamma_0)$ in Table 11 lie in the anticipated region around 1.250 or slightly smaller but appear to us as inconclusive as the dlog Padé approximant results reported in §2.1 with which they share the same sort of variation with lattice coordination number. For the leading subdominant singularity, Bessis et al[35] consider

$$0.08 < \gamma_1 < 0.52 \tag{2.8}$$

to be compatible with all lattices. This result is more in agreement with a regular subdominant singularity with exponent $\gamma-1 \simeq 0.25$ than with the RG value of $\gamma-\Delta_1 \simeq 0.75$. However not too much weight can be attached to this conclusion since the uncertainties in γ_1 must be quite large given the smallness of the amplitude C_1 and the number of series coefficients presently available. For the second subdominant singularity one finds from Table 10 and Bessis et al[35] that the highest-order estimates of γ_2 (i.e. utilising all the known coefficients) are

$$-1.1\ (\text{D}),\quad -1.15\ (\text{SC}),\quad -1.59(\text{BCC}),\quad 0.652(\text{FCC}). \tag{2.9}$$

Apart from the anomolously large result for the FCC lattice, the remaining values are once again in better agreement with a regular confluent singularity with exponent $\gamma-2 \simeq -0.75$ than with the exponent $\gamma-1 \simeq 0.25$ that is expected to follow the RG exponent $(\gamma-\Delta_1)$. However, this conclusion is probably fortuitous and little, if any, significance should be attached to it.

3. RATIO-BASED ANALYSIS

3.1 Method of Asymptotic Fits

According to RG theory, χ_0 is expected to exhibit close to the critical point a singularity of the form

$$\chi_0 \sim A_1\left(1 - \frac{K}{K_c}\right)^{-\gamma} + A_2\left(1 - \frac{K}{K_c}\right)^{-\gamma+1} + B\left(1 - \frac{K}{K_c}\right)^{-\gamma+\Delta_1} + \ldots \quad (3.1)$$

where, for the purposes of this subsection, we follow recent practice[28,29] and use the expansion variable $K = J/kT = \tanh^{-1}v$. However, entirely equivalent results[29] would be obtained using v. It follows[28] from (3.1) that the ratios r_n of successive coefficients behave like

$$r_n = K_c^{-1}\left[1 + (\gamma-1)n^{-1} + a\,n^{-1-\Delta_1} + bn^{-2}\right], \quad (n \to \infty) \quad (3.2)$$

where, for the moment, we have neglected higher-order terms. The constants a and b are related[28] to the amplitudes in (3.1) through

$$a = -\Delta_1\left(\frac{B}{A_1}\right)\frac{\Gamma(\gamma)}{\Gamma(\gamma-\Delta_1)}, \quad b = -\left(\frac{A_2}{A_1}\right)\frac{\Gamma(\gamma)}{\Gamma(\gamma-1)} \quad . \quad (3.3)$$

The method of asymptotic fits consists of fitting the known ratios to (3.2). The first application of this sort of analysis to high temperature χ_0 series seems to have been the work of Sykes et al[13]. Recently, McKenzie[29] has used this method to analyse the FCC lattice for which the series are known[12] through K^{15}. In some respects, her work extends the earlier analysis of Camp and Van Dyke[28] whose series were only through order K^{10}. On the other hand, Camp and Van Dyke did not confine their attention to spin ½ - hence, the use of K which is a more natural variable than v for the general spin problem.

Not surprisingly, it is not possible to obtain a sensible fit by leaving free all five parameters in (3.2). It is usual, therefore, to make certain assumptions about some of the parameters, determine the remaining parameters by fitting and then try, by assessing the quality of the fit, to test the validity of the assumptions. McKenzie[29] presented results for three different fits and these are given in Table 12. In part (a) of the Table, one assumes $\gamma = 1.250$ and $\Delta_1 = 0.5$, and then uses successive triplets of r_n to solve for K_c^{-1}, a and b. Precise agreement with Camp and Van Dyke[28] is obtained for $n \leq 10$ and shows why they were led to suggest that, in the limit, the correction-to-scaling term might be identically zero (i.e. a = 0 or B = 0). However, with more terms now available it is seen that the magnitude of a is still increasing quite rapidly

Table 12 McKenzie's asymptotic fit analysis, FCC lattice

(a) $\gamma = 1.250$, $\Delta_1 = 0.5$

n	K_c^{-1}	a	b
9	9.793837	-0.00145	-0.00963
10	9.794406	-0.00729	0.00307
11	9.794997	-0.01445	0.01951
12	9.795328	-0.01911	0.03079
13	9.795495	-0.02179	0.03756
14	9.795593	-0.02358	0.04228
15	9.795676	-0.02528	0.04695

(b) $\gamma = 1.250$, a = 0

n	K_c^{-1}	b
9	9.793695	-0.01278
10	9.793803	-0.01370
11	9.793969	-0.01543
12	9.794141	-0.01760
13	9.794229	-0.01998
14	9.794439	-0.02244
15	9.794564	-0.02499

(c) $\gamma = 1.241$, $\Delta_1 = 0.496$

n	K_c^{-1}	a	b
9	9.795658	0.03178	-0.04464
10	9.796027	0.02806	-0.03651
11	9.796457	0.02294	-0.02467
12	9.796657	0.02018	-0.01794
13	9.796713	0.01929	-0.01569
14	9.796718	0.01921	-0.01546
15	9.796721	0.01914	-0.01529

and shows no sign of converging even at $n = 15$. Table 12(b) is based on the assumption that the correction-to-scaling term is identically zero ($a = 0$) and $\gamma = 1.250$. Successive pairs of r_n are then used to solve (3.2) for K_c^{-1} and b. Again there is no evidence of convergence. In Table 12(c), the RG exponents $\gamma = 1.241$ and $\Delta_1 = 0.496$ were assumed and a three parameter fit performed. (Note that in both (a) and (c), it turns out[29] that the fits are relatively insensitive to small changes in Δ_1 of the order of ± 0.005.) In this case, there appears to be a marked improvement in the convergence, e.g. the last three values of a are constant to within 0.8%, and b to within 2.6%.

Of the three fits presented by McKenzie, there is a clear preference for the one using RG exponents. To study this indication further, I have investigated a number of other possibilities and some of the results are exhibited graphically in Figures 2 and 3. Curves labelled a, b, ... f in Figure 2 correspond to the assumptions $\gamma = 1.250$, $\Delta_1 = 0.5$, while those labelled A, B, ... F in Figure 3 correspond to $\gamma = 1.241$, $\Delta_1 = 0.496$. Plotted against n is the nth estimate of K_c^{-1} obtained by fitting the ratios r_n to the following asymptotic forms:

$$\text{a,A} \quad : \quad r_n = K_c^{-1}\left[1 + (\gamma-1)n^{-1}\right] \tag{3.4}$$

$$\text{b,B} \quad : \quad r_n = K_c^{-1}\left[1 + (\gamma-1)n^{-1} + bn^{-2}\right] \tag{3.5}$$

$$\text{c,C} \quad : \quad r_n = K_c^{-1}\left[1 + (\gamma-1)n^{-1} + bn^{-2} + dn^{-3}\right] \tag{3.6}$$

$$\text{d,D} \quad : \quad r_n = K_c^{-1}\left[1 + (\gamma-1)n^{-1} + an^{-1-\Delta_1} + bn^{-2}\right] \tag{3.7}$$

$$\text{e,E} \quad : \quad r_n = K_c^{-1}\left[1 + (\gamma-1)n^{-1} + bn^{-2} + cn^{-2-\Delta_1}\right] \tag{3.8}$$

$$\text{f,F} \quad : \quad r_n = K_c^{-1}\left[1 + (\gamma-1)n^{-1} + an^{-1-\Delta_1} + bn^{-2} + cn^{-2-\Delta_1}\right] . \tag{3.9}$$

Corresponding plots for the other parameters a,b,c,d may also be made. The form given by (3.4) is a one parameter fit, (3.5) is a two parameter fit, (3.6) to (3.8) are three parameter fits and (3.9) is a four parameter fit. The first three forms (3.4) to (3.6), contain no correction-to-scaling terms. In (3.9) we have included the first two contributions, namely $a/n^{1+\Delta_1}$ and $c/n^{2+\Delta_1}$, from correction-to-scaling terms. In (3.7) the second such term has been assumed to be negligibly small ($c = 0$) leaving only the dominant contribution, while in (3.8) the leading term accidentally vanishes ($a = 0$) leaving only the second contribution. The curves d, b and D correspond to the cases studied by McKenzie and reproduced in Table 12(a), (b) and (c), respectively.

Close inspection of these figures confirms that the best fit (curve D) is obtained using the asymptotic form (3.7) with the RG exponents $\gamma = 1.241$ and $\Delta_1 = 0.496$. Inclusion of the next correction

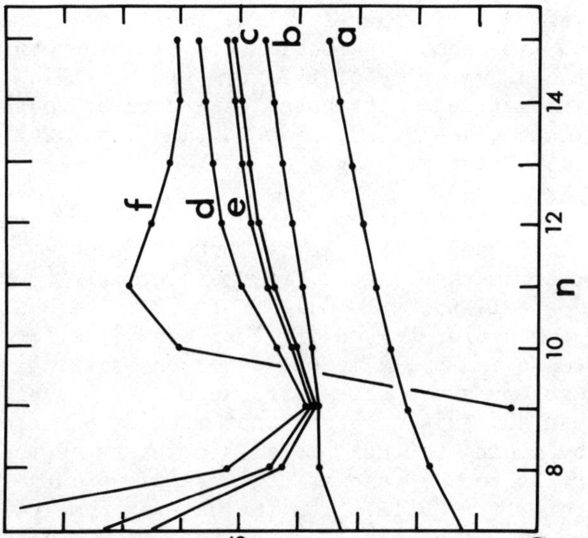

Fig.2 Asymptotic fit analysis, FCC lattice using $\gamma = 1.250$, $\Delta_1 = 0.5$.

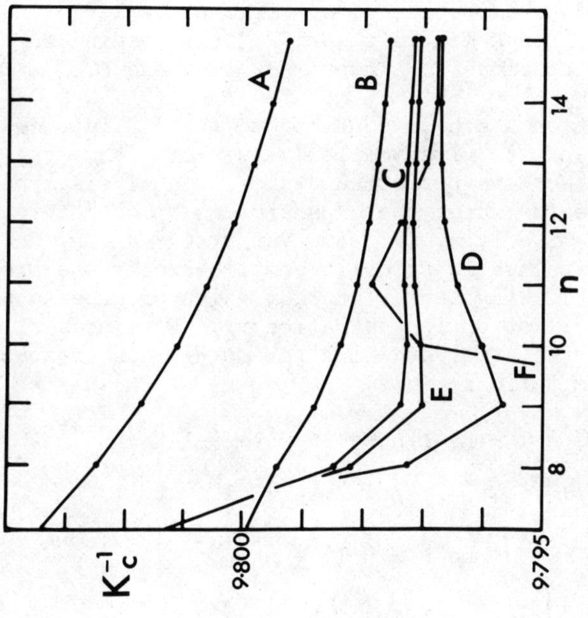

Fig.3 Asymptotic fit analysis, FCC lattice using $\gamma = 1.241$, $\Delta_1 = 0.496$.

term, as in (3.9), seems to give a very similar fit (curve F) provided, as would be expected, n is sufficiently large ($n \geq 14$). This analysis confirms McKenzie's work[29], thus lending further support to the RG predictions for the values of γ and Δ_1.

So far attention has been confined to the FCC lattice which, as we saw in §2.1, tends to give smaller values of γ than the loose-packed lattices. We have performed a similar analysis, therefore, for the SC lattice for which the expansion is known[14] through order K^{19}. As reported briefly by Camp and Van Dyke[28] and by McKenzie[29], analysis of loose-packed lattices is complicated by the presence of an antiferromagnetic singularity at $-K_c$, which produces a characteristic odd-even oscillation in the ratios r_n. Estimates of K_c^{-1}, calculated from three, four and five parameter fits, respectively, to the following asymptotic forms:

$$\text{a,A:} \quad r_n = K_c^{-1}\left[1 + (\gamma-1)n^{-1} + bn^{-2} + (-1)^n e\, n^{-\gamma+\alpha-1}\right] \qquad (3.10)$$

$$\text{b,B:} \quad r_n = K_c^{-1}\left[1 + (\gamma-1)n^{-1} + an^{-1-\Delta_1} + bn^{-2} + (-1)^n e\, n^{-\gamma+\alpha-1}\right] \qquad (3.11)$$

$$\text{c,C:} \quad r_n = K_c^{-1}\left[1 + (\gamma-1)n^{-1} + an^{-1-\Delta_1} + bn^{-2} + (-1)^n e\, n^{-\gamma+\alpha-1} + (-1)^n f\, n^{-\gamma+\alpha-1-\Delta_1}\right]. \qquad (3.12)$$

are plotted versus n in Figure 4. (Note that the vertical scale is much larger than was used in Figures 2 and 3.) In all three cases, it has been assumed[37,13] that the oscillatory behaviour arises from an antiferromagnetic singularity of the form $(1 + \mu v)^{1-\alpha}$. As before, curves labelled by lower and upper case letters refer to the use of the exponents $\gamma = 1.250$, $\alpha = 0.125$, $\Delta_1 = 0.5$ and $\gamma = 1.241$, $\alpha = 0.110$, $\Delta_1 = 0.496$, respectively.

The form (3.10) contains no correction-to-scaling terms and is equivalent to the form used in the early work of Sykes et al[13]. It appears to provide a remarkably good fit, particularly (curve a) with the exponents $\gamma = 1.250$, $\alpha = 0.125$. In (3.11) we have simply added the leading correction-to-scaling term $a/n^{1+\Delta_1}$. For both sets of exponents (curves b and B), a marked odd-even oscillation appears. However, the averages of successive pairs seem fairly constant for sufficiently large n and this suggests that most of the oscillation might be removed by adding the next oscillatory term as in (3.12). Indeed this appears to be the case as the results in curves c and C show. The asymptotic form (3.12) fits the data extremely well for $n \geq 16$ using either set of critical exponents, with no clear preference indicated. We have also experimented with replacing the second oscillatory term in (3.12) by one of order $n^{-\gamma+\alpha-2}$. The results are indistinguishable, to within graphical accuracy, from those given by curves c and C.

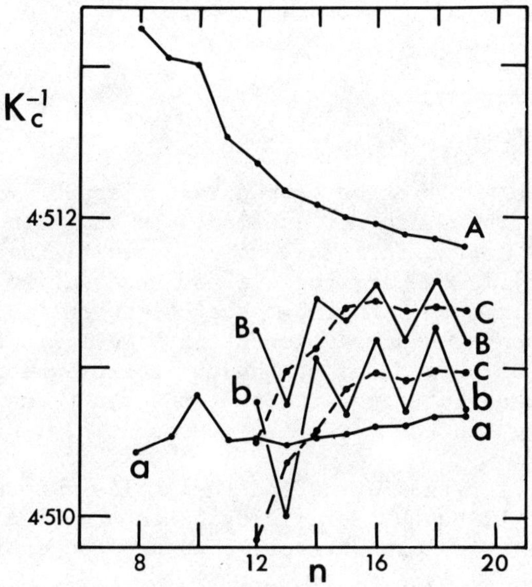

Fig.4 Asymptotic fit analysis, SC lattice. Curves a,b,c refer to the exponent set $\gamma = 1.250$, $\alpha = 0.125$, $\Delta_1 = 0.5$, and curves A,B,C to $\gamma = 1.241$, $\alpha = 0.110$, $\Delta_1 = 0.496$.

3.2 <u>Analysis of Zinn-Justin</u>

In his analysis of high temperature series, Zinn-Justin[38] utilised some novel extrapolation formulae markedly different from the standard expressions[25,27]. We will focus here on the susceptibility exponent γ although the exponents ν and α were also studied[38].

One begins by calculating the ratios of the ratios of coefficients and then defining the coefficients

$$b_n = 1 - (r_n/r_{n-1}) \ . \tag{3.13}$$

Using (3.2) it can be shown that

$$b_n \sim (\gamma-1)/n^2 \ , \quad (n \to \infty) \tag{3.14}$$

from which it follows that the quantities

$$\gamma_n = 1 + \frac{b_{n-1}\, b_n}{(\sqrt{b_{n-1}} - \sqrt{b_n})^2} \tag{3.15}$$

will approach γ as $n \to \infty$. The method is unbiased in that the value of the critical point v_c is not assumed. In addition, n does not appear explicitly in the formulae, which avoids[27] the problem of using n or (n + N). However, this is not necessarily an advantage in our opinion, since the knowledge that a sequence of estimates for a range of N values must all approach the same limit can sometimes be helpful. To reduce the effect of the antiferromagnetic singularity for the loose-packed lattices, the odd and even terms were first treated separately and then averaged appropriately.

Values of γ_n for the four standard lattices are plotted against 1/n in Figure 5. Although there is evidence of quite large oscillations for small n, the last few estimates are decreasing for all

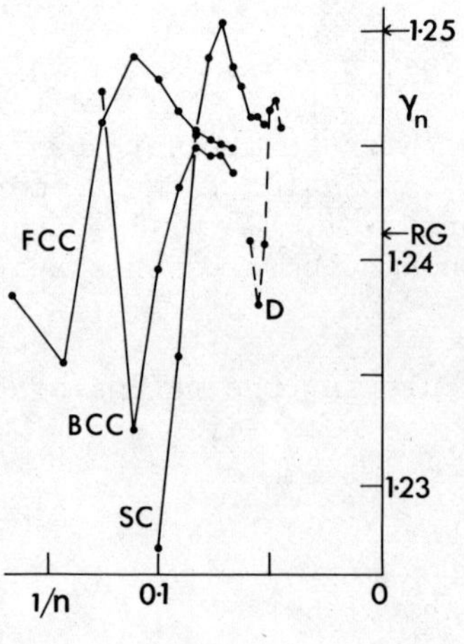

Fig.5 Zinn-Justin analysis. Estimates of γ plotted against 1/n.

lattices. However, the estimates have clearly not attained their true asymptotic behaviour which is either linear with 1/n for a regular confluence or, if there are non-zero correction-to-scaling terms, linear with $1/n^{\Delta_1}$. Our inclination, therefore, is to increase the uncertainties in Zinn-Justin's best estimate of[38]

$$\gamma = 1.245 \pm 0.002 \text{ or } 0.003 \qquad (3.16)$$

to around $\pm$ 0.005. The central value is intermediate between the traditional series estimate of $\gamma = 1.250$ and the RG estimate of $\gamma = 1.241$.

Zinn-Justin[38] also used a method of estimating γ that involved first estimating v_c. The biased estimates of γ so obtained are completely consistent with the direct estimate obtained above.

3.3 Analysis of Gaunt and Sykes

The work of Gaunt and Sykes[14] was the first in this recent series of papers which we are reviewing that have as their aim the resolution of the discrepancy between the traditional series estimates of critical exponents and RG estimates. Their approach involved only minor refinements to the classic ratio method as first used by Domb and Sykes[22] over twenty years ago for studying the exponent γ. Without these refinements, the ratio method yields the estimates plotted versus 1/n in Figure 6. Clearly, these plots are not extrapolable to a common limit - once again, the smaller the coordination number the larger the value of γ that is indicated.

The main refinement introduced by Gaunt and Sykes[14] was to work with the series $\chi_0(x)$ rather than $\chi_0(v)$, where the new variable x is defined by the Euler transformation

$$x = \frac{2v}{(1 + \frac{v}{v_c^*})} \quad . \qquad (3.17)$$

If v_c^* were the exact critical point, then the antiferromagnetic singularity at $v = -v_c^*$ would be mapped to infinity in the x-plane, while the ferromagnetic singularity at $v = v_c^*$ is a fixed point of the transformation (as is the origin $v = 0$). In practice, the exact critical point is not known and so v_c^* is simply a good estimate. The hope is that the antiferromagnetic singularity is mapped sufficiently far from the origin relative to the ferromagnetic singularity that it does not affect the extrapolations significantly.

Estimates of γ could be obtained by extrapolating successive estimates of

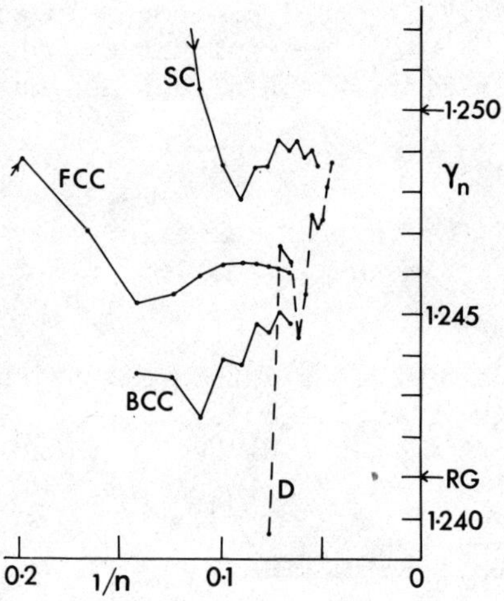

Fig.6 Ratio analysis. Estimates of γ plotted against 1/n.

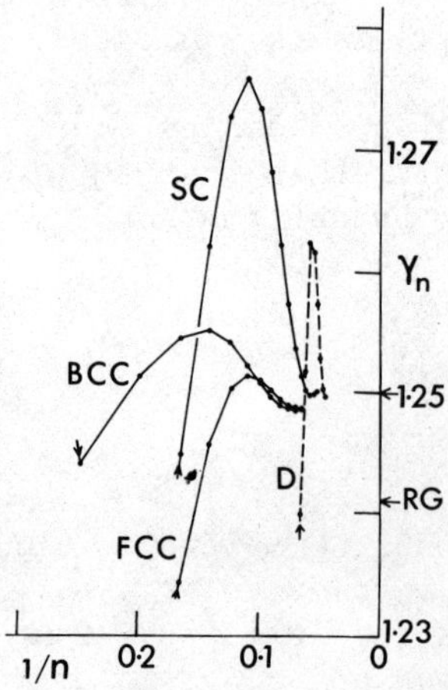

Fig.7 Modified ratio analysis of Gaunt and Sykes. Estimates of γ plotted against 1/n.

$$1 + n(\frac{\mu_n}{\mu} - 1) \ ,$$

where μ_n is the ratio of successive coefficients of the transformed series and $\mu = 1/x_c$. However, since x_c is not known exactly, we avoided the necessity of using an estimate by replacing μ with the nth order estimate

$$\mu'_n = n\mu_n - (n-1)\mu_{n-1} \ ,$$

which must approach μ as $n \to \infty$ and which corresponds to extrapolating successive ratios of the $\chi_0(x)$ series linearly against $1/n$. Thus, we form the sequence of estimates

$$\gamma'_n = 1 + n(\frac{\mu_n}{\mu'_n} - 1)$$

from which our final estimates γ_n of γ were obtained by linear extrapolation of adjacent γ'_n against $1/n$. These estimates are unbiased in the sense that they do not depend in a direct way upon an estimate of the critical point x_c of the transformed series. While it is true that a good estimate v_c^* was used in calculating the transformed series $\chi_0(x)$, that was simply to reduce the effect of the antiferromagnetic singularity and could have been achieved quite well with a much less accurate value of v_c^*. In fact, γ_n turns out to be very stable with respect to small changes in v_c^*; for example, the uncertainties in the estimates (2.2) of v_c do not affect the last estimate of γ_n until at least the sixth decimal place.

The estimates γ_n are plotted versus $1/n$ in Figure 7. The oscillations apparent in these plots are smoothly varying with steadily decreasing amplitudes. They are probably caused[14,38] by unknown singularities which have been moved closer to the critical point by the Euler transformation. We have investigated their location by forming dlog Padé approximants to the $\chi_0(x)$ series. For the BCC lattice, for example, one finds non-physical singularities lying behind x_c, off the real axis, at a distance of about 1.6 x_c from the origin. The effect of such singularities should therefore decay exponentially fast. In summary, what particularly impresses us about Figure 7 is the essentially consistent picture that it conveys, all four lattices being in good accord with the view that they have a common limit close to $\gamma = 1.250$[†].

[†]A critique of this analysis will be found in the Lecture Notes of B. G. Nickel.

4. Summary and Conclusions

We have reviewed recent attempts† at reanalysing high temperature susceptibility expansions for the simple, spin $\frac{1}{2}$ Ising model in the hope of shedding some light on the apparent discrepancy that has arisen between traditional series estimates of critical exponents and RG estimates.

All of the Padé-based results (see §2) exhibit a characteristic variation of the apparent value of γ with coordination number q. Lattices with small q (D, SC, tetrahedron) give estimates of γ close to the traditional series estimate of $\gamma = 1.250$, whilst lattices with larger q (BCC, FCC) tend to give smaller values of γ between 1.240 and 1.248. This lattice dependence is almost certainly caused by convergence problems arising from relatively short series and should not be seen as indicating a breakdown of the universality hypothesis. Unfortunately, it is not at all clear, a priori, which of the lattices should give estimates lying closest to the exact result. We conclude that with the number of terms presently available none of the Padé-based methods described in §2 is able to resolve the discrepancy.

The ratio-based methods in §3 appear to give conflicting evidence. Thus,

(i) The method of asymptotic fits applied to the FCC lattice indicates a definite preference for the RG predictions of γ and Δ_1. For the SC lattice, equally excellent fits to the data can be obtained using either the traditional series exponents or the RG estimates. No clear preference is apparent.

(ii) An analysis of the type used by Zinn-Justin yields an intermediate value for γ around $\gamma = 1.245$ for all lattices. However, our uncertainties are large enough to embrace the RG estimate as well as the traditional series estimate.

---

† The only other recent work of which I am aware is the ratio-type analysis of Sheludyak and Rabinovich[39] which yields $\gamma = 1.244$ with unrealistically small uncertainties of $\pm$ 0.0005. I am grateful to J. V. Sengers for drawing my attention to this paper whilst we were in Cargèse.

(iii) The Gaunt and Sykes analysis also indicates a common limit for all four standard lattices. However, an objective choice would seem to lie within the confidence limits of the traditional series estimate $\gamma = 1.250 \pm 0.003$. If the true value of γ turns out to be closer to the RG estimate, then as pointed out by Gaunt and Sykes[14] further research will be required to explain why exponent error assessments have always been far too optimistic.†

References

1. M.E. Fisher, Rep Prog Phys 30, 615 (1967)
2. G.A. Baker, B.G. Nickel, M.S. Green and D.I. Meiron, Phys Rev Letts 36, 1351 (1976)
3. G.A. Baker, B.G. Nickel and D.I. Meiron, Phys Rev B 17, 1367 (1978)
4. J.C. Le Guillou and J. Zinn-Justin, Phys Rev Letts 39, 95 (1977)
5. R.F. Chang, H. Burstyn, J.V. Sengers and A.J. Bray, Phys Rev Letts 37, 1481 (1976)
6. R. Hocken and M.R. Moldover, Phys Rev Letts 37, 29 (1976)
7. J.W. Essam and D.L. Hunter, J Phys C 1, 392 (1968)
8. M.F. Sykes, D.L. Hunter, D.S. McKenzie and B.R. Heap, J Phys A 5 667 (1972)
9. M.A. Moore, D. Jasnow and M. Wortis, Phys Rev Letts 22, 940 (1969)
10. G.A. Baker, Phys Rev B 15, 1552 (1977)
11. B.G. Nickel and B. Sharpe, J Phys A 12, 1819 (1979)
12. S. McKenzie, J Phys A 8, L102 (1975)
13. M.F. Sykes, D.S. Gaunt, P.D. Roberts and J.A. Wyles, J Phys A 5, 640 (1972)
14. D.S. Gaunt and M.F. Sykes, J Phys A 12, L25 (1979)
15. D.S. Gaunt and M.F. Sykes, J Phys A 6, 1517 (1973)

† A plausible explanation requiring the existence of significant correction-to-scaling terms is proposed by B.G. Nickel in his Lecture Notes. He concludes on the basis of a ratio-type analysis that reasonable fits to the series can be obtained using $\gamma = 1.237 \pm 0.005$.

16. D.L. Hunter, Thesis, University of London (1967)

17. M.F. Sykes, J Math Phys 2, 52 (1961)

18. C. Domb and B.J. Hiley, Proc Roy Soc A268, 506 (1962)

19. M. Wortis, D. Jasnow and M.A. Moore, Phys Rev 185, 805 (1969)

20. E. Block, Ark Fys 24, 79 (1963)

21. J.W. Essam and M.F. Sykes, Physica 29, 378 (1963)

22. C. Domb and M.F. Sykes, Proc Roy Soc A240, 214 (1957)

23. C. Domb and M.F. Sykes, J Math Phys 2, 63 (1961)

24. D.C. Rapaport, J Phys A 7, 1918 (1974)

25. D.L. Hunter and G.A. Baker, Phys Rev B 7, 3346 (1973)

26. G.A. Baker and D.L. Hunter, Phys Rev B 7, 3377 (1973)

27. D.S. Gaunt and A.J. Guttmann, Asymptotic Analysis of Coefficient in: "Phase Transitions and Critical Phenomena 3", C. Domb and M.S. Green, eds., Academic, London: New York (1974)

28. W.J. Camp and J.P. Van Dyke, Phys Rev B 11, 2579 (1975)

29. S. McKenzie, J Phys A 12, L185 (1979)

30. J.A. Leu, D.D. Betts and C.J. Elliott, Can J Phys 47, 1671 (1969)

31. J. Ho-Ting-Hun and J. Oitmaa, J Phys A 8, 1920 (1975)

32. J. Oitmaa and C.J. Elliott, Can J Phys 48, 2383 (1970)

33. J. Oitmaa and J. Ho-Ting-Hun, J Phys A 12, L281 (1979)

34. C. Domb, Ising Model, in: "Phase Transitions and Critical Phenomena 3", C. Domb and M.S. Green, eds., Academic, London: New York (1974)

35. D. Bessis, P. Moussa and G. Turchetti, J Phys A 13, 2763 (1980)

36. T.T. Wu, B. McCoy, C. Tracy and E. Barouch, Phys Rev B 13, 316 (1976)

37. M.E. Fisher, Physica 25, 521 (1959)

38. J. Zinn-Justin, J Physique 40, 63 (1979)

39. Yu. E. Sheludyak and V.A. Rabinovich, High Temp 17, 40 (1979)

DERIVATION OF HIGH TEMPERATURE SERIES EXPANSIONS: ISING MODEL

Sati McKenzie

Wheatstone Physics Laboratory
King's College
Strand London WC2R 2LS UK

INTRODUCTION

In this talk, I shall discuss the derivation of high temperature series expansions for the Ising model, with special reference to the spin $\frac{1}{2}$ case. In the second lecture, we shall see how these methods of derivation can be extended to the classical vector model.

The basic aim, in all series expansion work, is to express the free energy (or any other thermodynamic property) as a power series in the field and temperature variables. The coefficients of the series are calculated by considering graphs which can be embedded on the lattice. Depending on the types of graph considered, one is led to different methods of calculation, all of which must give the same series. The two main approaches are (A) Expansions in excluded volume graphs and (B) Expansions in free graphs.

There are further subcategories within each approach, such as connected cluster expansions and star graph expansions in (A) and unrenormalized, vertex renormalized and bond renormalized forms of (B). We shall discuss some of these developments and their applications.

Graph Definitions

At this stage, I would like to introduce some graphical terminology that will be used in what follows. Most of the concepts are either well-known or intuitively obvious and are best illustrated by the examples in Fig. 1.

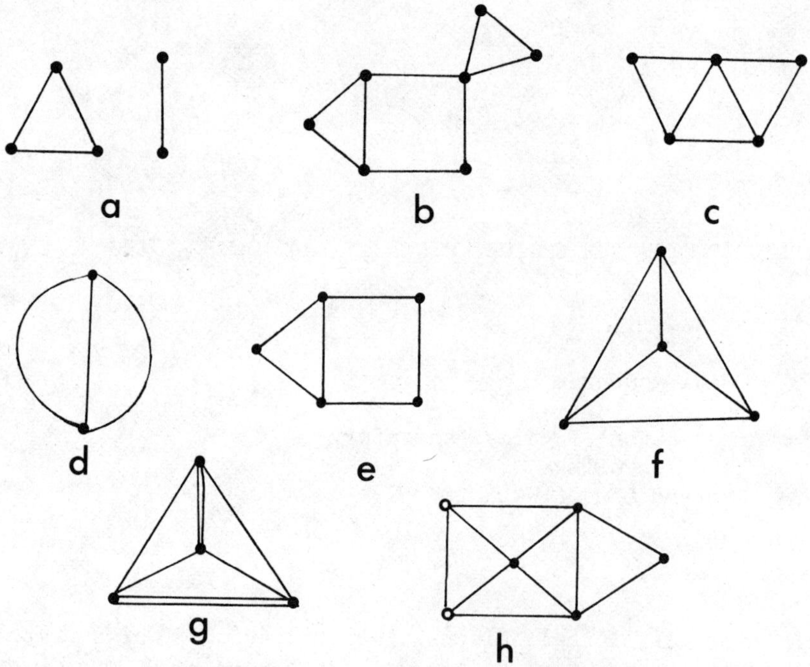

Fig. 1. Examples of graphs.

Graphical terms

Graph G(N,L): Is defined by N vertices and L edges or lines, each edge joining a pair of vertices. (Fig. 1: a,b,c.)

Cycle index: $C(G) = L - N + 1$.

Degree or valence of a vertex: The number of edges incident at a vertex.

Odd (even) vertices: Vertices of odd (even) valence.

Connected and separated graphs (Fig. 1: b,a).

1-reducible (articulated) graphs and irreducible (star) graphs (Fig. 1: b,c).

Articulation vertex: Deletion of edges incident at the vertex produces a separated graph (Fig. 1 b).

Topology of a graph (Fig. 1: d).

Realization of a topology (Fig. 1: e).

Simple and multigraphs (Fig. 1: f,g).

Embedding of a graph on another graph or lattice: A way of

identifying the vertices of the graph with a subset of sites of the lattice such that connectivity is preserved.

Free (weak) embeddings: Embeddings without (with) the excluded volume restriction of only one graph vertex (edge) being allowed to occupy a particular lattice site (bond).

Rooted and unrooted graphs (Fig. 1: h,g).

Lattice constant: Number of distinct embeddings per site.

Free multiplicity: Total number of free embeddings per site.

Expansions in Excluded Volume Graphs [1]

We consider a system of N spins with L spin pair interactions, represented by G(N,L). The Ising model Hamiltonian for such a system is given by,

$$\mathcal{H} = -\frac{J}{S^2} \sum_{<ij>}^{L} S_i^z S_j^z - \frac{mH}{S} \sum_{i}^{N} S_i^z , \tag{1}$$

where J is the interaction strength, m the magnetic moment of the spins and H the magnetic field which acts in the z direction. The first sum is over all L nearest neighbour pairs and the second is over all N spins. The coupling between spins is confined to their z-components.

The partition function Z of such a system is defined in the usual manner,

$$Z = \mathrm{tr}\,\exp(-\beta\mathcal{H}) = \mathrm{Tr} \prod_{<ij>}^{L} \exp(K\sigma_i\sigma_j) \prod_{i}^{N} \exp(\beta mH\sigma_i) , \tag{2}$$

where $\beta = 1/kT$, $K = \beta J$, $\sigma_i = S_i^z/S_i$. For the spin $\frac{1}{2}$ model, the variables σ_i assume values of ± 1 and the trace consists in summing over all σ_i for those values. Using the van der Waerden transformation [2], one can write

$$Z = \cosh^L K \cosh^N \beta mH\, \mathrm{Tr} \prod_{<ij>}^{L} (1 + v\sigma_i\sigma_j) \prod_{i}^{N} (1 + \tau\sigma_i) , \tag{3}$$

with $v = \tanh\beta J$, $\tau = \tanh\beta mH$.

The product of $(1 + v\sigma_i\sigma_j)$ corresponds to taking all possible subsets of edges of the lattice G, each such subset forming a subgraph of G. Since the trace over an odd power of σ_i is zero, any odd vertices of the subgraph must be combined with the appropriate $\tau\sigma_i$ term to ensure a nonzero trace. Thus, in terms of subgraphs of G, the partition function is simply,

$$Z = \cosh^L K \cosh^N(\beta mH) 2^N \left[1 + \sum_{\ell,m} \tau^{2m} v^{\ell} P_{\ell,2m} \right], \tag{4}$$

where $P_{\ell,2m}$ denotes the number of subgraphs of G with ℓ edges and 2m odd vertices. The graphs may be connected or separated, but no multigraphs are allowed. The zero field free energy, Z_0 and the zerofield susceptibility χ_0 are very simply related to the $P_{\ell,0}$ and $P_{\ell,2}$ terms respectively.

The next step is to express (4) as an expansion in terms of connected graphs only. This leads to the finite cluster development. Going further, it is possible to express the contribution of articulated graphs in terms of star graphs and obtain the star cluster expansion.

Before going into details, I would like to introduce an important result about graph weights which will be used frequently in what follows. This result defines a sufficient condition for a particular type of graph to make no contribution to an expansion.

Define a property A(G) associated with a graph G, such that,

$$A(G) = \sum_{g \leq G} \phi(g) . \tag{5}$$

The sum is over all subgraphs g of G. A subgraph, here, is defined by a subset of the edge set of G. $\phi(g)$ is the contribution or weight of g and does not depend on G. We are not concerned about the precise form of ϕ at this stage.

If G can be decomposed into G_1 and G_2, such that for $g_1 \leq G_1$ and $g_2 \leq G_2$,

$$A(g_1 \cup g_2) = A(g_1) + A(g_2) , \tag{6}$$

it can be shown that $\phi(G) = 0$.

To see this, we invert (5), using the principle of inclusion-exclusion [3], to derive

$$\phi(G) = \sum_{g \leq G} (-)^{L-\ell} A(g) , \tag{7}$$

where ℓ denotes the number of edges of g.

Subgraphs g can be of three types:

(a) $g = g_1(\ell_1) \leq G_1(L_1)$

(b) $g = g_2(\ell_2) \leq G_2(L_2)$

(c) $g = g_1 \cup g_2$ with $g_1 \leq G_1$, $g_2 \leq G_2$.

Substituting in (7), we obtain,

$$\phi(G) = \sum_{g_1 \leq G_1} (-)^{L-\ell_1} A(g_1) + \sum_{g_2 \leq G_2} (-)^{L-\ell_2} A(g_2)$$

$$+ \sum_{\substack{g=g_1 \cup g_2 \\ g_1 \leq G_1 \\ g_2 \leq G_2}} (-)^{L-\ell_1-\ell_2} \left[A(g_1) + A(g_2) \right]$$

$$= \sum_{\ell_2=0}^{L_2} (-)^{-\ell_2} \binom{L_2}{\ell_2} \sum_{g_1 \leq G_1} (-)^{L-\ell_1} A(g_1) + \sum_{\ell_1=0}^{L_1} (-)^{-\ell_1} \binom{L_1}{\ell_1} \sum (-)^{L-\ell_2} A(g_2)$$

$$= 0 \ . \tag{8}$$

This result will be used in two ways. To show the existence of a connected graph expansion, we identify $A(G)$ with $\ell n\, Z$ and show that $\phi(G) = 0$ if G is disconnected. To demonstrate the existence of a star graph expansion, we have to show that $\phi(G) = 0$ for articulated graphs. This is found to be the case when the free energy is expanded as a function of magnetization and temperature.

Connected Graph Expansions

For a separated graph $G = G_1 \cup G_2$, it is clear that

$$\ell n\, Z(G) = \ell n\, Z(G_1) + \ell n\, Z(G_2) \ , \tag{9}$$

and only connected graphs contribute to the free energy expansion. One can thus write,

$$\ell n\, Z(G) = \sum_{g \leq G} \phi(g) \ , \tag{10}$$

and calculate $\phi(G)$ by considering the set of connected graphs in order of increasing size. Thus,

$$\begin{aligned} \ell n\, Z(/) &= \phi(/) \\ \ell n\, Z(\wedge) &= \phi(\wedge) + 2\phi(/) \\ \ell n\, Z(\mathsf{N}) &= \phi(\mathsf{N}) + 2\phi(\wedge) + 3\phi(/) \end{aligned} \tag{11}$$

$$\ln Z(\lambda) = \phi(\lambda) + 3\phi(\wedge) + 3\phi(/) \qquad (11)\ \text{cont'd}$$

$$\ln Z(\Delta) = \phi(\Delta) + 3\phi(\wedge) + 3\phi(/) \quad .$$

On inverting (11), one obtains $\phi(G)$ for all the graphs on the LHS. It should be noted that since (11) involves only connected graphs, eq.(7) is no longer valid.

To derive an expansion for the free energy of a lattice $\mathcal{L}$, correct to order v^n (say), we consider all connected graphs G with up to n edges that can be embedded on $\mathcal{L}$, and write,

$$\ln Z(\mathcal{L}) = \sum_G (G;\mathcal{L})\phi(G) \quad , \tag{12}$$

where the $\phi(G)$ are calculated from the partition functions of G, using (11). The $\phi(G)$ are expanded as power series in v and τ and truncated at the appropriate power. $(G;\mathcal{L})$ denotes the weak lattice constant per site of G on $\mathcal{L}$. The free energy is also defined per site.

Alternatively, one can derive expansions for particular properties of interest, such as the zerofield free energy ($\ln Z_0$), the zero field susceptibility (χ_0) and higher field derivatives of the free energy ($\chi_0^{(2n)}$). Thus,

$$\ln Z_0(\mathcal{L}) = \sum_G (c_i;\mathcal{L})\ k_i(v) \quad , \tag{13}$$

where c_i is the set of all connected graphs. The weights k_i are calculated from the zero field partition functions of c_i and involve only the $P_{\ell,0}$ term in (4). This suffices to derive the specific heat series,

$$C_0 = k\beta^2 \frac{\partial^2}{\partial\beta^2} \ln Z_0 \quad , \tag{14}$$

which can be used to obtain estimates for the exponent α.

For the zero field susceptibility χ_0, the expansion is of the form,

$$\frac{kT\chi_0}{m^2} = \sum_{c_i} (c_i;\mathcal{L})h_i(v) \quad . \tag{15}$$

For a finite cluster c_i, the susceptibility can be related to pair correlations $<\sigma_i\sigma_j>$, using the relation,

$$\frac{kT\chi_N}{m^2} = N + 2 \sum_{<ij>} <\sigma_i\sigma_j> \quad . \tag{16}$$

The sum is over all pairs of vertices $<ij>$. To calculate $<\sigma_i\sigma_j>$, one joins the vertices i and j by a pseudo bond v^* to obtain a new graph G_{ij}. Then,

$$<\sigma_i\sigma_j> = \lim_{v^*\to 0} \frac{\partial \ln Z_0(G_{ij})}{\partial v^*} \quad . \tag{17}$$

For the spin ½ Ising model, (17) merely enumerates all subgraphs of G with exactly two odd vertices, i and j.

Higher field derivatives like $\chi_0^{(2)}$, $\chi_0^{(4)}$ ... can be treated similarly in terms of multiple correlations, but the procedure gets very complicated.

The connected graph expansion is undoubtedly a great improvement on the "primitive method" in terms of all graphs. However, it still entails dealing with articulated graphs, some of which have very large lattice constants.

Star Graph Expansions

One is therefore led to consider the feasibility of obtaining expansions in terms of 1-irreducible graphs or stars. We first note that a connected cluster expansion for A(G) becomes a star cluster expansion if the weight $\phi(G)$ is zero for all articulated graphs. This would be the case if for an articulated graph G with star blocks G_1, G_2,

$$A(G) = A(G_1) + A(G_2) \quad . \tag{18}$$

This property is seen to hold for $A = \ln Z_0 - N\ln 2$. For instance the graph shown in Fig. 2,

$$Z_0(G) = \tfrac{1}{2}Z_0(G_1)Z_0(G_2) = 2^6(\cosh K)^8(1+3v^3+v^4+2v^6+v^7) \quad . \tag{19}$$

The factor of ½ comes from the loss of a vertex when G_1 and G_2 are joined at the articulation point.

Thus the zerofield free energy of the Ising model has a star graph expansion. Incidentally, this result is also true for the more general classical vector model as we shall see later.

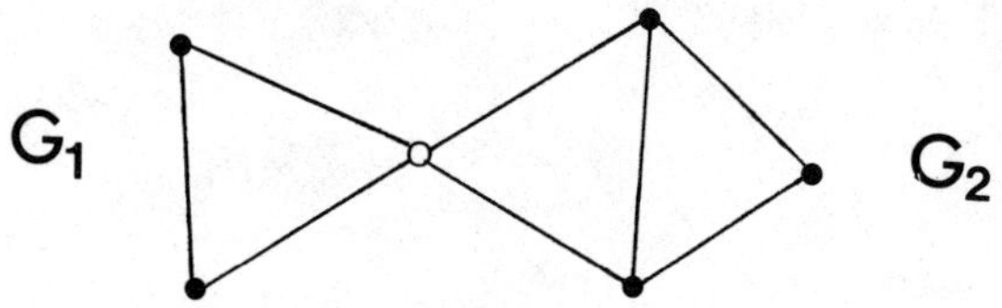

Fig. 2. Articulated graph $G = G_1 \circ G_2$.

The next problem is to generalize the star cluster approach to include the calculation of susceptibility and higher field derivatives as star graph expansions. Ideally, one would like to be able to calculate the energy in a field as a star graph expansion. To do this, we consider the partition function of an articulated graph $G_1 \circ G_2$, when the magnetization m at the articulation point is fixed. It can be seen from (19) that $Z(G_1 \circ G_2)$ now becomes,

$$Z(G_1 \circ G_2) = A\, Z(G_1)Z(G_2) \ , \tag{20}$$

where A depends on the spin at the articulation point but is otherwise independent of G_1 and G_2. This suggests that $\ell n\ Z(M,v)$ can be developed as a star graph expansion, as can all the derivatives of $\ell n\ Z$ with respect to the magnetization M [4].

A more rigorous justification for this approach comes from the application of the Mayer cluster integral theory to the problem of interacting spins on a crystal lattice. This was pursued by several workers such as Fuchs, Yvon, Rushbrooke and Scoins, and Domb and Hiley [5-9]. The activity expansion in terms of reducible cluster integrals is interpreted as the connected graph expansion for $\ell n\ Z(H,T)$. The cluster integrals over the volume v of the system are now replaced by sums over lattice sites, which are related to the lattice constants. The virial expansion in terms of irreducible cluster integrals is obtained by a change of variable from activity to density. This transformation, for a magnetic system, involves a change of variable from H to the overturned spin density α, which is closely related to the magnetization M. The irreducible cluster integrals are now interpreted in terms of star lattice constants. Details of the derivation can be found in literature and would be too lengthy to discuss here.

Uhlenbeck and Ford [10] obtained the same results by considering generating functions for the three types of graph, namely, the set of all N-vertex graphs G_N, the set of ℓ-vertex connected graphs c_ℓ and the set of m-vertex star graphs S_m. Thus,

$$
\begin{aligned}
F(x) &= \sum_{N} F_N \frac{x^N}{N!} & F_N &= \sum_{G_N} W(G_N) \\
f(x) &= \sum_{\ell=1}^{\infty} f_\ell \frac{x^\ell}{\ell!} & f_\ell &= \sum_{C_\ell} W(C_\ell) \\
S(y) &= \sum_{m=2}^{\infty} \frac{r_m}{m!} y^m & r_m &= \sum_{S_m} W(S_m) \; .
\end{aligned}
\tag{21}
$$

Making two assumptions regarding the weights W, they showed that the generating functions could be related to each other.

$$
\begin{aligned}
1 + F(x) &= \exp f(x) \\
T(z) &= z \frac{df}{dz} = z \exp \frac{dS(T)}{dT} \; ,
\end{aligned}
\tag{22}
$$

where T(z) is the generating function for 1-rooted connected graphs.

The density expansion for the free energy thus involves only star graphs. As originally formulated [8,9], the method yields a low temperature expansion in the variables α (density of overturned spins) and z (a temperature variable related to the Mayer f function). The graphs that enter into the expansion are star graphs which can be embedded in the "strong" sense [11]. The low temperature expansion can then be transformed into the high temperature expansion in the variable v [12].

It is also possible, knowing the existence of a star graph expansion for $\ln Z(\alpha,T)$, to derive directly, high temperature expansions for χ_0^{-1}, $\chi_0^{(2)}/\chi_0^4$ and higher derivatives. This approach has proved quite successful in deriving long series [13-16], and it would be interesting to discuss it in some detail.

It proves convenient to rewrite the free energy of a graph G in the form

$$
\begin{aligned}
\ln Z_N(G) &= N\beta mH + \frac{Nq}{2}\beta J + N\ln 2 - N\ln(1+\tau) - \frac{Nq}{2}\ln(1+v) \\
&\quad + \ln\left[1 + \sum_{\ell,m} \tau^{2m} v^{\ell}{}_{\ell,2m}\right] \\
&= N\beta mH + \frac{Nq}{2}\beta J + \ln \Lambda_N \; .
\end{aligned}
\tag{23}
$$

The first two terms represent the ground state energy, when all the spins are aligned parallel to the magnetic field. The fraction of overturned spins, denoted by α, is given by,

$$\alpha = \frac{n}{N} = \tfrac{1}{2}\left[1 - \frac{M}{Nm}\right] = -\frac{1}{2Nm\beta}\frac{\partial \ln\Lambda}{\partial H} \quad . \tag{24}$$

A change of variable to α enables us to write,

$$\frac{1}{N}\ln\Lambda_N(G) = \sum_{S\leq G} H_s(\alpha,v) \ , \tag{25}$$

where the sum is over all star subgraphs S. The weights H_s must be calculated from the partition functions of star subgraphs.

In the zero field ($\alpha = \frac{1}{2}$) limit, we obtain, for the susceptibility of a finite graph G,

$$\frac{2Nm^2\beta}{\chi_0} = \lim_{\alpha\to\frac{1}{2}}\frac{\partial \ln\Lambda_N}{\partial\alpha} = \sum_{S\leq G} W_S(v) \ , \tag{26}$$

where W_S represents the inverse susceptibility weight of S and is now a function of v alone. If W_S is expanded as a polynomial in v, it can be shown that the lowest possible power of v is v^{ℓ}, where ℓ is the number of edges of S.

The calculation of inverse susceptibility weights involves writing down the high temperature, zero field susceptibility of star graphs, taking the reciprocal and subtracting off the weights of all star subgraphs. This is readily done, by considering star graphs with increasing number of edges, so that at any stage, the weights of all subgraphs have been previously calculated.

Additional complications arise when one is dealing with inhomogeneous star graphs. The magnetization (and hence α) varies from site to site, as the environments of the sites are not all the same. We now replace α by a set $\{\alpha_i\}$ and τ by $\{\tau_i\}$, $i = 1,2,\ldots N$. (Since most graphs possess little or no symmetry, it proves simpler to ignore the symmetry altogether and treat all vertices as being distinct). The inverse susceptibility is now given by,

$$\frac{2Nm^2\beta}{\chi_0} = \sum_{i=1}^{N}\frac{\partial \ln\Lambda_N}{\partial\alpha_i} \ , \quad \alpha_i = -\frac{1}{2Nm\beta}\frac{\partial \ln\Lambda_N}{\partial H_i} \ . \tag{27}$$

To calculate the partial derivatives with respect to α_i, we write,

$$\frac{\partial \ln\Lambda_N}{\partial H_i} = \sum_j \frac{\partial \ln\Lambda_N}{\partial\alpha_j}\frac{\partial\alpha_j}{\partial H_i} \ , \tag{28}$$

with,

$$\frac{\partial \ln\Lambda_N}{\partial H_i} = -\beta m ,$$

$$\frac{\partial \alpha_i}{\partial H_i} = -\frac{\beta m}{2N} , \tag{29}$$

$$\frac{\partial \alpha_j}{\partial H_i} = -\frac{\beta m}{2N}\frac{\partial^2 \ln\Lambda_N}{\partial \tau_i \partial \tau_j} = -\frac{\beta m}{2N}\frac{\sum_\ell P^{(ij)}_{\ell,2} v^\ell}{1+\sum_\ell P_{\ell,0} v^\ell} .$$

The term $P^{(ij)}_{\ell,2}$ denotes the number of subgraphs with ℓ edges and exactly two odd vertices at i and j, while $P_{\ell,0}$ denotes the number of subgraphs with ℓ edges and no odd vertices.

We define a matrix $\underline{M}$, with elements,

$$M_{ii} = 1, \quad M_{ij} = \frac{\partial^2 \ln\Lambda_N}{\partial \tau_i \partial \tau_j} , \tag{30}$$

and a vector $\underline{X}^T = \{x_1, x_2 \ldots x_N\}$, with

$$x_i = \frac{\partial \ln\Lambda_N}{\partial \alpha_i} . \tag{31}$$

The inverse susceptibility is now given by,

$$\frac{m^2\beta}{\chi_0} = \frac{1}{2N}\underline{1}^T\underline{x} = \underline{1}^T\underline{M}^{-1}\underline{1} . \tag{32}$$

For purposes of deriving series expansions, we can write,

$$\underline{M}^{-1} = (\underline{I} + \underline{X})^{-1} = (\underline{I} - \underline{X})(\underline{I} + \underline{X}^2)(\underline{I} + \underline{X}^4)\ldots , \tag{33}$$

and use the fact that $\underline{X} \sim v$, $\underline{X}^n \sim v^n$. We only need compute $\underline{M}^{-1}$ to the order required.

Having calculated $\chi_0^{-1}(G)$, we use (26) to obtain $W_G(v)$. When all G with up to ℓ edges have been considered, we let $G = \mathcal{L}$, and write, correct to v^ℓ,

$$\frac{m^2\beta}{\chi_0} = \sum_G (G;\mathcal{L})\, W_G(v) , \tag{34}$$

where $(G;\mathcal{L})$ denotes weak lattice constants and χ_0^{-1} is defined per lattice site.

In table I are given a list of star lattice constants and inverse susceptibility weights. The data is sufficient to calculate χ_0^{-1} to order v^7 on the FCC lattice. The weights, however, are independent of the lattice.

Table I: FCC lattice, χ_0^{-1} expansion

G	$(G;\mathcal{L})$	$W_G(v)$
	1	1
	6	$-2v + 2v^2 - 2v^3 + 2v^4 - 2v^5 + 2v^6 - 2v^7$
	8	$6v^3 - 12v^4 + 12v^5 - 6v^6 + 0v^7$
	33	$8v^4 - 16v^5 + 16v^6 - 16v^7$
	168	$10v^5 - 20v^6 + 20v^7$
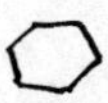	970	$12v^6 - 24v^7$
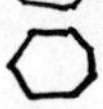	6168	$14v^7$
	36	$-8v^5 + 4v^6 + 28v^7$
	384	$-8v^6 + 2v^7$
	36	$-8v^6 + 0v^7$
	2400	$-8v^7$
	966	$-8v^7$
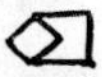	600	$-8v^7$
	2	$48v^7$
	24	$4v^7$
	192	$0v^7$
	48	$0v^7$

$$\chi_0^{-1}\,(\mathrm{FCC}) = 1 - 12v + 12v^2 + 36v^3 + 180v^4 + 948v^5 + 5556v^6 + 36132v^7$$

Higher Derivatives

In the last section, we saw that the first derivative of the free energy with respect to α corresponds to χ_0^{-1}. The third derivative is given by,

$$\lim_{\alpha\to\frac{1}{2}} \frac{\partial^3 \ln\Lambda_N}{\partial\alpha^3} = -8\,\frac{\dfrac{\partial^4 \ln\Lambda_N}{\partial H^4}}{\chi_0^4} = \sum_S (S;\mathcal{L})\, h_S(v) \ , \tag{35}$$

and can be developed as a star graph expansion.

Calculation of the third α-derivative for a finite star cluster is a little more involved; the algebra is tedious but straightforward, and one derives, for a star graph S with N sites,[14]

$$\lim_{\alpha\to\frac{1}{2}} \frac{\partial^3 \ln\Lambda_N}{\partial\alpha^3} = -2\sum_{i=1}^{N} B_i^3 + \sum_{j,k} B_k B_j\, \underline{1}^T \left[\underline{B}\ \underline{C}_{jk}\ \underline{B}\right] \underline{1} \ , \tag{36}$$

where $\underline{B} = \underline{A}^{-1}$, and $A_{ij} = \partial\alpha_j/\partial\tau_i$. The matrix $\underline{A}$ is similar to the matrix $\underline{M}$ used in the calculation of χ_0^{-1}. We also have,

$$B_i = \sum_{j=1}^{N} B_{ji}, \quad \underline{C} = 2\underline{A} - \underline{I} \ . \tag{37}$$

The matrix $\underline{C}_{ij}$ is derived by differentiating all the elements $\underline{C}$ with respect to τ_i, τ_j before imposing the zero field limit. The elements of $\underline{C}_{ij}$ are therefore easily written down in terms of subgraphs of S with four odd vertices.

All these manipulations are easily performed on a computer. The star subgraph subtraction is exactly the same as for the χ_0^{-1} problem and use of the weights and lattice constant data in (35) yields the required series. An example of such a calculation to a low order in v is given in Table II, again for the FCC lattice.

Excluded volume graph expansions have, on the whole, proved most successful with thermodynamic properties of the Ising model. I shall now discuss another powerful method which has been particularly useful in the study of correlation functions. This is the linked cluster expansion which involves free graphs rather than excluded volume graphs.

Linked Cluster Expansion[17-20]

The main features of this approach are:

(a) A many variable Taylor expansion of the free energy about the non-interacting limit yields an expansion in the variables $v(=\beta J)$

Table II: FCC Lattice, $-\chi_0^{(2)}/\chi_0^4$ series

S	(S;ℓ)	$W_S(v)$						
•	1	2						
	6		$-12v^2$	$+32v^3$	$-60v^4$	$+96v^5$	$-140v^6$	$+192v^7$
	8			-72	432	-1080	1332	-216
	33				-144	768	-1824	3072
	168					-240	1200	-2760
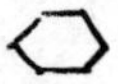	970						-360	1728
	6168							-504
	36					384	-1560	240
	384						480	-1800
	36						480	-1152
	2400							576
	966							576
	600							576
	24							-1680
	192							-384
	2						-384	-3456
	48							-384

$$-\chi_0^{(2)}/\chi_0^4 = 2 - 72v^2 - 384v^3 - 1656v^4 - 9216v^5 - 53304v^6 - 374{,}400v^7$$

and the semiinvariants M_n^o, which are functions of spin and magnetic field.

(b) The coefficients of the series are made up of contributions from multiline, free graphs, and in some cases, rooted graphs.

(c) Successive renormalizations are possible to sum over contributions from selected subsets of graphs. One can thereby reduce the number of contributing graphs, though at the expense of increased algebraic complexity. Two such schemes are used, namely, vertex renormalization and bond renormalization. The former gives an expansion in 1-irreducible graphs while the latter involves only 2-irreducible graphs.

I shall confine myself to a discussion of bare and vertex renormalized expansions for the free energy, magnetization and pair correlation function for the spin-S Ising model.

Notation:

We shall start with the Hamiltonian for a system of N Ising spins,

$$\mathcal{H} = -\frac{1}{S^2}\sum_{<ij>} J_{ij}\, S_i^z\, S_j^z - \frac{m}{S}\sum_{i}^{N} S_i^z\, H_i \; . \tag{38}$$

The field H_i and the interaction constant J_{ij} are allowed to vary. This is done for convenience and eventually we shall set all $H_i = H$ and $J_{ij} = J$.

Defining the variables,

$$\mu_i = S_i^z/S, \quad v_{ij} = \beta J_{ij} \quad \text{and} \quad h_i = \beta m H_i \; ,$$

we rewrite (38) in the form (with $J_{ii} = 0$),

$$-\beta\mathcal{H} = \tfrac{1}{2}\sum_{i,j} v_{ij}\, \mu_i \mu_j + \sum_i h_i\, \mu_i \; . \tag{39}$$

The free energy W and other properties are defined in the usual way. Denoting the magnetization by M_1, the susceptibility per site by $\chi(i)$ and the pair correlation function by $\mathcal{M}_2(ij)$, we obtain,

$$W = \ell n\, \mathrm{Tr}\, \exp(-\beta\mathcal{H}) \tag{40a}$$

$$M_1(i) = \langle\mu_i\rangle = \frac{\delta W}{\delta h_i} \tag{40b}$$

$$\mathcal{M}_2(ij) = \langle\mu_i\mu_j\rangle - \langle\mu_i\rangle\langle\mu_j\rangle$$

$$= \frac{\delta^2 W}{\delta h_i\, \delta h_j} = \frac{\delta W}{\delta v_{ij}} - \frac{\delta W}{\delta h_i}\,\frac{\delta W}{\delta h_j} \tag{40c}$$

$$\chi(i) = \sum_j \mathcal{M}_2(ij) \; . \tag{40d}$$

The linked cluster expansion is obtained by developing W as a Taylor expansion about the non-interacting limit, $v_{ij} = 0$. In this limit,

$$W(h,v{=}0) \equiv W_0(h) = \ell n\, \mathrm{Tr}\, \exp \sum_i h_i \mu_i = \sum_i M_0^o(i) \; ,$$

where,

$$M_0^o(i) = \ell n \sum_{\mu_i=-1}^{1} \exp h_i\mu_i = \ell n \frac{\text{Sinh}\,(\frac{2S+1}{2S})h_i}{\text{Sinh}\,(\frac{h_i}{2S})} . \qquad (41)$$

Derivatives of $M_0^o(i)$ with respect to h_i are called bare semiinvariants and defined as follows,

$$\frac{\delta^n M_0^o(i)}{\delta h_k \delta h_\ell \ldots \delta h_s} = M_n^o(i)\ \delta(i,k,\ell\ldots s) \quad . \qquad (42)$$

The $\delta(i,k,\ell\ldots s)$ term arises because $M_0^o(i)$ involves only the ith lattice site. Another property of the $M_n^o(i)$ is that differentiation with respect to h_i merely increases the order of the semiinvariant by one.

Unrenormalized Expansions

The expansion for W takes the form,

$$W = W_0 + \sum_{<ij>} v_{ij} \frac{\delta W}{\delta v_{ij}} \Big|_{v=0} + \frac{1}{2!} \sum_{\substack{<ij>\\<k\ell>}} v_{ij}v_{k\ell} \frac{\delta^2 W}{\delta v_{ij}\delta v_{k\ell}}\Big|_{v=0}$$

$$+ \ldots \quad . \qquad (43)$$

The terms v_{ij}, $v_{ij}v_{k\ell}$ ..., correspond to choosing subsets of lattice bonds in different ways and each such choice corresponds to a subgraph of the lattice. Since a particular bond may be chosen more than once, multigraphs occur in the expansion. The coefficients of the πvij terms can be related to bare semiinvariants by the use of (40) and (42). We derive,

$$\lim_{v=0} \frac{\delta W}{\delta vij} \lim_{v=0} \left[\frac{\delta^2 W_0}{\delta h_i \delta h_j} + \frac{\delta W_0}{\delta h_i}\frac{\delta W_0}{\delta h_j}\right] = M_1^o(i)M_1^o(j) \quad . \qquad (44)$$

The quadratic term in $v_{ij}v_{k\ell}$ gives a contribution,

$$\lim_{v=0} \frac{\delta^2 W}{\delta v_{ij}\delta v_{k\ell}} = \{\delta(ik)\delta(j\ell)+\delta(i\ell)\delta(jk)\}M_2^o(i)\ M_2^o(j)$$

$$+ M_1^o(i)M_1^o(k)M_2^o(j)\delta(j\ell) + 3 \text{ similar terms.} \qquad (45)$$

The $\delta(ik)$ terms are necessary to ensure nonvanishing derivatives of W_0 with respect to h_i, h_k as seen in (42). It is the localized property of the $M_0^o(i)$ that ensures a connected graph expansion for W.

To second order in v, we derive,

$$W = \sum_i M_0^o(i) + \tfrac{1}{2} \sum_{i,j} M_1^o(i) M_1^o(j) v_{ij} + \tfrac{1}{4} \sum v_{ij}^2 M_2^o(i) M_2^o(j)$$

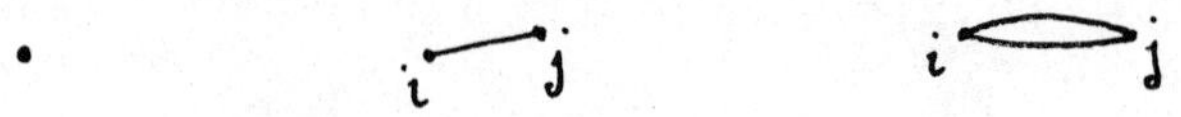

$$+ \tfrac{1}{2} \sum_{i,j,k} v_{ij} v_{jk} M_1^o(i) M_2^o(j) M_1^o(k) + \ldots \qquad (46)$$

The graphical interpretation and weighting for the W expansion are now becoming clear. Rigorous derivations can be found in literature. The free energy W is given by the weighted sum of all unrooted connected graphs, including multigraphs. The weight of a graph G is given by,

$$\frac{FM(G)}{S(G)} \prod_i M_{d_i}^o \prod_{<ij>} v_{ij} \; , \qquad (47)$$

where FM denotes the free multiplicity (or number of free embeddings per site), S denotes the symmetry number of the graph; the first product is over all vertices, where d_i is the degree of the vertex i. The second product is over all edges of the graph. The free multiplicity of a multigraph is the same as that of the basic graph, where each set of multiple edges is replaced by a single edge. The symmetry number, however, depends on the multiedges.

Expansions for $M_1(i)$ and $\mathcal{M}_2(ij)$ are obtained by differentiating the expansion for W with respect to h_i and h_i, h_j respectively. The effect of differentiation is to fix each vertex in turn and change the corresponding semiinvariant $M_n^o(i)$ to $M_{n+1}^o(i)$. Since singling out vertices for differentiation can be looked upon as rooting the graph, expansions for $M_1(i)$ and $\mathcal{M}_2(ij)$ involve 1- and 2-rooted graphs respectively. Thus,

$$M_1(i) = \frac{\delta}{\delta h_i}$$

$$= M_1^o(i) + \sum_j M_1^o(j)v_{ij}M_2^o(i) + \tfrac{1}{2}\sum_j M_3^o(i)M_2^o(j)v_{ij}^2$$

$$+ \sum_{j,k} M_2^o(i)M_2^o(j)M_1^o(k)v_{ij}v_{jk} + \tfrac{1}{2}\sum_{j,k} M_3^o(i)M_1^o(j)M_1^o(k)v_{ij}v_{ik} + \ldots \tag{48}$$

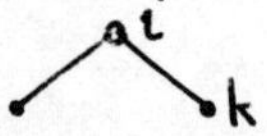

The magnetization can therefore be expanded in terms of all 1-rooted graphs, with the root vertex being assigned a factor of $M_{n+1}^o(i)$ if it is of degree n. Also, the free sum over lattice sites (free embeddings) is now performed with the root point fixed at site i. The symmetry number is now that of the rooted graph which is in general smaller than that of the unrooted graph. Apart from these changes, the rules for weighting are the same as those for the W expansion (47).

The two spin correlation function $\mathcal{M}_2(ij)$ has an expansion in terms of 2-rooted graphs. Thus (for i ≠ j),

$$\mathcal{M}_2(ij) = v_{ij}M_2^o(i)M_2^o(j) + \tfrac{1}{2}v_{ij}^2M_3^o(i)M_3^o(j) + \sum_k v_{ik}v_{jk}M_2^o(k)M_2^o(i)M_2^o(j)$$

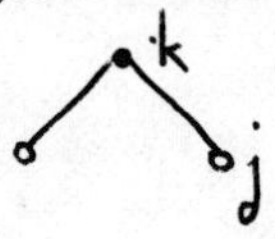

$$+ \sum_k v_{ij}v_{ik}M_1^o(k)M_3^o(i)M_2^o(j) + \sum_k v_{ij}v_{jk}M_1^o(k)M_3^o(j)M_2^o(i) + \ldots$$

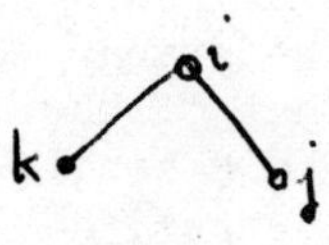

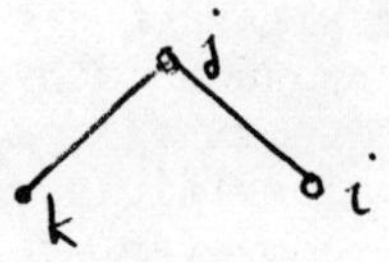

(49)

The weight of a two rooted graph is given by,

$$\frac{FM(G)_{ij}}{S(G)_{ij}} \prod_i M_{d_i}^o \prod_{<ij>} v_{ij}\; M_{r_1+1}^o(i)\; M_{r_2+1}^o(j)\;, \tag{50}$$

where $FM(G)_{ij}$ denotes the free multiplicity of G with root vertices fixed at sites i,j and $S(G)_{ij}$ denotes the symmetry of the rooted graph. The first product is over all internal (non-root) vertices and d_i denotes the degree of the vertex i. r_1 and r_2 denote the degrees of the root vertices at i and j.

Higher correlations can similarly be interpreted in terms of 3-, 4-rooted graphs.

Vertex Renormalized Expansions

We saw that the bare expansion for W, M_1 and $\mathcal{M}_2$ involves all connected multiline graphs which are numerous. To make the linked cluster expansion reasonably tractable at higher orders of v, one has to find a way of reducing the number of contributing graphs. This is done by going over to the vertex renormalized form of the expansion.

The basic idea behind vertex renormalization is similar to the transformation from reducible graphs to star graphs in excluded volume expansions. Any rooted graph can be decomposed into a unique irreducible skeleton together with several unrooted 1-insertions. The procedure is to take each articulation point in turn, clip away all edges incident at that point and then reconnect any disconnected parts containing root points. All the unrooted disconnected parts are 1-insertions, the point of insertion being treated as being distinct from the other vertices (Fig. 3). The reason for starting with a rooted graph is now clear - only a rooted graph possesses a unique skeleton.

Such a decomposition suggests that it should be possible to write down expansions for $M_1(i)$ and $\mathcal{M}_2(ij)$ in terms of the 1-irreducible skeletons and take account of all possible 1-insertions implicitly. This is done by modifying the M_n^o associated with the skeletal vertices. Formally, we make the change,

$$M_n^o(i) \rightarrow M_n(i) \ , \qquad (51)$$

where $M_n(i)$ is the renormalized semi-invariant, and allows for all possible ways of decorating the vertex with 1-insertions. The only problem is to calculate $M_n(i)$.

As we are working with free graphs, the free multiplicity of any reducible graph is simply the product of the free multiplicity of the skeleton and those of all the 1-insertions. That part of the calculation is therefore trivial. Any 1-insertion of degree ℓ at a vertex i must change the $M_n^o(i)$ to $M_{n+\ell}^o(i)$. Identical 1-insertions at a vertex increase the symmetry number of the graph, while

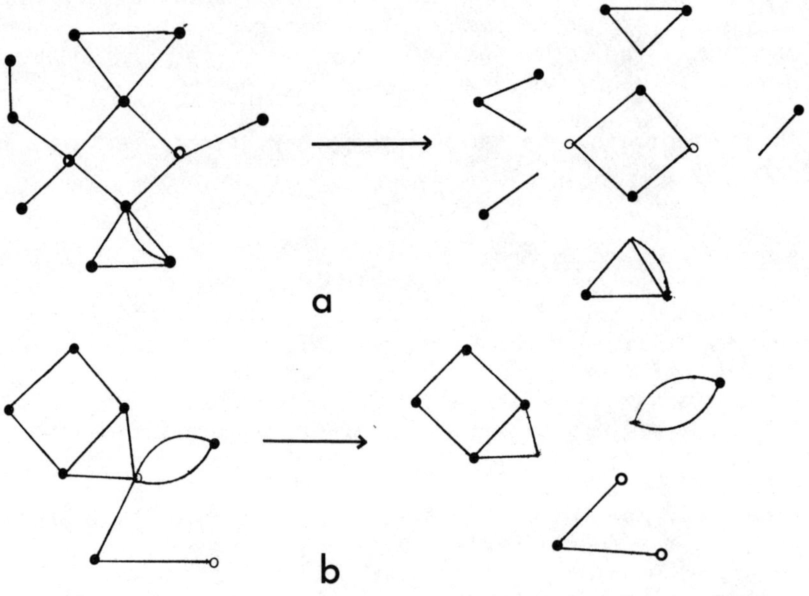

Fig. 3. Decomposition of a rooted, 1-reducible graph.

distinct decoration of equivalent skeletal vertices reduces the symmetry number. These changes must be allowed for in calculating the renormalized semiinvariants.

We first classify 1-insertions according to the valence of the insertion vertex (i). Let G_ℓ denote the sum of all ℓ-valent 1-insertions weighted as follows:

$$\frac{FM(G)_i}{S(G)_i} \prod_i M^o_{d_i} \prod_{<ij>} v_{ij} \,, \tag{52}$$

where $FM(G)_i$ denotes the free multiplicity of G with the insertion vertex fixed at i and $S(G)_i$ denotes the symmetry number of G rooted at i. The first product is over all vertices except i and the second is over all edges. The insertion vertex is weighted with a factor of 1, because it will be lost when the insertion is joined to the skeleton. The valence of the insertion vertex modifes the M^o_n associated with the skeletal vertex $[M^o_n(i) \to M^o_{n+\ell}(i)]$. Defined with the $M^o_{d_i}$, (52) defines the bare self fields G_ℓ. They will be renormalized eventually.

Once the M_n are known in terms of h and v, all the bare expansions (except W) can be written down in renormalized form, involving

$$M_n = M_n^o + \sum_{\ell=1}^{\infty} G_\ell M_{n+\ell}^o + \frac{1}{2!} \sum_{\ell,m=1}^{\infty} G_\ell G_m M_{n+\ell+m}^o + \ldots$$

$$= \left[\exp \sum_\ell G_\ell \frac{\delta^\ell}{\delta h^\ell}\right] M_n^o \quad . \tag{53}$$

It can be easily verified that (53) takes account of all possible decorations of the vertex. The factors $\frac{1}{2!}$, $\frac{1}{3!}$ allow for the increased symmetry due to identical insertions at the vertex. Reduction of symmetry due to distinct decorations at equivalent vertices is compensated for by equivalent cross terms for the product of M_n at those vertices. Thus, by replacing M_n^o by M_n at all skeletal vertices, one takes account of contributions due to all reducible graphs that can be derived from the 1-irreducible skeleton.

Any unrenormalized expansion involving rooted graphs can be converted into the vertex renormalized form involving only 1-irreducible graphs by replacing all M_n^o by M_n. This applies to the self fields G_ℓ as well, and the renormalized form $G(\underline{M})$ involves only 1-irreducible 1-insertions. The first few contributions are given by,

G_1 =

G_2 = $\frac{1}{2}$  + $\frac{1}{2}$  + ... (54)

G_3 = $\frac{1}{6}$  + ...

To calculate $M_n(\underline{M}^o)$, we iterate (53) and (54) simultaneously, as shown in the following scheme:

$M_n = M_n^o \longrightarrow G_\ell(v)$ (correct to v)

$M_n(v) \longleftarrow$

$\longrightarrow G_\ell(v^2)$ (correct to v^2)

$M_n(v^2) \longleftarrow$ (55)

$\longrightarrow G_\ell(v^3)$ (correct to v^3)

\- - - - - - - - - - -

$M_n(v^n) \longleftarrow G_\ell(v^n)$ (correct to v^n) .

In terms of M_n^o and G_ℓ, the renormalized semiinvariants are given by only 1-irreducible graphs. Also, M_1 is simply the magnetization, as it enumerates all ways of decorating a single root vertex. All field derivatives of the free energy (such as χ_0, $\chi_0^{(2)}$ ...) can be obtained by differentiating M_1 with respect to h.

The free energy expansion (46) involves unrooted graphs which do not possess a unique skeleton. To calculate W in renormalized form, one integrates M_1, subject to the boundary condition:

$$h(1) = \infty \Rightarrow M_1^o = 1, \quad M_n^o = 0, \quad n > 1 \quad . \tag{56}$$

We shall not pursue the derivation in detail, as one rarely calculates W explicitly. One finally obtains,

$$W = \sum_i M_o(i) + \phi - \sum_i \sum_{n=1}^{\infty} G_n(i)\, M_n(i) \quad , \tag{57}$$

where ϕ is the weighted sum of all 1-irreducible o-rooted graphs as in (47), with the M_n^o replaced by M_n.

Pair Correlations in Zero Field Limit

To calculate $\mathcal{M}_2(ij)$ in vertex renormalized form, all 2-rooted 1-irreducible graphs must be considered. In the zero field limit, only a subset of these graphs contribute. To see this, we note that, for n odd,

$$h(i) = 0 \Rightarrow M_n^o(i) = G_n(i) = M_n(i) = 0 \quad . \tag{58}$$

Since n-valent internal vertices are weighted with a factor of $M_n(i)$ and root vertices with $M_{n+1}(i)$, (see 50), the only graphs that contribute to the renormalized expansion of $\mathcal{M}_2(ij)$ in the zero field limit are 2-rooted 1-irreducible graphs with root vertices of odd degree ($M_{n+1} \neq 0$) and all other vertices of even degree ($M_n \neq 0$). All such graphs can be built up from star graphs with exactly two odd vertices, by rooting the odd vertices and joining the stars at their root vertices. Denoting the star graphs by A, B, C... and their odd vertices by i,j,k,..., we can represent this schematically as follows:

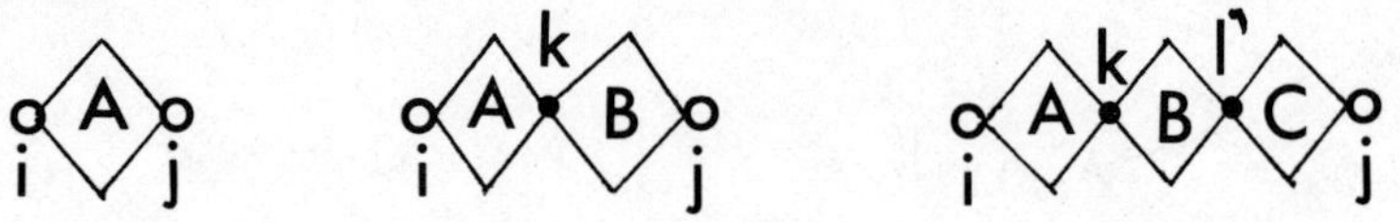

(59)

where the open circles (i and j) denote the roots of the correlation graph.

Define $I_{mn}^{(ij)}$ as the sum of all star graphs with two odd vertices of degree m, n at sites i and j, weighted as follows

$$\frac{FM(G)_{ij}}{S(G)_{ij}} \prod_{<k\ell>} v_{k\ell} \prod_i M_{d_i} \quad (h = 0) \quad , \tag{60}$$

where the free multiplicity FM is defined with the root vertices fixed at sites i and j, and the symmetry number S is that of the rooted graph. The first product is over all edges and the second is over all internal vertices. The M_{d_i} are calculated in the h = 0 limit. We can now relate $\mathcal{M}_2(ij)$ to the $I_{mn}^{(ij)}$. For i ≠ j, we obtain,

$$\begin{aligned}\mathcal{M}_2(ij) = & \sum_{m,n} M_{m+1}(i)\, I_{mn}^{(ij)}\, M_{n+1}(j) \\ & + \sum_{\substack{m,n,p,\gamma \\ k}} M_{m+1}(i)\, I_{mp}^{(ik)} M_{p+r}(k)\, I_{rn}^{(k,j)} M_{n+1}(j) \\ & + \sum_{\substack{m,n,p,r,t,q \\ k,\ell}} M_{m+1}^{(i)}\, I_{mp}^{(ik)} M_{p+r}(k) I_{rt}^{(k\ell)} M_{t+q}(\ell) I_{qn}^{(\ell,j)} M_{n+1}(j) \\ & + \ldots \quad . \end{aligned} \tag{61}$$

The problem is thus reduced to computing the $I_{mn}^{(ij)}$ from star graphs with two odd vertices and combining them in all possible ways according to (61). The semiinvariants M_n have already been calculated in (55), and it is a simple matter to set h = 0.

Conclusion

In this lecture, we discussed two basic approaches commonly used in deriving series expansions, with special reference to the Ising model. In the next talk, I shall show how some of these methods generalize to the classical vector model.

References

1 C. Domb in "Phase Transitions and Critical Phenomena" (eds. C. Domb and M.S. Green) Vol. 3, Ch. 1, 6 (1974).

2 B.L. Van der Waerden, Z.Phys. 118, 473 (1941).

3 J. Riordan in "An Introduction to Combinatorial Analysis", (J. Wiley, N.Y.) (1958).

4 C. Domb, J.Phys.A 7, L45 (1974).

5 K. Fuchs, Proc.Roy.Soc. A, 179, 340 (1942).

6 J. Yvon, Cah. Phys. No. 28 (1945).

7 G.S. Rushbrooke and H.I. Scoins, Proc.Roy.Soc. A 230, 74 (1955).

8 " " J.Math.Phys. 3, 176 (1962).

9 C. Domb and B.J. Hiley, Proc.Roy.Soc. A 268, 506 (1962).

10 G.E. Uhlenbeck and G.W. Ford in "Studies in Statistical Mechanics", Vol.1 p.123 (J. de Boer and G.E. Uhlenbeck eds: North Holland Co.) (1962).

11 M.F. Sykes, J.W. Essam, B.R. Heap and B.J. Hiley, J.Maths.Phys. 7, 1557 (1966).

12 G.A. Baker Jr., Phys.Rev. B, 15, 1552 (1977).

13 S. McKenzie, J.Phys.A 8, L102 (1975).

14 S. McKenzie, J.Phys.A 13, 1007 (1980).

15 S. McKenzie and D.S. Gaunt, J.Phys.A 13, 1015 (1980).

16 D.S. Gaunt and M.F. Sykes, J.Phys.A 12, L25 (1979).

17 M. Wortis in "Phase Transitions and Critical Phenomena" (eds. C. Domb and M.S. Green: N.Y. Academic), Vol.3, Ch.3 (1974).

18 F. Englert, Phys.Rev. 129, 567 (1963).

19 G. Horwitz and H.B. Callen, Phys.Rev. 124, 1757 (1961).

20 M. Wortis, D. Jasnow and M.A. Moore, Phys.Rev. 185, 805 (1969).

SERIES EXPANSIONS FOR THE CLASSICAL VECTOR MODEL

Sati McKenzie

Department of Physics
King's College
Strand, London WC2R 2LS, UK

INTRODUCTION

In the last lecture, we dealt with the spin-$\frac{1}{2}$ Ising model, with the Hamiltonian,

$$\mathcal{H}_I = -J \sum_{<ij>} \sigma_i \sigma_j - mH \sum \sigma_i , \tag{1}$$

where the coupling between the spins is confined to the z-components, with $\sigma_i = \pm 1$. We shall now consider a more general Hamiltonian of the form,

$$\mathcal{H} = -\frac{1}{S^2} \sum_{<ij>} J_{ij} \underline{S}_i . \underline{S}_j - \frac{m}{S} \sum_{i=1}^{N} \underline{S}_i . \underline{H}_i . \tag{2}$$

The spins $\underline{S}_i$ are now D-dimensional vectors given by,

$$\underline{S}_i = \{S_i^1 \; S_i^2 \; \ldots \; S_i^D\} . \tag{3}$$

The interaction between the spins takes the form of a scalar product $\underline{S}_i . \underline{S}_j$ and (2) defines the D-vector model.

The spins $\underline{S}_i$ can be treated either as quantum mechanical operators or as classical vectors. In the former case, one obtains the quantum Heisenberg model [1] (D = 3) and the X-Y model [2] (D = 2, spin-spin coupling along the z-axis being zero). On allowing S to tend to infinity, so that the $\underline{S}_i/S$ become classical D-dimensional unit vectors, one realises the classical D-vector model [3] .

For special values of D, the classical vector model reduces to certain well studied problems:

(1) $D = 1$, $\underline{S}_i.\underline{S}_j = S_i^z S_j^z$ (Ising model)

(2) $D = 2$, $\underline{S}_i.\underline{S}_j = S_i^x S_j^x + S_i^y S_j^y$ (Planar model)

(3) $D = 3$, $\underline{S}_i.\underline{S}_j = S_i^x S_j^x + S_i^y S_j^y + S_i^z S_j^z$ (Classical Heisenberg model)

(4) $D = 0$ (Excluded volume problem)

(5) $D = \infty$ (Spherical model)

Other variations that can be incorporated into the Hamiltonian include (a) lattice anisotropy, (b) spin anisotropy and (c) next nearest neighbour interactions. All these problems can be studied and have been studied by exact series expansion methods. In this talk, however, I shall confine myself to a discussion of the isotropic, nearest neighbour, classical D-vector model.

Formalism

We rewrite (2) in the form,

$$= - J \sum_{<ij>} \underline{\sigma}_i . \underline{\sigma}_j - m H \sum_{i=1}^{N} \sigma_i^z . \qquad (4)$$

The first sum is over all nearest neighbour spin pairs and the second is over all spins σ_i. The $\underline{\sigma}_i$ are classical vectors which can assume all orientations on the surface of a D-dimensional unit hypersphere. The magnetic field H is assumed to act in the z direction. The partition function Z for a system of N spins is now given by,

$$Z_N = \mathrm{tr}\ \exp(-\beta \mathcal{H}) = \mathrm{tr} \prod_{<ij>} \exp(K \underline{\sigma}_i . \underline{\sigma}_j) \prod_{i=1}^{N} \exp(h \sigma_i^z) , \qquad (5)$$

where $K = \beta J$, $h = \beta m H$ and $\beta = 1/kT$. The trace for a classical system is obtained by integrating over the angular variables Ω_i associated with the spin vectors. Thus,

$$<A> = \mathrm{tr}\ A = \frac{\int \ldots \int A \prod_{i=1}^{N} d\Omega_i}{\int \ldots \int \prod_{i=1}^{N} d\Omega_i} = S_D^{-N} \int \ldots \int A \prod_{i=1}^{N} d\Omega_i . \qquad (6)$$

S_D is simply the surface area of a unit hypersphere in D dimensions, given by,

$$S_D = 2\pi^{D/2}/\Gamma(D/2) . \qquad (7)$$

The free energy F_N and other thermodynamic properties can now be written down in the usual way. The free energy is given by

$$F_N = - kT \ln Z, \tag{8}$$

and the internal energy U_N, the specific heat C_H, and magnetization M are given by,

$$U_N = - \frac{\partial \ln Z_N}{\partial \beta}$$

$$C_H = \frac{\partial U}{\partial T} \tag{9}$$

$$M = m \sum_i \langle \sigma_i^z \rangle = \frac{1}{\beta} \frac{\partial \ln Z}{\partial H} .$$

The susceptibility χ_N is defined by,

$$\chi_N = \frac{1}{\beta} \frac{\partial^2 \ln Z_N}{\partial H^2} = \frac{m^2}{kT} \sum_i \sum_j \langle \sigma_i^z \, \sigma_j^z \rangle = \frac{m^2}{DkT} \sum_i \sum_j \langle \underline{\sigma}_i \cdot \underline{\sigma}_j \rangle . \tag{10}$$

The zero field susceptibility per spin, χ_0, is given by,

$$\frac{DkT\chi_0}{m^2} = 1 + \frac{2}{N} \sum_{i>j} \langle \underline{\sigma}_i \cdot \underline{\sigma}_j \rangle \Big|_{h=0} . \tag{11}$$

For an N-spin system represented by a graph G(N,L) the zero field correlation $\langle \underline{\sigma}_i \cdot \underline{\sigma}_j \rangle$ between spins at i and j can be calculated very simply by joining the vertices i and j by a pseudo edge K^* to get a graph $G_{ij}(N,L+1)$. In terms of G_{ij}, we can write,

$$\langle \underline{\sigma}_i \cdot \underline{\sigma}_j \rangle \Big|_{h=0} = \frac{\partial \ln Z_0(G_{ij})}{\partial K^*} \Big|_{K^*=0} \tag{12}$$

where Z_0 represents the partition function in zero magnetic field.

The derivation of series expansions for the D-vector model parallels that for the Ising model. The problem of calculating traces is more complicated in this case. As for the Ising model, derivation of series for the free energy and susceptibility in the zero field limit proves considerably simpler than the calculation of the entire free energy as a function of h and K.

Summary of Methods Available for Deriving Series Expansions

Some of the commonly used methods include

(a) the moment method [4] which yields an expansion in terms of all graphs including disconnected ones;

(b) the cumulant approach [5,6] which involves only connected graphs;

(c) the finite cluster development [7] which is very closely related to the cumulant expansion;

(d) the linked cluster expansion [8] in terms of free graphs and

(e) the star graph expansion method [9,10,11] in terms of 1-irreducible or star graphs.

Moment and Cumulant Expansions

It is convenient to rewrite the Hamiltonian (4) in the form

$$-\beta\mathcal{H} = -\beta\mathcal{H}_0 - \beta\mathcal{H}_1 = h\sum_i^N \sigma_i^z + K\sum_{<ij>} \underline{\sigma}_i\cdot\underline{\sigma}_j \ . \tag{13}$$

The partition function is given by,

$$Z = Z_1 Z_2 \ , \tag{14}$$

where

$$Z_1 = \mathrm{tr}\ \exp(-\beta\mathcal{H}_0) = \left[\frac{1}{S_D}\int d\Omega\ e^{h\sigma^z}\right]^N \ . \tag{15}$$

The spin-spin interactions contribute to the Z_2 term which is defined by,

$$Z_2 = \langle \exp K \sum_{<ij>} \underline{\sigma}_i\cdot\underline{\sigma}_j \rangle_1 = \frac{\mathrm{Tr}\ \exp(-\beta\mathcal{H})}{\mathrm{Tr}\ \exp(-\beta\mathcal{H}_0)} \ . \tag{16}$$

On expanding the exponential in (16), one obtains the moment expansion. Thus,

$$Z_2 = \sum_{\ell=0}^{\infty} \frac{K^\ell}{\ell!} \langle (\sum \underline{\sigma}_i\cdot\underline{\sigma}_j)^\ell \rangle_1 = \sum_{\ell=0}^{\infty} \frac{K^\ell}{\ell!} M_\ell \ . \tag{17}$$

The terms in (17) can now be interpreted in terms of interaction graphs on the lattice. The ℓth moment M_ℓ includes contributions from all graphs with ℓ edges, including disconnected and multigraphs. We can write,

$$M_\ell = \sum_{g(n,\ell)} \langle g \rangle \ , \tag{18}$$

where the sum is over all such graphs $g(n,\ell)$ and $\langle g\rangle$ represents the moment trace of g. Since g is in general a multigraph, one can think of it in terms of a basic skeleton with m edges, and assign multiplicities $\ell_1, \ell_2, \ldots \ell_m$ to the edges. The moment trace $\langle g\rangle$ can now be written in the form,

$$\langle g\rangle = \frac{\ell!}{\prod_i \ell_i!} \frac{1}{Z_1} \int \ldots \int \frac{1}{S_D^N} \prod_e \underline{\sigma}_i \cdot \underline{\sigma}_j \prod_{i=1}^{N} \exp h\, \sigma_i^z \, d\,\Omega_i \,, \tag{19}$$

where the combinatorial factor takes account of the distinct permutations of the edge labels. The first product is over all edges and the second is over all vertices.

We shall postpone calculation of $\langle g\rangle$ for the present, merely observing that for a separated graph $g = (g_1 \cup g_2)$,

$$\langle g\rangle = \langle g_1\rangle\langle g_2\rangle \,. \tag{20}$$

The expansion (17), (18) is the moment expansion in terms of all graphs, including separated ones. To derive the cumulant expansion, we write,

$$\ln Z_2 = \sum_{n=1}^{\infty} \frac{K^n}{n!} C_n \,, \quad \text{with} \quad C_n = \sum_g [g] \,. \tag{21}$$

The quantity $[g]$ denotes the cumulant weight of the graph g, and is related to $\langle g\rangle$ in the usual way. Namely,

$$[g] = \sum_{k=1}^{n} (-)^{k-1}(k-1)! \sum_{P(n,k)} \langle g_1\rangle\langle g_2\rangle \ldots \langle g_k\rangle \,. \tag{22}$$

The sum over $P(n,k)$ includes all partitions of the n edges of g into k subsets, represented by subgraphs $g_1, g_2 \ldots g_k$.

It can be readily shown (by induction, for instance), that the cumulant trace $[g]$ vanishes if g is disconnected. This is a consequence of (20). The expansion (21) thus involves only connected graphs. Rewriting (22) in the form

$$[g] = \langle g\rangle - \sum_{k=2}^{n} \sum_{P(n,k)} [g_1][g_2] \ldots [g_k] \,, \tag{23}$$

one can calculate the cumulant trace of a graph from its moment trace, provided the cumulant averages of its subgraphs have been calculated previously. The only remaining problem is to calculate $\langle g\rangle$.

For the Heisenberg model (D = 3), the $\underline{\sigma}_i$ are classical three dimensional unit vectors, and we can write,[5]

$$\sigma_i^z = \cos\Theta_i \,, \tag{24}$$

$$\underline{\sigma}_i . \underline{\sigma}_j = \cos\Theta_{ij} = \cos\Theta_i \cos\Theta_j + \sin\Theta_i \sin\Theta j \cos(\phi_i - \phi_j) \,.$$

In terms these angular variables, (15), (19) become,

$$Z_1 = \left[\frac{1}{4\pi} \int_0^{2\pi} \int_0^{\pi} e^{h\cos\Theta} \sin\Theta d\Theta d\phi\right]^N \text{ and,}$$

$$\langle g\rangle = \frac{\ell!}{\underset{i}{\Pi}\ell_i!}\, \frac{1}{Z_1} \int\ldots\int \frac{1}{(4\pi)^N} \underset{e}{\Pi} (\cos\Theta_i\cos\Theta_j + \sin\Theta_i\sin\Theta_j\cos(\phi_i-\phi_j) \times$$

$$\prod_{i=1}^{N} \exp(h\cos\Theta_i)\sin\Theta_i d\Theta_i d\phi_i \,. \tag{25}$$

The integrals in (25) can be decomposed into contributions from integrals of the form

$$\int_0^{2\pi} \cos^m\phi \sin^n\phi d\phi \quad\text{and}\quad \int_0^{\pi} \cos^m\Theta \sin^n\Theta\, e^{h\cos\Theta}\, d\Theta \,, \tag{26}$$

which can be evaluated, thus yielding an expansion for ℓnZ in K and h.

In practice, it is possible, by using various theorems and lemmas concerning traces, to reduce considerably, the number of graphs for which explicit calculation of traces is necessary.

Finite Cluster Approach

The development in the last section in no way assumed an infinite regular lattice. The expansions for Z and ℓnZ are equally valid for a finite cluster represented by a graph G(N,L). For such a system, we can write,

$$\ell nZ(G) = \sum_{g\leq G} \phi(g) = \ell n\left[\sum_{g'\{\ell_i\}} \frac{K^{\Sigma\ell_i}}{(\Sigma\ell_i)!} \langle g'\rangle\right] , \tag{27}$$

where $g'\{\ell_i\}$ denotes multigraphs formed by assigning multiplicites $\{\ell_i\}$ to the edges of G, zero multiplicities being allowed. The sum over g is over all simple subgraphs g of G, including G itself. $\phi(g)$ denotes the weight of g and is a function of K and h.

For any finite graph G, one can explicitly consider all g' and obtain Z(K,h). The logarithm of Z can then be expanded as a series in K and h. To calculate the weight $\phi(G)$, one has to subtract off the weights of all subgraphs g. This can be done by considering finite clusters in order of increasing size, so that at any stage, the $\phi(g)$ of subgraphs have already been calculated.

To illustrate the procedure, we consider the first few graphs that arise:

$$
\begin{aligned}
\ln Z(\bullet) &= \phi(\bullet) \\
\ln Z(/) &= \phi(/) + 2\phi(\bullet) \\
\ln Z(\wedge) &= \phi(\wedge) + 2\phi(/) + 3\phi(\bullet) \\
\ln Z(\mathcal{N}) &= \phi(\mathcal{N}) + 2\phi(\wedge) + 3\phi(/) + 4\phi(\bullet) \\
\ln Z(\curlywedge) &= \phi(\curlywedge) + 3\phi(\wedge) + 3\phi(/) + 4\phi(\bullet) \\
\ln Z(\Delta) &= \phi(\Delta) + 3\phi(\wedge) + 3\phi(/) + 3\phi(\bullet) \, .
\end{aligned}
\tag{28}
$$

By inverting (28), it is possible to calculate $\phi(G)$ in terms of K and h. We also note that for a separated graph $G = (G_1 \cup G_2)$, traces of subgraphs in the two components are independent of each other and the product property of Z is realized. Thus, $\ln Z(G)$ is simply the sum of the free energies of G_1 and G_2 and this condition together with the basic lemma introducted in the first talk, is sufficient for $\phi(G)$ to be zero. Thus the left hand side of (28) involves only connected graphs.

To extend the above treatment to a lattice $\mathcal{L}$, we limit ourselves to calculating a series expansion correct to say K^n. By considering all graphs G with up to n edges which can be embedded on the lattice, and calculating ϕ correct to K^n, we obtain,

$$\ln Z(\mathcal{L}) = \sum_G (G;\mathcal{L}) \, \phi(G) \, , \tag{29}$$

where $(G;\mathcal{L})$ denotes the weak lattice constant of G on $\mathcal{L}$, defined per site. The free energy is also defined per site.

At this stage, it would be interesting to consider in a little more detail, the calculation of the zero field free energy ($\ln Z_0$) and the zero field susceptibility (χ_0). These quantities are considerably easier to derive than the complete partition function and are very useful in their own right.

The Zero Field Limit

In calculating the zero field partition function Z_0, we make use of two results regarding traces in the no field limit.

(i) If an interaction graph g has an articulation point (o), so that we can write

$$g = g_1 \circ g_2 \, , \tag{30}$$

where g_1 and g_2 are the two star blocks joined at o,

$$\langle g\rangle_o = \langle g_1\rangle_o \langle g_2\rangle_o \quad , \tag{31}$$

where the subscript denotes trace in the zero field limit. This result follows from the fact that when h = 0, the trace involves only the relative orientations of neighbouring spin vectors. On fixing the orientation of the spin at the articulation point, the trace integral over the remaining spins factorises as in (31). Integration over the spin at the articulation point yields a numerical factor which cancels on taking the reduced trace.

This leads us to the very important result that only star graphs contribute to the expansion for $\ell n\, Z_0$.

(ii) $\langle g\rangle_o = 0$ unless every vertex of g is of even degree. This result can be derived [1] by considering the S = ∞ limit of the quantum model. An alternative approach [12] involves expanding Z_0 in terms of Bessel functions and spherical harmonics, and will be discussed shortly.

The above results can also be used to simplify the calculation of the zero field susceptibility χ_0 [7]. For a finite cluster G, we write,

$$\frac{DkT\chi_N}{m^2} = N + \sum_{i\neq j} \langle \underline{\sigma}_i . \underline{\sigma}_j \rangle \big|_{h=0} = N + \sum_{g\leq G} w(g) \quad , \tag{32}$$

where w(g) represents the susceptibility weight of g. The correlations $\langle \underline{\sigma}_i . \underline{\sigma}_j \rangle$ can be calculated very simply (see eq.12) from the zero field partition functions of finite clusters. Thus the finite cluster development for χ_0 follows naturally from that for $\ell n\, Z_0$. It can be shown that only two classes of graph contribute to the susceptibility expansion, namely, star graphs and articulated graphs derived from star graphs by deleting one edge.

Star Graph Expansions

We have already seen that the finite cluster approach leads to a star graph expansion for the zero field free energy. The susceptibility expansion, however, involves articulated graphs such as "chains" which have large lattice constants and are best avoided. This can be done by working with χ^{-1} rather than χ_0. If the reciprocal susceptibility is develOped as a finite cluster expansion, only star graphs contribute with non-zero weight.

The justification for this approach has come chiefly from pursuing the analogy with the Ising model [13]. The development for the Ising model suggests that for the classical vector model, at any rate, the product property of Z for articulated graphs can be recovered by fixing the magnetization at the articulation point. In other words, $\ell n\, Z(M,K)$ has a star graph expansion. It then follows that all derivatives of the free energy with respect to M must have star graph expansions, χ_0^{-1} being the first derivative.

An alternative method [14] for showing the existence of a star graph expansion for χ_0^{-1} is to use the fact that for an articulated graph $g = (g_1 o g_2)$, the correlation between spins at sites i and j on g_1 and g_2 factorizes in the form,

$$\langle \underline{\sigma}_i \cdot \underline{\sigma}_j \rangle = \langle \underline{\sigma}_i \cdot \underline{\sigma}_o \rangle \langle \underline{\sigma}_o \cdot \underline{\sigma}_j \rangle \; . \tag{33}$$

It can then be shown that

$$\psi(g) = \chi_0^{-1}(g) - N = \psi(g_1) + \psi(g_2) \; , \tag{34}$$

which together with the basic lemma is sufficient to establish that the weight of g is zero.

As for the Ising model, calculation of χ_0^{-1} for an inhomogeneous graph requires the inversion of the matrix of correlations. The reasons for this additional complication arise from the need to distinguish between inequivalent vertices of the graph. Details of the calculation will be discussed at a later stage.

To summarize, we have now seen that $\ell n\, Z_0$ and χ_0^{-1} can be developed as star graph expansions. Provided one can calculate Z_0 for finite star clusters, one can relate the susceptibility of star graphs to the zero field partition function of star graphs of higher cycle index. The most important problem at this stage, is to find an efficient method for calculating Z_0, a method that can be readily programmed for the computer and one which can be generalized to any value of D.

Zero Field Partition Function of Finite Star Clusters

The method I now describe is due to Joyce [12] and involves expanding the interaction term in the partition function in terms of Bessel functions and spherical harmonics. The results obtained from this approach together with certain modifications due to Domb [9,10] and coworkers, provide an almost completely automatic procedure for calculating $Z_0(G)$ for all star graphs.

In the absence of a magnetic field, the partition function Z_0 is given by

$$Z_0(G) = \int \ldots \int \prod_i^N \frac{d\Omega_i}{4\pi} \prod_{\langle ij \rangle} \exp K\, \underline{\sigma}_i \cdot \underline{\sigma}_j \, . \tag{35}$$

We now expand $\exp K\, \underline{\sigma}_i \cdot \underline{\sigma}_j$ as follows,

$$\exp K\, \underline{\sigma}_i \cdot \underline{\sigma}_j = \exp K \cos\Theta_{ij} = \left(\frac{\pi}{2K}\right)^{\frac{1}{2}} \sum_{\ell=0}^{\infty} (2\ell+1) I_{\ell+\frac{1}{2}}(K)\, P_\ell(\cos\Theta_{ij}) \, , \tag{36}$$

with

$$P_\ell(\cos\Theta_{ij}) = 4\pi(2\ell+1)^{-1} \sum_{m=-\ell}^{\ell} Y^*_{\ell m}(\Theta_j \phi_j) Y_{\ell m}(\Theta_i, \phi_i) \, .$$

The $I_{\ell+\frac{1}{2}}(K)$ are modified Bessel functions of the first kind and the $P_\ell(\cos\Theta)$ are expanded using the addition theorem for spherical harmonics. We thus derive,

$$Z_0(G) = (4\pi)^{C(G)-1} \sum_{\{\ell_{ij}\}} \prod_{\langle ij \rangle} \lambda_{\ell_{ij}}(K) \sum_{\{m_{ij}=-\ell_{ij}\}}^{\{+\ell_{ij}\}}$$

$$\int \ldots \int \prod_{i=1}^{N} d\Omega_i \prod_{\langle ij \rangle} Y^*_{\ell_{ij} m_{ij}}(\Theta_j \phi_j)\, Y_{\ell_{ij} m_{ij}}(\Theta_i \phi_i) , \tag{37}$$

with $\lambda_\ell = \left(\frac{\pi}{2K}\right)^{\frac{1}{2}} I_{\ell+\frac{1}{2}}(K)$.

A slight rearrangement of (37) enables us to express Z_0 in terms of vertex function. Denoting the set of edges incident at vertex i by δ_i, we write,

$$Z_0(G) = (4\pi)^{C(G)-1} \sum_{\{\ell_{ij}\}} \prod_{\langle ij \rangle} \lambda_{\ell_{ij}}(K) \sum_{\{m_{ij}\}} \prod_i^N \int d\Omega_i \prod_{\delta_i} Y_{\ell_{\delta_i} m_{\delta_i}}(\Theta_i \phi_i)$$

$$= \sum_{\{\ell_{ij}\}} C\{\ell_{ij}\} \prod_{\langle ij \rangle} \lambda_{\ell_{ij}}(K) \, . \tag{38}$$

The $\lambda_\ell(K)$ are easily expanded in powers of K. Calculation of the "weight" $C\{\ell_{ij}\}$ is not trivial, as it involves integration over products of spherical harmonics. Closed form expressions in terms of the Wigner 3 n-j symbols have been derived for graphs of cycle index up to four [12]. However, each integral has to be treated separately and there is no obvious way of automating the procedure to handle products of any given number of spherical harmonics. Also, to deal with values of D greater than three, one has to generalize the

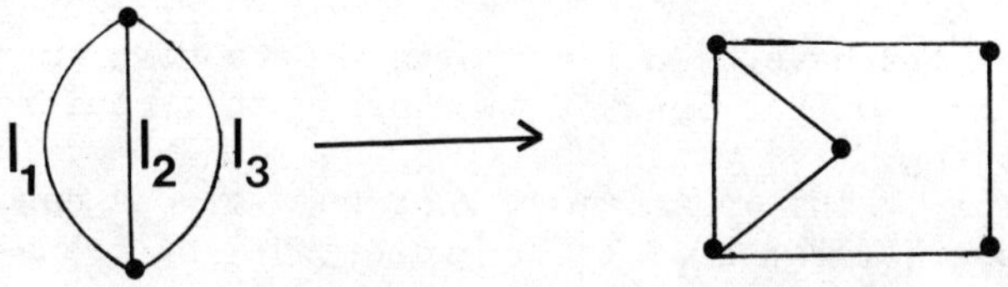

Fig. 1. Graph of theta topology.

treatment in terms of hyperspherical harmonics.

Nevertheless, several important results have emerged from this work which have been incorporated into later work. I shall briefly summarize these results before discussing more recent developments.

(i) If a graph G has vertices of degree two, $Z_0(G)$ can be calculated from the partition function of the topology obtained by suppressing all vertices of degree two.

Thus, the theta graph 1-2-3 (Fig 1) can be obtained from the topology by associating bridges 2 and 3 with two and three edges respectively. The partition function of the graph is obtained by replacing λ_{ℓ_2}, λ_{ℓ_3} in the partition function of the topology by $\lambda^2_{\ell_2}$, $\ell^3_{\ell_3}$ respectively.

This result is valid in all cases, and implies that we only need calculate Z_0 for topologies which are far fewer than the total number of graphs.

(ii) Any choice of $\{\ell_i\}$ in (38) can be represented by a "bonding" of the topology. Thus Fig 2 represents the bonding 112211 of the "alpha" topology. Corresponding to each "bonding", is a "weight" C, which is a number.

(iii) Only a subset of bondings have non-zero weight. There are several constraints on the multiplicities of bridges incident at a vertex. These can be conveniently summarized by saying that only those bondings which correspond to a superposition of simple polygons of the topology can have nonzero weignt.

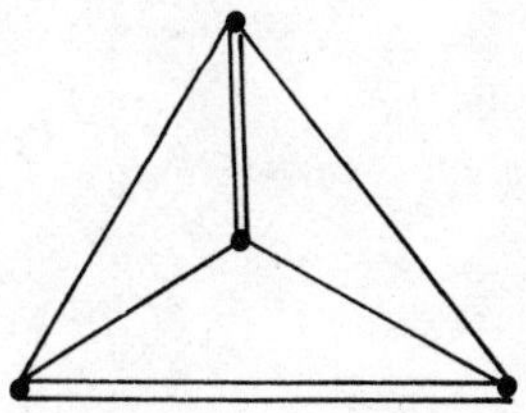

Fig. 2. Bonding of a topology.

All these results have been incorporated into work currently being done at King's College London, together with a modified procedure for calculating $C\{\ell_i\}$ which obviates the need to evaluate the integrals explicitly. The vast majority of the weights can be obtained from those of graphs of lower cycle index by means of the "ladder transformation" [9]. The few that remain can be dealt with using other techniques which have been developed. Generalization to higher values of D presents no difficulty.

The Ladder Transformation

At this stage, we need to distinguish between ladder topologies and non ladder topologies. A ladder topology is one with at least one pair of nodes connected by at least two bridges. A non ladder topology is one which has no pair of nodes joined by more than one bridge. Thus Fig 3a represents a ladder topology while Fig 3b represents a non ladder topology.

The ladder transformation relates the $C\{\ell_i\}$ for a ladder topology to the $C'\{\ell_i\}$ of the subtopology derived by deleting a multibridg Thus, the weights for the theta topology (Fig 1) can be derived from those of the polygon.

The transformation consists in replacing a portion K^* of an edge K by a pair of parallel edges K', K" (Fig 4). The problem is to relate $\lambda_\ell(K'+K'')$ to $\lambda_r(K')\ \lambda_s(K'')$. This can be done (for any value of D), by using well known properties of Bessel functions.

Generalizing the definition of λ_ℓ, we write [15],

$$\lambda_\ell = 2^\nu \Gamma(\nu+1) K^{-\nu}\ I_{\nu+\ell}(K) \tag{39}$$

with $\nu = D/2 - 1$.

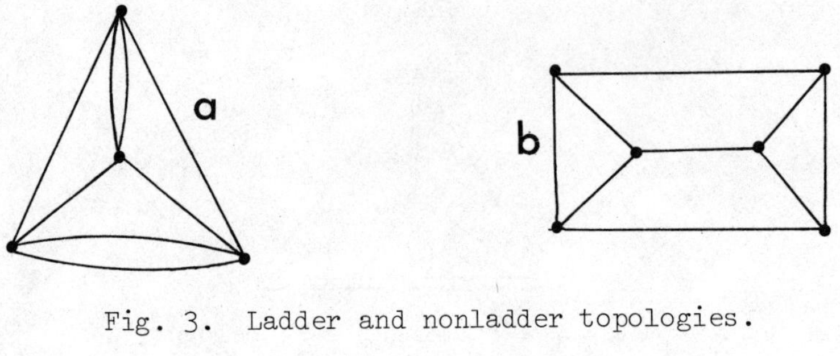

Fig. 3. Ladder and nonladder topologies.

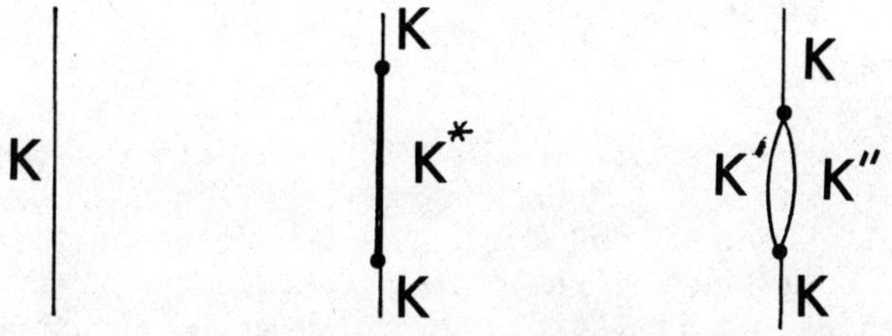

Fig. 4. Laddering.

Using the difference formulae,

$$I'_\Theta = \tfrac{1}{2}(I_{\Theta-1} + I_{\Theta+1})$$
$$\Theta I_\Theta = \frac{K}{2}(I_{\Theta-1} + I_{\Theta+1}) \; , \tag{40}$$

we can relate $\lambda_{\ell+1}$ to $\chi_{\ell-1}$, λ_ℓ. Thus,

$$\lambda'_\ell = \tfrac{1}{2}\left(\frac{\ell}{\nu+\ell}\,\lambda_{\ell-1} + \frac{2\nu+\ell}{\nu+\ell}\,\lambda_{\ell+1}\right) . \tag{41}$$

We use an expansion for $\lambda_\ell(K'+K")$ of the form

$$\lambda_\ell(K' + K") = \sum_{r,s} d^{(\ell)}_{r,s}\,\lambda_r(K')\lambda_s(K") \; , \tag{42}$$

where the $d^{(\ell)}_{rs}$ can be calculated recursively, using,

$$\frac{D+\ell-2}{D+2\ell-2}\,d^{(\ell+1)}_{rs} = \frac{r+1}{D+2r}\,d^{(\ell)}_{r+1,s} + \frac{D+r-3}{D+2r-4}\,d^{(\ell)}_{r-1,s} - \frac{\ell}{D+2\ell-2}\,d^{(\ell-1)}_{r,s} . \tag{43}$$

The simplest case, $\ell = 0$, corresponds to calculating polygon weights from a pair of isolated vertices. Since the partition function of the polygon is known exactly, we derive

$$d^{(0)}_{rr} = \frac{(D+2r-2)(D+r-3)!}{\gamma!(D-2)!} \; , \quad \text{all others zero.} \tag{44}$$

Using (43) to calculate $d^{(1)}_{rs}$, we derive,

$$d^{(1)}_{r-1,r} = d^{(1)}_{r,r-1} = \frac{(D+r-3)!}{(D-2)!(r-1)!} \; , \quad \text{all others zero.} \tag{45}$$

This enables us to calculate C(1,r,s) for the theta topology, namely,

$$C(1,2,1) = D(D-1)$$
$$C(1,2,3) = D^2(D-1)/2 \; . \qquad (46)$$

In practice, a large number of possible (r,s) are ruled out because of the constraints on $\{\ell_i\}$. Only a very small number of coefficients $d_{rs}^{(\ell)}$ are actually needed, and for any given value of D, can be generated by a computer program. To calculate the free energy series on the face centred cubic lattice to order K^{13}, only about 27 coefficients are needed.

Non-ladder Topologies, Special Methods

The weights $C\{\ell_i\}$ are of two types, those which correspond to complete terms (no $\ell_i=0$) and those which correspond to incomplete terms ($\ell_i=0$ for some i). The latter present no difficulty. They correspond to topologies of lower cycle index and can be assumed known. The problem is to calculate $C\{\ell_i\}$ when no ℓ_i is zero. Two techniques are available for dealing with this situation.

(i) Collapse of a bond

A partition interaction is allowed to tend to infinity, so that the ends of that bond coalesce, and one obtains a new topology whose weights can be related to those of the original topology. On allowing the bridge K_1 of G (Fig 5) to collapse, one derives a new topology G^* which is a ladder topology whose weights are known. It can be shown that,

$$C(\ell_2,\ell_3\ldots\ell_6)_{G^*} = \sum_{\ell_1=0}^{\infty} C(\ell_1,\ell_2,\ell_3\ldots\ell_6)_G \; . \qquad (47)$$

Since only the first few terms on the right hand side of (47) are nonzero, they can usually be calculated unambiguously. In any case, by a careful choice of K_1, one can obtain at least a relation between the unknown $C\{\ell_i\}$, and use other methods to resolve the ambiguity.

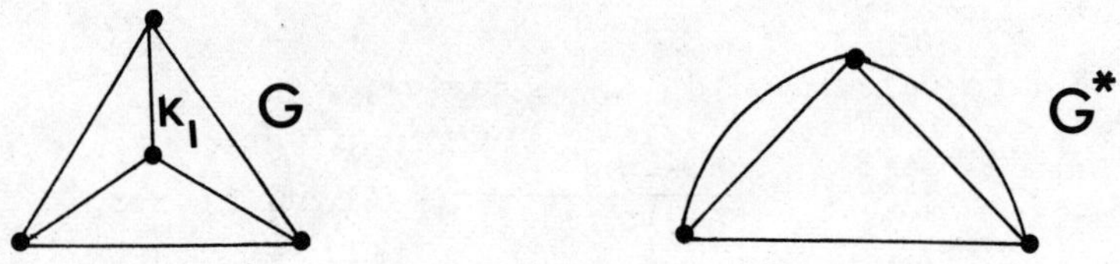

Fig. 5. Collapse of a bond.

(ii) Direct averaging

We use the fact that the following two definitions of $Z_0(G)$ are equivalent

$$\left\langle \prod_{<ij>} \sum_{r=0}^{\infty} \frac{K_{ij}^{r}}{r!} \left(\sum_{\alpha=1}^{D} \sigma_{\alpha i} \sigma_{\alpha j} \right)^{r} \right\rangle = \sum_{\{\ell_i\}} C(\ell_{ij}) \prod_{<ij>} \lambda_{\ell_{ij}}(K), \quad (48)$$

where α denotes the component of spin.

A term corresponding to $K_1^a K_2^b \ldots K_r^h$ on the LHS is the average of a product,

$$\frac{(\sigma_{\alpha i}\sigma_{\alpha j})^{a}}{a!} \frac{(\sigma_{\beta_i}\sigma_{\beta_k})^{b}}{b!} \cdots \frac{(\sigma_{\nu_m}\sigma_{\nu_p})^{h}}{h!}, \quad (49)$$

which can be treated as a colouring problem on a bonding of G, a choice of D colours being allowed for each bond. For each colouring, one averages over each vertex and takes the product (49). It can be shown that a vertex of degree two has an average of the form A/D, a vertex of degree four has an average of the form B/D(D+2) and so on, whatever the colouring. Working with the form of the averages and treating A, B ... as unknowns, one can obtain a relation between the two sides of (49). Using known results for particular values of D, one can usually calculate the required coefficients unambiguously.

In practice, a combination of all these techniques proves adequate for all cases. This completes the procedure for all star graphs. To summarize,

(a) Ladder topologies: starting with the polygon (n=1), build up the set of all $C(\ell_i)$ for topologies of cycle index n from those of cycle index n-1, using the transformation coefficients $d_{rs}^{(\ell)}$.

(b) Non ladder topologies: calculate complete terms by using the techniques described above. Incomplete terms correspond to bondings of topologies of lower cycle index.

(c) To calculate Z_0 for a particular graph G of a given topology, make the substitution

$$\prod_{<ij>} \lambda_{\ell_{ij}}(K) \to \prod_{<ij>} \lambda_{\ell_{ij}}^{m_{ij}}(K), \quad (50)$$

where m_{ij} represents the number of edges of G corresponding to the bridge <ij> of the topology.

(d) Expand the $\lambda_{\ell_{ij}}$ in powers of K to obtain,

$$Z_0(G) = \sum_{r=0}^{rmax} a_r K^r \; . \tag{51}$$

To calculate the zero field free energy, we use the relation,

$$\ell n\, Z_0(G) = \sum_{G'\leq G} \phi(G') \; , \tag{52}$$

to compute $\phi(G)$ correct to order K^{rmax}. Considering all G with up to rmax edges which can be embedded on a lattice $\mathcal{L}$, we write, for the free energy of $\mathcal{L}$,

$$\ell n\, Z_0(\mathcal{L}) = \sum_G (G;\mathcal{L})\phi(G) \; . \tag{53}$$

The zero field specific heat C_0 is obtained by differentiating (53) with respect to K (eq.9).

Zero-field Susceptibility

As indicated before, the reciprocal susceptibility χ_0^{-1} can be developed as a finite cluster expansion,

$$\chi_0^{-1}(\mathcal{L}) = \sum_G (G;\mathcal{L})h_G(K) \; , \tag{54}$$

where $h_G(K)$ is nonzero only for star graphs. For a homogeneous star cluster G, we use (11), (12) to relate $\chi_0(G)$ to the zerofield partition function of star clusters of higher cycle index, and calculate $h_G(K)$ using (54) in the form,

$$\chi_0^{-1}(G) = \sum_{G'\leq G} h_{G'}(K) \; . \tag{55}$$

For inhomogeneous clusters, we need to distinguish between different types of vertex and we follow a procedure very similar to that used for the Ising model. Defining a matrix M of spin correlations, we write,

$$M_{ij} = 1/D \frac{1}{Z_0(G)} \frac{\partial Z_0(G_{ij})}{\partial K^*_{ij}} \Big|_{K^*_{ij}=0} = \langle\sigma_i^z \sigma_j^z\rangle \; , \qquad M_{ii} = 1 \; . \tag{56}$$

G_{ij} denotes a graph obtained from G by joining vertices i and j with an edge K^*. The differentiation with respect to K^*_{ij} in the limit $K^*_{ij} = 0$, corresponds to selecting from the bondings of G_{ij}, only those terms with $\ell_{ij} = 1$. On expanding the λ_ℓ in K, one obtains M_{ij} as polynomials in K, correct to K^{rmax}. The inverse susceptibility is then given by,

$$\chi_0^{-1}(G) = \underline{1}^T \underline{M}^{-1} \underline{1} = \sum_{G' \leq G} h_{G'}(K) \ . \qquad (57)$$

Use of (54) gives the reciprocal susceptibility series for the lattice.

Computational Details

In practice, it proves more convenient to work with the variable $w_1 = \lambda_1/\lambda_0$, rather than K. Defining $w_r = \lambda_r/\lambda_0$, we note that $w_r \simeq w_1^r + O(w_1^{r+2}) \simeq K^r + O(K^{r+2})$. Another small departure from (56), (57) is to define,

$$M'_{ii} = Z_0, \quad M'_{ij} = Z_0 \ M_{ij} \ ,$$
$$\chi_0^{-1} = Z_0 \ \underline{1}^T(\underline{M}')^{-1} \ \underline{1} \ . \qquad (58)$$

Worked Examples (D=3)

We first consider the bond (Fig 6a) for which

$$Z_0 = 1, \quad Z_0(G_{12}) = 1 + 3w_1w_1^* + 5w_2^*w_2 + \ldots$$

and $M_{12} = M_{21} = w_1$.

$$\chi_0^{-1}(/) = 2 - 2w_1 + 2w_1^2 - 2w_1^3 + 2w_1^4 - \ldots$$
$$-2h(\cdot) = -2. \qquad (59)$$
$$h(/) = -2w_1 + 2w_1^2 - 2w_1^3 + 2w_1^4 - \ldots$$

For the triangle (Fig 6b), we have,

$$Z_0 = 1 + 3w_1^3 + 1.08w_1^6 + \ldots$$
$$M_{12} = M_{13} = M_{23} = w_1 + w_1^2 + 1.2w_1^4 + \ldots$$
$$\chi_0^{-1}(\Delta) = Z_0(\underline{1}^T\underline{M}^{-1}\underline{1}) = 3 - 6w_1 + 6w_1^2 - 1.2w_1^4 + \cdots$$
$$-3h(\cdot) - 3h(/) \qquad = -3 + 6w_1 - 6w_1^2 + 6w_1^3 - 6w_1^4 + \cdots$$
$$h(\Delta) = 6w_1^3 - 7.2w_1^4 + \cdots \qquad (60)$$

Finally, for the square (Fig 6c),

$$Z_0 = 1 + 3w_1^4 + \ldots$$

$$M_{12} = M_{14} = M_{23} = M_{34} = w_1 + w_1^3 + \ldots$$

$$M_{13} = M_{24} = 2w_1^2 + \ldots$$

$$\chi_0^{-1}(\square) = 4 - 8w_1 + 8w_1^2 - 8w_1^3 + 16w_1^4 + \ldots$$

$$-4(\bullet) - 4(/) = -4 + 8w_1 - 8w_1^2 + 8w_1^3 - 8w_1^4 + \ldots$$

$$h(\square) = 8w_1^4 + \ldots \; . \qquad (61)$$

For the FCC lattice,

$$(\bullet;\mathcal{L}) = 1; \quad (/;\mathcal{L}) = 6$$

$$(\triangle;\mathcal{L}) = 8; \quad (\square;\mathcal{L}) = 33.$$

Thus,

$$\chi_0^{-1}(\mathrm{FCC}) = 1 - 12w_1 + 12w_1^2 + 36w_1^3 + 218.4w_1^4 + \ldots$$

and

$$\chi_0 = 1 + 4K + 14\tfrac{2}{3}K^2 + 51.7333K^3 + 178.459259K^4 + \ldots \; .$$

Only three graphs (excluding the single site) have been considered to this order. The advantage of using the star graph expansion becomes more apparent at higher orders, as was seen in the first lecture. The method outlined here has been used to extend the susceptibility series on various three and four dimensional lattices and the new series coefficients will be published when extrapolation studies have been completed.

Conclusion

The main object of these lectures was to give an account of some of the commonly used methods of deriving high temperature series

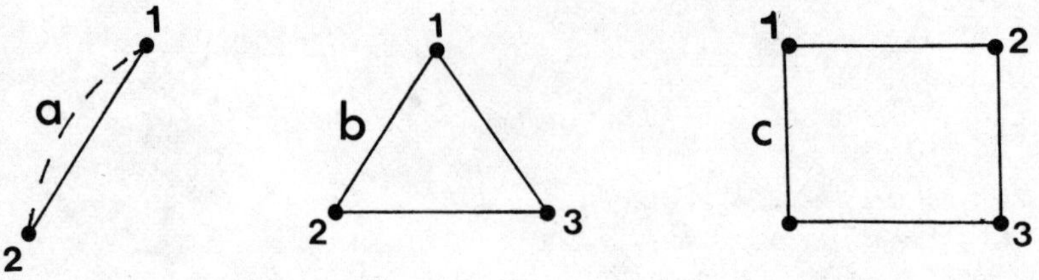

Fig. 6. Star graphs for χ_0^{-1}.

expansions. The treatment has been essentially pedagogical and I have not tried to ensure a strictly historical development. The list of references has been kept to a minimum, but the interested reader should find all relevant papers cited in those references.

References

1 G.S. Rushbrooke, G.A. Baker Jr. and P.J. Wood in "Phase Transitions and Critical Phenomena", Vol.3, Eds. C. Domb and M.S. Green (London: Academic Press) Ch.5 (1974).

2 D.D. Betts, in "Phase Transitions and Critical Phenomena", Vol.3, Eds. C. Domb and M.S. Green (London: Academic) Ch.8 (1974).

3 H.E. Stanley, in "Phase Transitions and Critical Phenomena" Vol.3, Eds. C. Domb and M.S. Green (London: Academic) Ch.7 (1974)

4 G.S. Rushbrooke and P.J. Wood, Mol.Phys. 1, 257 (1958).

5 R.L. Stephenson and P.J. Wood, Phys.Rev. 173, 475 (1968).

6 H.E. Stanley, Phys.Rev. 158, 537, 546 (1967).

7 G.S. Joyce and R.G. Bowers, Proc.Phys.Soc.(London) 88, 1053 (1966).

8 M. Ferer, M.A. Moore and M. Wortis, Phys.Rev.B 4, 3954 (1971).

9 C. Domb, J.Phys.C 5, 1417 (1972).

10 C. Domb, J.Phys.A 9, 983 (1976).

11 P.S. English, D.L. Hunter and C. Domb, J.Phys.A 12, 2111 (1979).

12 G.S. Joyce, Phys.Rev. 155, 478 (1967).

13 C. Domb, J.Phys.A 7, L45 (1974).

14 D.C. Rapaport, J.Phys.A 7, 1918 (1974).

15 H.E. Stanley, Phys.Rev. 179, 570 (1969).

THE PROBLEM OF CONFLUENT SINGULARITIES

Bernie G. Nickel†

Physics Department
University of Guelph
Guelph, Ontario, Canada N1G 2W1

I INTRODUCTION

The numerical verification of the validity of universality and hyperscaling from series expansions of model systems cannot be separated from the question of the existence of confluent singularities. Confluent singular behavior in the critical region is expected on the basis of a formal analysis[1] of Wilson's renormalization group picture but the problem that remains from the point of view of the series mechanician is that series coefficients represent "experimental" data and the exponential fitting of data is known to be horribly unstable. Even if the correlation function F of interest is believed to be well approximated near the critical point $x = x_c$ by $F(x) = C_1(x_c - x)^{-\gamma_1} + C_2(x_c - x)^{-\gamma_2}$ many values of C_1, γ_1, C_2, and γ_2 will fit perfectly well if the data on F(x) is noisy and does not cover a sufficiently wide range of x. For lattice models we can choose x as the correlation length ξ and then note the accessible range is bounded below by a lattice spacing and above by the maximum distance sampled by the lattice graph embeddings contributing to the series. At best this represents a decade in ξ and one and one-half decades in temperature for three dimensional models. Furthermore, since the theoretical data is "noisy", corresponding to the observed irregular behavior of lattice embedding constants with order, I do not find it surprising that the debate concerning the correct values for critical exponents is continuing.

In these lecture notes I will discuss three separate topics

†Alfred P. Sloan Fellow

that relate to the question of confluent singularities. The first is an analysis, by standard ratio and Padé methods, of recently obtained high temperature series for the body-centered cubic (B.C.C.) lattice for both the susceptibility χ and second moment correlation length squared M_2/χ, for spin ½, 1, 2, and ∞. There is clear evidence for a confluent singularity in the spin ½ model and reasonable evidence for universality, both among the various spin models and with the field theoretic ϕ^4 model, of the susceptibility and correlation length exponents γ and ν. Although the analysis of these series is not yet completed, it appears unlikely that very precise exponent estimates can be given without extra information as input. This simply confirms the idea that exponential fitting is basically unstable against correlated changes in the parameters.

Second, since it now appears likely that universality and hyperscaling are satisfied it is useful to define a renormalized coupling constant analog y as described by Nickel and Sharpe.[2] High temperature series can then be analysed with y as the independent variable either by Dlog Padé or, and probably more correctly, by second order homogeneous differential approximants as described by Rehr.[3] In either case, y plane analysis has the virtue that the leading divergences in the correlation functions used in the definition of y have been eliminated and one can therefore concentrate directly on the terms responsible for corrections to scaling. This analysis will yield both exponents and amplitudes which can be fed back into conventional temperature plane analysis to eliminate much of the instability problem.

Third, because it appears likely that the "hidden" confluent singularities in short high temperature series have led to erroneous exponent estimates one can legitimately ask whether a similar problem plagues the field theoretic ϕ^4 model analysis. In an attempt to answer this question I present an analysis, as described by Le Guillou and Zinn-Justin,[4] of the ϕ^4 model in one dimension where confluent singularities are known to be present and indeed to be large. I am almost certain that confluent singularities, even if present with significant amplitude in the three dimensional model, will never be detected by series expansion methods. It seems that if the exponents as obtained from the continuum ϕ^4 model are indeed correct, this is the result of an accident and is not indicative of the utility of field theoretic expansions.

I do not mean to imply in these lectures that the question of universality and hyperscaling has been resolved. One should continue to check for violations but these checks can only be realistically made with much longer series than are presently available. In particular there is an urgent need for extended series for the second field derivative of the susceptibility χ'' on the B.C.C., preferably for general spin S, but certainly for spin ½. Extension of series data for lattices with low coordination number, e.g. simple cubic

(S.C.) and tetrahedron, is also important since Padé estimates for γ for these lattices have consistently been closer to 1.25[5,6] than the estimates for the B.C.C. and face-centered cubic (F.C.C.).

II HIGH TEMPERATURE SERIES: TEMPERATURE ANALYSIS

Second order homogeneous differential approximants[7] will in principle handle a simple confluence of singularities, i.e. $F(x) = (x_c - x)^{-\gamma_1} A_1(x) + (x_c - x)^{-\gamma_2} A_2(x)$ where $A_i(x)$ are analytic at $x = x_c$. However, because of the expected instability of exponential fitting I believe it is preferable to start with simpler methods of analysis and ask whether there is evidence for the existence of confluent singular behavior without attempting to determine specific details. Thus in the following I apply both ratio and Dlog Padé methods to extended high temperature series; notably for the B.C.C. for which I have recently obtained χ and ξ^2 series to order 21.

As an aside, to give some feeling for the effort required for the new series generation an outline is given below; specific details are of limited interest and will be published elsewhere. The order 21 B.C.C. terms were generated by the Wortis free embedding scheme[8] which starts with a table of elementary 2-rooted 2-skeletons generated by the Heap rules.[9] To obtain this table required about 12 hrs CPU on the University of Guelph Amdahl V/5 and corresponds to slightly over one-half of the CPU time for the entire project. The second stage in the calculation is the replacement of every bond in the 2-skeletons by the full 2-point correlation function, and the evaluation of the embedding constants of the resulting graphs on the lattice. This is normally the most time consuming part of all series calculations but the combination of free embedding rules and B.C.C. lattice reduces this problem to about 3 hrs CPU. The essential simplification is that the unrestricted summation over nearest neighbor vectors, $\sum_{\vec{\delta}_i}$, factorizes into $\left(\sum_{\delta_i^x = \pm 1}\right)\left(\sum_{\delta_i^y = \pm 1}\right)\left(\sum_{\delta_i^z = \pm 1}\right)$ and hence any graph embedding constant is simply the cube of the embedding constant on a one dimensional chain. The remainder of the calculation is dominated by the sorting of large arrays of graphical entries rather than the algebraic manipulations required to generate the nodal and ladder contributions to the 2-point correlation function. In fact, the principle limiting factor in the present calculation is the sheer volume of data storage which amounts to nearly three million entries at the final vertex renormalized stage and several times this number at intermediate stages. An order 23 calculation is expected to require about 20 times the resources of the present order 21 and is probably not feasible. As a final but very important comment: although the calculation is totally automated one must guard against subtle coding errors and thus some kind of checking procedure is vital. With the present method, because the

B.C.C. is only one of a family of lattices in various dimension d, complete sum rule checks involving every term in the graph table are possible by simply setting d = 0 (spin pair), d = 1 (linear chain), and d = 2 (simple quadratic (S.Q.)).

II-A Ratio Tests

The classic ratio method remains a very useful tool for analysing high temperature lattice series. I will first describe below the particular variation of the method used by Gaunt and Sykes[10] for directly determining critical exponents. I will then discuss two modifications which help disentangle the asymptotic behavior of the estimates, that arise specifically from possible confluent singularities, from the short series "noise" and from the expected analytic corrections. Finally I test the modified method on the 34 term spin ½ susceptibility series for the S.Q. lattice and then apply it to an analysis of the 15 term F.C.C. spin ½ susceptibility and the 21 term B.C.C. general spin S susceptibility and correlation length series.

Suppose a function F has the expansion

$$F = \sum f_n x^n \tag{2.1}$$

$$\approx A\left(1 - \frac{x}{x_c}\right)^{-\gamma}\left\{1 + A_1\left(1 - \frac{x}{x_c}\right) + A_2\left(1 - \frac{x}{x_c}\right)^2 + \ldots + A_\Delta\left(1 - \frac{x}{x_c}\right)^\Delta + \ldots\right\}$$

where the last form is valid for x near the unknown x_c. We assume in the following that all other singularities of F lie outside the circle $|x| = x_c$. The ratios $R_n \equiv f_n/f_{n-1}$ will give estimates of x_c to $O(n^{-1})$ but this can be improved by the use of the extrapolated ratio $\hat{R}_n$ defined by

$$\hat{R}_n \equiv n\, R_n - (n - 1)\, R_{n-1} \tag{2.2}$$

$$\approx \frac{1}{x_c}\left\{1 + (\gamma - 1)\, A_1\, n^{-2} + \ldots + \Delta^2\, A_\Delta \frac{\Gamma(\gamma)}{\Gamma(\gamma - \Delta)}\, n^{-(\Delta + 1)} + \ldots\right\}$$

Estimates of the exponent γ can be obtained from the normalized slope

$$S_n \equiv \hat{R}_n^{-1}\left(R_n - R_{n-1}\right)\Big/\left(n^{-1} - (n - 1)^{-1}\right) \tag{2.3a}$$

$$\approx (\gamma - 1)\left(1 - 2A_1\, n^{-1} + \ldots\right) - \Delta(1 + \Delta)\, A_\Delta \frac{\Gamma(\gamma)}{\Gamma(\gamma - \Delta)}\, n^{-\Delta} + \ldots$$

or the extrapolated slope

$$\hat{S}_n \equiv n\, S_n - (n-1)\, S_{n-1}$$

$$\approx (\gamma - 1)\left\{1 - \left[2A_1(\gamma - 3) - 6A_2(\gamma - 2) + 3A_1^2(\gamma - 1)\right] n^{-2} + \ldots \right.$$

$$\left. - \Delta(1 - \Delta^2)\, A_\Delta \frac{\Gamma(\gamma)}{\Gamma(\gamma - \Delta)}\, n^{-\Delta} + \ldots \right. \qquad (2.3b)$$

in which the $O(n^{-1})$ corrections have been eliminated.

In practice any attempt to obtain γ from successive slopes via eqn. 2.3 is complicated because $n^{-\Delta}$ is a slowly varying function if $\Delta \approx \frac{1}{2}$ as expected from renormalization group calculations. It is quite possible that interference from the analytic corrections proportional to n^{-k} will lead one to misinterpret all or part of the confluent term as a constant and thus lead to erroneous γ. Furthermore, the first few terms of all lattice expansions are rather erratic; whether these fluctuations correspond to singularities in F in the complex plane or just to polynomial corrections is immaterial for ratio analysis since in either case one is stuck with something resembling noise. The modifications to the ratio method we describe below are attempts to deal with both these problems.

If the function F does not have any zeros for $|x| \leq x_c$ then the effect of the analytic correction terms involving A_1, A_2, ... can very nearly be eliminated by simply applying the ratio method to the function F^{1/γ_a} where γ_a is a good guess for the true exponent γ. Our n^{th} order guess for the exponent is then either

$$\gamma_n \equiv \gamma_a\left(1 + S_n\left(F^{1/\gamma_a}\right)\right)$$

$$\approx \gamma - \frac{\Delta(1 + \Delta)}{\Gamma(1 - \Delta)} A_\Delta\, n^{-\Delta} + \ldots \qquad (2.4a)$$

or

$$\hat{\gamma}_n \equiv \gamma_a\left(1 + \hat{S}_n\left(F^{1/\gamma_a}\right)\right)$$

$$\approx \gamma - \frac{\Delta(1 - \Delta^2)}{\Gamma(1 - \Delta)} A_\Delta\, n^{-\Delta} + \ldots \qquad (2.4b)$$

In the asymptotic expressions for large n in eqn. 2.4 we have dropped all corrections proportional to $\gamma - \gamma_a$; in practice estimates of γ are insensitive to changes in γ_a of the order of the uncertainty in γ. Note that use of the extrapolated $\hat{\gamma}_n$ does not eliminate the

$n^{-\Delta}$ correction term but does reduce its magnitude by a factor $1-\Delta$ relative to γ_n. Thus which form of eqn. 2.4 we use depends on the application; if we wish to determine γ then it is important to reduce the correction terms and hence 2.4b is appropriate. If γ is known and we wish to estimate the corrections then 2.4a might be more useful.

The series "noise" problem is more difficult to treat. Since we do not understand the origin of the fluctuations in the series coefficients we cannot do better than to assume the fluctuations are random from term to term. Then a reasonable, and hopefully unbiased, procedure is to average over several consecutive terms. The choice of width of averaging distribution must be subjective; if too narrow the noise remains, if too broad we have discarded useful information. A final constraint on the averaging procedure results in that we are dealing with a finite number of terms and this suggests that a binomial weighting function be used.

The particular family of averages we use below is generated by Euler transformations dependent on a parameter α. Let

$$y = x/(1 + \alpha x/x_c) , \tag{2.5}$$

then

$$F = \sum_n f_n x^n = \sum_n \left\{ \sum_m \binom{n-1}{n-m} \left(\frac{\alpha}{x_c}\right)^{n-m} f_m \right\} y^n \tag{2.6}$$

and the coefficient of y^n depends only on f_m with $m \leq n$. Since for large n, $f_n \simeq x_c^{-n}$, the coefficient of y^n is dominated by f_m with m centered at $\bar{m} \approx n/(1 + \alpha)$ and lying in a range $\Delta m \approx \pm \bar{m}\sqrt{\alpha/n}$. As observed by Gaunt and Sykes,[10] the choice $\alpha = 1$ very effectively reduces the series noise but from the above discussion it is also clear that they used effectively only one-half the available series coefficients. Furthermore, under Euler transformation the asymptotic estimate for $\hat{\gamma}_n$ becomes

$$\hat{\gamma}_n(\alpha) \approx \gamma - \frac{\Delta(1 - \Delta^2)}{\Gamma(1 - \Delta)} A_\Delta \left(\frac{1 + \alpha}{n}\right)^{\Delta} + \ldots \quad . \tag{2.7}$$

Thus to minimize the correction terms we want $1 + \alpha$ as small as possible and the choice $\alpha = 1$ is not optimal from this point of view.

We have tried two procedures to obtain "optimal" results. In the first we simply make the usual $\hat{\gamma}_n$ vs n^{-1} plots for fixed α and reduce α to the point where the noise just becomes tolerable. For the F.C.C. with 15 series coefficients one can apparently use $\alpha \gtrsim -0.2$ although the interpretation of the results for negative α is obscure. For the B.C.C. with 21 series coefficients we require $\alpha \gtrsim 0.3$ to move the antiferromagnetic singularity minimally away from the origin. The second procedure is based on the observation that the effective inverse number of terms after averaging is $\bar{m}^{-1} = (1 + \alpha)\, n^{-1}$. Thus we fix n at the maximum available series coefficients, plot $\hat{\gamma}_n(\alpha)$ vs $1 + \alpha$, and obtain γ by extrapolation to the point $1 + \alpha = 0$. The advantage of this second method lies in the coupling of the effective number $\bar{m} \approx n/(1 + \alpha)$ and the dispersion $\Delta m \approx \sqrt{\alpha n}/(1 + \alpha)$. That is, at least in the range $\alpha < 1$, the averaging width $\sqrt{\alpha n}/(1 + \alpha)$ has the useful qualitative feature of being broad in the large α (small $\bar{m}$) regime where the series noise is typically large and narrow in the small α (large $\bar{m}$) regime where the noise is small.

Extended series for the S.Q. 2-point function can be obtained fairly simply by the use of the integral formulae given by Wu, et al.[11] The first 35 coefficients of the expansions of $\chi = \sum_i \langle S_o S_i \rangle$ and $M_2 = \sum_i (r_i/a)^2 \langle S_o S_i \rangle$ in $v = \tanh K$ are as follows.

χ(S.Q.): 1, 4, 12, 36, 100, 276, 740, 1972, 5172, 13492, 34876, 89764, 229628, 585508, 1486308, 3763460, 9497380, 23918708, 60080156, 150660388, 377009364, 942106116, 2350157268, 5855734740, 14569318492, 36212402548, 89896870204, 222972071236, 552460084428, 1367784095156, 3383289570292, 8363078796612, 20656054608404, 50987841944612, 125771030685740

M_2(S.Q.): 4, 32, 164, 704, 2708, 9696, 32948, 107648, 340916, 1052960, 3185188, 9468480, 27729316, 80168352, 229179140, 648697984, 1820052468, 5066498144, 14004100644, 38461119936, 105017024900, 285226504608, 770910141140, 2074298686976, 5558291406068, 14837038975136, 39464550391748, 104623143422144, 276505869141556, 728657760246752, 1914976819498948, 5019905137386368, 13127567836074788, 34252265131016096, 89179462091122052 (2.8)

The estimates of $\hat{\gamma}_n$ for χ from eqn. 2.4b are shown as a function of Euler transform parameter $1 + \alpha$ in Fig. 1 for various n. Series "noise" prevents any estimate of γ to much better than 0.01 for

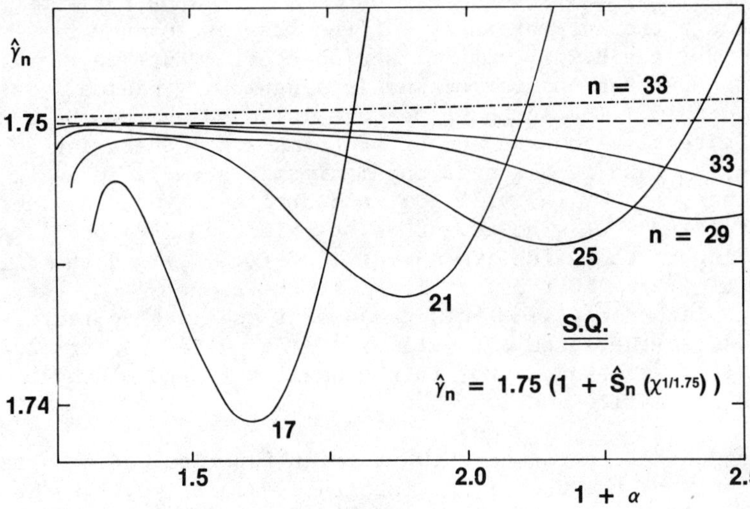

Fig. 1. Ratio method susceptibility exponent estimates from 2.4b as a function of Euler transform parameter for the S.Q. lattice keeping a varying number of series terms. The chain curve is the asymptotic estimate 2.7 for n = 33 using the data of Ref. 12.

$n \lesssim 17$. Only for $n \gtrsim 21$ are there relatively flat regions in the curves which could be interpreted as "asymptotic" and used to extrapolate the estimates to $1 + \alpha = 0$. The confluent singular corrections are believed to be small in this case; if we accept the functional form for χ guessed by Sykes, et al[12] then $\Delta = \gamma = 1.75$, $A_\Delta \approx -0.095$ and the correction to γ given by the asymptotic form 2.7 with n = 33 is less than 0.0006 for $1 + \alpha \lesssim 2$.

The application of the method to the F.C.C. χ series obtained by McKenzie[13] is shown in Figs. 2 and 3. There is a dramatic reduction in the "noise" in the Euler transform plot as compared with the standard n^{-1} plot although the valley and peak at $n \approx 6$ and 9 in Fig. 2 are still discernable in Fig. 3 at a slightly shifted $\bar{m} \approx 7$ and 10. Furthermore, an additional valley and peak at $\bar{m} \approx 13$ and 17 can be seen in Fig. 3 superimposed on what looks like a uniform downward drift with increasing $\bar{m}$. An entirely reasonable fit to the asymptotic form 2.7 is obtained assuming $\Delta = 0.5$ and $\gamma = 1.235$. The fit is certainly subjective; even with Δ fixed at 0.5 one could obtain equally plausible fits with $\gamma = 1.24$ or 1.23. The choice $\gamma = 1.25$ is unreasonable for any choice of Δ as concluded also by McKenzie.[14]

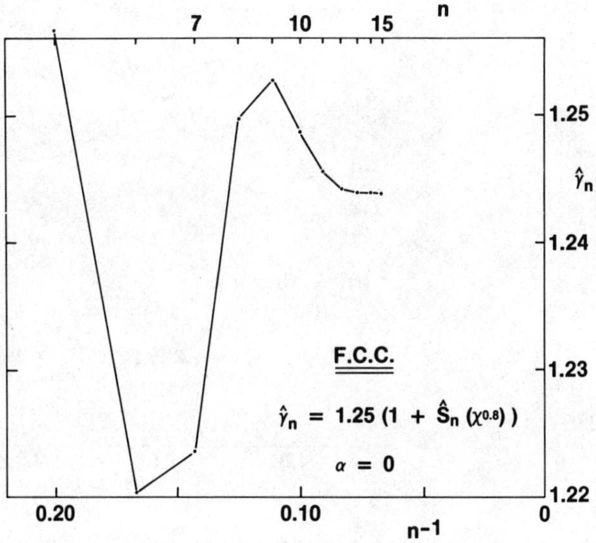

Fig. 2. Ratio method γ estimates from 2.4b for the F.C.C. χ series as a function of inverse number of terms with no Euler transformation.

All analysis for the B.C.C. given below has been performed with inverse temperature K as the independent variable since Wortis free embedding expansions[8] yield directly, for general spin S, series in K. These series are available from the author; for the spin ½ model the series in v = tanh K are

χ(B.C.C.): 1, 8, 56, 392, 2648, 17864, 118760, 789032, 5201048, 34268104, 224679864, 1472595144, 9619740648, 62823141192, 409297617672, 2665987056200, 17333875251192, 112680746646856, 731466943653464, 4747546469665832, 30779106675700312, 199518218638233896

M_2(B.C.C.): 8, 128, 1416, 13568, 119240, 992768, 7948840, 61865216, 470875848, 3521954816, 25965652936, 189180221184, 1364489291848, 9757802417152, 69262083278152, 488463065172736, 3425131086090312, 23896020585393152, 165958239005454632, 1147904794262960384, 7910579661767454248 (2.9)

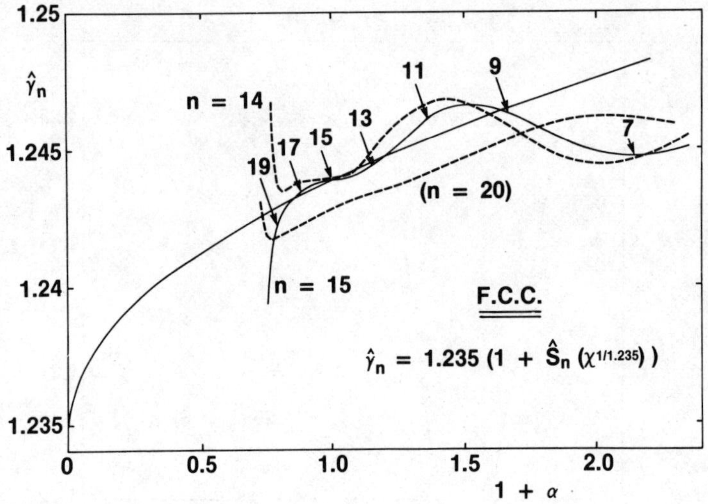

Fig. 3. Ratio method γ estimates from 2.4b for the F.C.C. χ series as a function of Euler transform parameter. The n = 15 curve is labelled by the effective number $\bar{m} = n/(\alpha + 1)$. The dash curve labelled n = 20 is based on a "fictitious" model series of 20 terms generated from the F.C.C. χ series by a particular approximant in the recursion scheme of Ref. 7 for which $v_c = 0.101716$, $\gamma = 1.2346$, $\Delta = 0.4366$. Note the near absence of "noise". The solid smooth curve is the asymptotic estimate 2.7 with $\Delta = 0.5$.

The data for the B.C.C. spin ½ χ is shown in Fig. 4 as a $\hat{\gamma}_n$ vs n^{-1} plot with $\alpha = 0.4$ just large enough to damp the anti-ferromagnetic oscillations. In this case the clear drift in the data suggests that with $\Delta = 0.5$, $\gamma = 1.237$ is a reasonable estimate. This estimate can be made more convincing if we force universality in a comparison of the general spin S models as shown in Fig. 5. Again, the fits are subjective and in drawing the asymptotic family of parabolas I have assumed the small α regime is the most significant as noted for the S.Q. test case in Fig. 1. Complete consistency cannot be achieved; on the basis of curve fitting alone a larger $\gamma \approx 1.239$ would be the better choice but eqn. 2.7 also predicts how the curves should collapse with increasing n and the observed drifts between n = 20 and 21 suggests $\gamma \approx 1.235$. Note also that I have assumed $\Delta = 0.5$ although a comparison of the predicted collapse rate with increasing n with that observed at $\alpha = 0.5$ would have suggested this value.

Fig. 6 displays the data for $(\xi/a)^2 \propto M_2/\chi$ similar to Fig. 5 for χ. Any unbiased estimate for 2ν is rather uncertain but if we rely on the universality of correction to scaling amplitude ratios[15] then the choice $\gamma = 1.237$ forces $2\nu = 1.259$ as shown. Note in this case the rate of collapse of the curves in going from n = 19 to 20 suggests $\Delta \approx 0.65$ rather than the assumed 0.5.

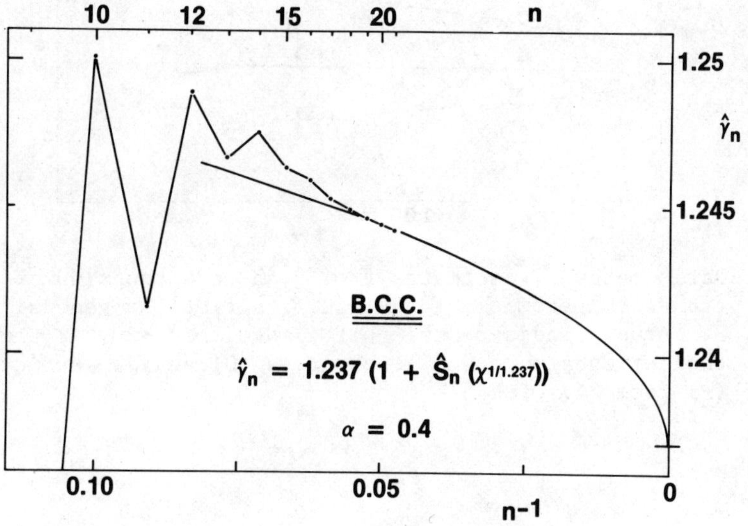

Fig. 4. Ratio method γ estimates from 2.4b for B.C.C. χ series and asymptotic parabola from 2.7 with Euler transform parameter fixed at $\alpha = 0.4$.

I conclude on the basis of ratio analysis that very precise estimates of γ and ν cannot be given but that at least within the ranges $\gamma = 1.237 \pm 0.005$ and $2\nu = 1.259 \pm 0.005$ reasonable fits to the series can be obtained. Similarly Δ is very uncertain; $\Delta = 0.5$ is reasonable, $\Delta \approx 0.6$ cannot be excluded. On the other hand, with the assumption of universality we can conclude rather precisely the ratio

$$A_\Delta(\xi)/A_\Delta(\chi) \;=\; 1.0 \tag{2.10}$$

which is larger by a factor 2 than the order ε prediction.[15] It is also larger than the order ε^2 result[16] and the high temperature series estimate[17] of 0.7 based on 12 terms. We can also deduce, rather precisely, that if $\Delta = 0.5$ then

$$A_{0.5}(\chi, S=\infty) - A_{0.5}(\chi, S=\tfrac{1}{2}) \;=\; 0.16 \tag{2.11a}$$

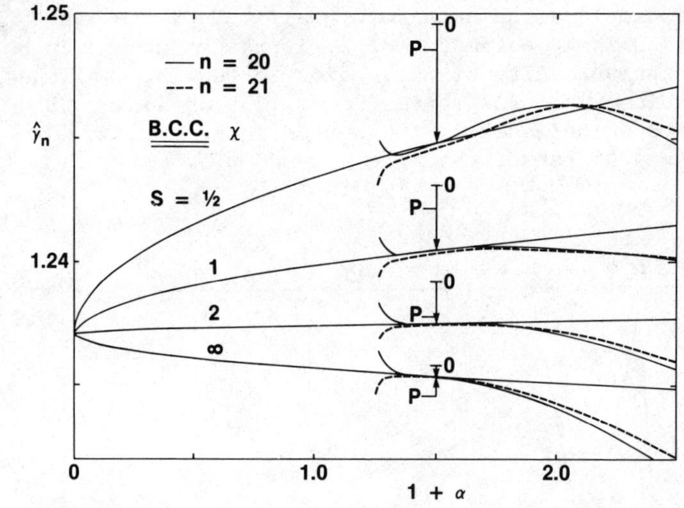

Fig. 5. Ratio method γ estimates from 2.4b as a function of Euler transform parameter for B.C.C. χ series for general spin S. Arrows indicate the shift, magnified ×20, in the last two curves at $\alpha = 0.5$ as observed (O) and as predicted (P) from 2.7 with $\Delta = 0.5$.

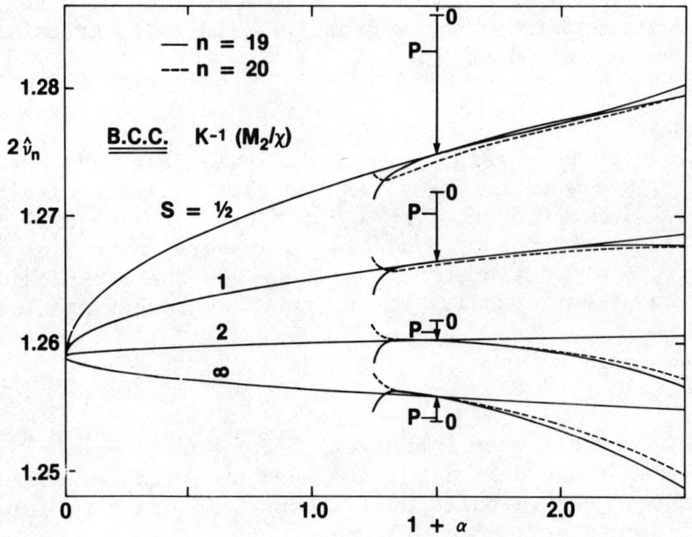

Fig. 6. As in Fig. 5 for 2ν for B.C.C. K^{-1} M_2/χ.

or if $\Delta = 0.6$

$$A_{0.6}(\chi, S = \infty) - A_{0.6}(\chi, S = \tfrac{1}{2}) = 0.26 \quad . \tag{2.11b}$$

It is both in the determination of Δ and in the specific assignment of an $A_\Delta(S = \tfrac{1}{2})$ or $A_\Delta(S = \infty)$ that one is very uncertain and it is here that I believe longer χ'' series and the y plane analysis as outlined in the following section will play a crucial role. Finally, note that our estimates 2.11 are in the same range as those given in the original work of Saul, et al.[18] The essential difference in the two calculations is that with only 12 terms Saul, et al assumed $A_\Delta(S = \tfrac{1}{2}) = 0$ and attributed the entire difference in 2.11 to $A_\Delta(S = \infty)$. I would prefer to ascribe most of the amplitude to the $S = \tfrac{1}{2}$ model and guess

$$A_{0.5}(\xi/a) \approx A_{0.5}(\chi) \approx -0.13 \quad . \tag{2.12}$$

II-B Padé Analysis

The basic idea behind Dlog Pade analysis is that if the function F of interest has the form 2.1, then the logarithmic derivative

$$\frac{d}{dx} \ell nF \approx \frac{\gamma}{x_c - x} - \frac{A_1}{x_c} - \frac{2A_2 - A_1^2}{x_c}\left(1 - \frac{x}{x_c}\right) - \ldots - \frac{\Delta\, A_\Delta}{x_c}\left(1 - \frac{x}{x_c}\right)^{\Delta - 1} - \ldots \tag{2.13}$$

will be dominated by the pole term with residue $-\gamma$ for x near x_c. Thus if the logarithmic derivative is written as a Padé approximant, i.e. a polynomial ratio, there should exist a zero of the denominator that can be interpreted as x_c and a simply calculable residue that can be interpreted as $-\gamma$.

Unfortunately, if the amplitude A_Δ is non-zero, the Padé will also try to simulate the confluent branch point singularity by a sequence of poles terminating near x_c. Thus the Padé pole nearest x_c will be an approximation both to the true pole and to a segment of the branch cut from $x = x_c$ to $x = x_c + \ell$ where ℓ is some fraction of the distance to the next nearest pole. The integrated discontinuity across the cut

$$D = 2i \frac{\Delta\, A_\Delta}{x_c} \operatorname{Sin}\pi\Delta \int_{x_c}^{x_c + \ell} dx \left(\frac{x}{x_c} - 1\right)^{\Delta - 1}$$

$$= 2i\, A_\Delta \operatorname{Sin}\pi\Delta \left(\frac{\ell}{x_c}\right)^{\Delta} \tag{2.14}$$

is $2\pi i$ times the effective residue contribution from this cut segment. Thus we expect the observed residue R at the Padé pole to be given by

$$R = -\gamma + A_\Delta \frac{\mathrm{Sin}\pi\Delta}{\pi}\left(\frac{\ell}{x_c}\right)^{\Delta} . \qquad (2.15)$$

If we assume $\Delta = 0.5$ and use as a reasonable estimate for the spin ½ model $A_\Delta \approx -0.1$ (cf. eqn. 2.12), then even the relatively low estimate of $\ell/x_c \approx 0.1$ would lead to shift in the observed Padé residue away from the true γ of order 0.01. Since we hope to achieve much better accuracy for critical exponents than 0.01 it is clear that some extrapolation from finite order Padé estimates to infinite order must be made.

The following example, appropriate for $\Delta = 0.5$, is intended to suggest an extrapolation procedure based on the nature of the convergence of the Padé estimates with order. The simple cut function $(\omega^2 - 1)^{-\frac{1}{2}}$ has the series and Padé representations

$$\frac{1}{\sqrt{\omega^2 - 1}} = \sum_{n=0}^{\infty} \frac{\Gamma(n + \frac{1}{2})}{\Gamma(\frac{1}{2})\Gamma(n+1)} \left(\frac{1}{\omega}\right)^{2n+1}$$

$$= \frac{U_{n-1}(\omega)}{T_n(\omega)} + 0\left(\left(\frac{1}{\omega}\right)^{2n+1}\right) \qquad (2.16)$$

where T_n and U_n are Chebyshev polynomials of the first and second kind.[19] The polynomial of the first kind has the explicit representation $T_n(\mathrm{Cos}\theta) = \mathrm{Cos}n\theta$ from which we find the zeros closest to $\omega = 1$ are located at approximately $\omega \approx 1 - \frac{\pi^2}{8n^2},\ 1 - \frac{9\pi^2}{8n^2},\ \ldots$. Thus we expect the length ℓ in eqn. 2.14 to scale as n^{-2} and the asymptotic corrections to the residue given by 2.15 to scale as n^{-1}. I leave it as an exercise for the reader to attempt to make this heuristic argument more precise or to work out the expected behavior if $\Delta \neq 0.5$!

Plots of Padé residue estimates vs n^{-1} for extrapolation purposes for the B.C.C. χ and M_2/χ series are shown in Figs. 7 and 8. Here n specifies the number of terms in the power series on which the Padés were based; the estimates shown are those of the four Padés nearest the diagonal (n/2, n/2) form. Straight lines converging to the ratio estimates $\gamma = 1.237$ and $\nu = 0.6295$ at $n^{-1} = 0$ have been drawn to show these estimates are reasonable. The large scatter in the residues and the uncertainty in the precise form of the asymptotic approach to $n^{-1} = 0$ precludes estimates based on Padé analysis alone to better than the $\gamma = 1.238 \pm 0.003$ and $\nu = 0.631 \pm 0.003$ shown in Figs. 7 and 8.

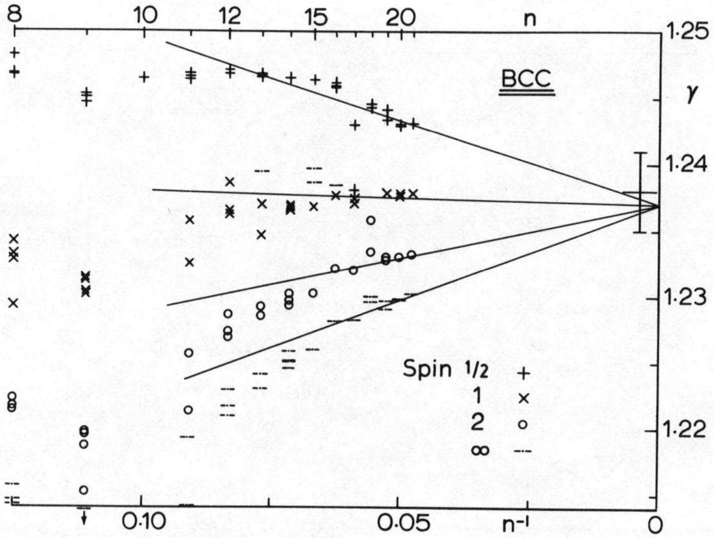

Fig. 7. Dlog Padé estimates of γ for B.C.C. χ series for general spin S. Points mark estimates based on n terms from four near diagonal (n/2, n/2) Padé approximants. Solid straight lines are a crude fit to the Padé estimates and the ratio value of $\gamma = 1.237$.

One observation regarding the spin ½ model should be noted. For the χ series, Padés using less than 15 terms show an isolated pole on the real axis. In this case the Padé pole must simulate the true pole in 2.13 and a fixed segment of the cut resulting in a false plateau of very stable residue estimates. It is only when one exceeds 15 terms that a second pole appears near x_c and drifts with increasing n leading to a corresponding drift in the Padé estimate of γ. A similar false plateau is observed in the F.C.C. χ series except that the break occurs at n = 12. There is no corresponding observable break in the S.C. χ series although most Padés based on 17 to 19 terms are defective and could be signalling the appearance of a second pole.

III High temperature series: Coupling constant analysis

Since all series analysis methods are just extrapolation procedures it is essential that as many distinct methods as possible be tried and checked for consistency. In this way one might hope to uncover the implicit biases to which all methods are prone. For example, the methods described in the preceding section assume that observable correlation functions are smooth functions of temperature

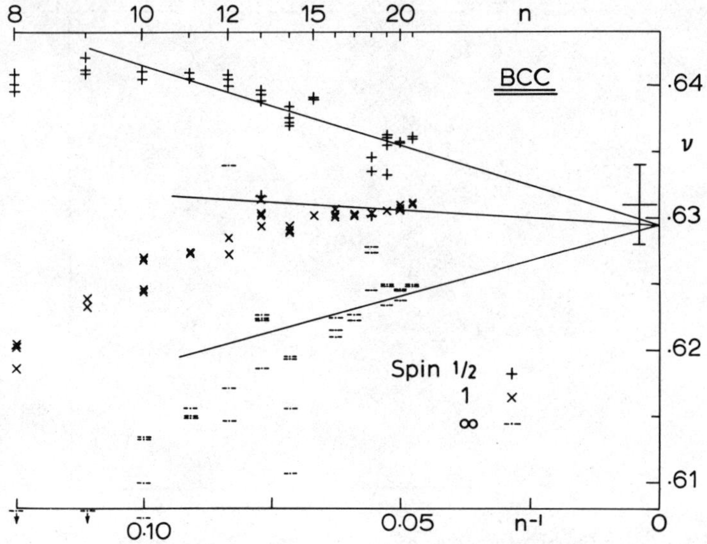

Fig. 8. As in Fig. 7 for 2ν for B.C.C. M_2/χ. Extrapolated 2ν = 1.259 is the ratio method estimate.

with very well defined asymptotic behavior as the temperature tends to the critical temperature. A very distinct method I will describe below[2] originates in the renormalization procedures of field theory; namely, one defines a large enough number of physical observables that between them the "bare" parameters defining the model can be eliminated. Then one specifically looks for the behavior of some observables as functions of others assuming, of course, that one is again dealing with smooth functions that can be extrapolated into regions not accessible to direct calculation.

As a specific example consider the (second moment) definition of the correlation length squared:

$$(\xi/a)^2 = (2d)^{-1} \left(\sum_i \left(\vec{r}_i/a\right)^2 \langle S_o\, S_i\rangle_c\right) \Big/ \left(\sum_i \langle S_o\, S_i\rangle_c\right) \qquad (3.1)$$

where a is the nearest neighbor spin separation, $\vec{r}_i$ is the displacement from the origin to the site i, and $\langle\ \rangle_c$ denotes thermodynamic (cumulant) average. Because $(\xi/a)^2$ is a monotonically increasing function of inverse temperature K and is linear in K for small K, 3.1 can be reverted order by order to obtain K as a power series in $(\xi/a)^2$. Any other correlation function which is normally calculated as a series in K can thus be reexpressed as a series in $(\xi/a)^2$.

Treating ξ/a as independent variable is possible but not particularly useful because of the infinite range $0 \leq (\xi/a)^2 \leq \infty$. A more useful independent variable is an analog of the renormalized coupling constant defined by Nickel and Sharpe[2] for the spin ½ model in dimension d as

$$y = (\xi/a)^2 \left(\langle -\frac{1}{2} \sum_{ijk} S_0 S_i S_j S_k \rangle_c \Big/ \langle \sum_i S_0 S_i \rangle_c^2 \right)^{-2/d} \tag{3.2}$$

In the limit $K \to 0$, the expression in brackets approaches unity so that y is linear in $(\xi/a)^2$ or K for small K. In the critical region $(\xi/a)^2$ diverges as $(K_c - K)^{-2\nu}$, the 4-point average, χ'', as $(K_c - K)^{-\gamma-2\Delta}$, and the 2-point average, χ, as $(K_c - K)^{-\gamma}$. Thus $y \propto (K_c - K)^{-2(d\nu + \gamma - 2\Delta)/d}$ and since $d\nu + \gamma - 2\Delta \geq 0$,[20] y either diverges if hyperscaling fails or tends to a finite limit if hyperscaling is satisfied. Note that at present the case for failure of hyperscaling is weak. With our new estimates $\gamma = 1.237$, $2\nu = 1.259$, and the old estimate $\Delta = 1.563$,[21] $3\nu + \gamma - 2\Delta$ is precisely zero although one cannot exclude the possibility that with longer χ'' series there will result a downward revision of Δ. In any case the numerical series evidence indicates y is a monotonic increasing function of K and hence 3.2 can be reverted to yield K as a function of y. Then, for example, we can express $(\xi/a)^2$ as a series in y and if hyperscaling fails $(\xi/a)^2$ should not diverge for any finite y. On the other hand, if hyperscaling is satisfied as I will assume in the following, then the physical range for y, $0 \leq y \leq y^*$, is finite and y becomes a particularly useful expansion parameter. Furthermore, the leading divergences of all the functions used in the definition 3.2 cancel so that y is a very direct indicator of the corrections to scaling.

For the function $(\xi/a)^2$ of y our expectations on the basis of renormalization group calculations can be worked out fairly simply. If in eqn. 2.1 we choose $F = \xi/a$ and $x = K$, then on reversion we obtain

$$1 - \frac{K}{K_c} \approx B(\xi/a)^{-1/\nu} \Big\{ 1 + B_1(\xi/a)^{-\Delta_1/\nu} + B_2(\xi/a)^{-2\Delta_1/\nu} + B_3(\xi/a)^{-1/\nu} + B_4(\xi/a)^{-\Delta_2/\nu} + \ldots \Big\} . \tag{3.3}$$

Here we have added one extra correction to scaling term not explicitly given in 2.1 since the renormalization group estimates of Golner and Reidel,[22] $\Delta_1/\nu = 0.884$, $1/\nu = 1.608$, and $\Delta_2/\nu = 1.883$, suggest the last three terms in 3.3 are comparable. The function y will have a structure similar to 3.3 except for the leading divergence; explicitly

$$y = y^* \left\{ 1 + C_1(\xi/a)^{-\Delta_1/\nu} + \ldots \right\} . \tag{3.4}$$

Finally, on reverting 3.4 we obtain

$$(\xi/a)^2 \simeq D(y^* - y)^{-2\nu/\Delta_1} \left\{ 1 + D_1(y^* - y) + D_2(y^* - y)^{(1 - \Delta_1)/\Delta_1} \right.$$

$$\left. + D_3(y^* - y)^{(\Delta_2 - \Delta_1)/\Delta_1} + \ldots \right\} \tag{3.5}$$

which shows explicitly how the leading y plane divergence of $(\xi/a)^2$ is related to the exponent Δ_1. Also, the amplitude D can be related back to the A_Δ in eqn. 2.1. We find

$$D = (- y^* C_1)^{2\nu/\Delta_1} , \tag{3.6a}$$

$$C_1 = \frac{2}{d} \left\{ d\, A_{\Delta_1}(\xi/a) + 2A_{\Delta_1}(\chi) - A_{\Delta_1}(\chi'') \right\} B^{\Delta_1} \tag{3.6b}$$

so that when longer χ'' series become available consistency between K plane and y plane amplitudes can be checked.

Two analyses of 3.5 for the B.C.C. spin ½ model have been reported in the literature. Nickel and Sharpe[2] ignored the confluent singularities and from Padé estimates of the inverse logarithmic derivative

$$\gamma(y) \equiv \left[\frac{d}{dy} \ln(\xi/a)^2 \right]^{-1} \approx \frac{\Delta_1}{2\nu} (y^* - y) \tag{3.7}$$

estimated $y^* = 0.1594$, $\Delta_1/2\nu = 0.59$ ($\Delta_1 \approx 0.74$). Such a large value for Δ_1 leads to a confluent correction in eqn. 3.5 proportional to $D_2(y^* - y)^{0.35}$ so that the simple Padé analysis is unlikely to be correct. Rehr[3] analysed 3.5 using second order homogeneous approximants and concluded $y^* = 0.1602$ and $\Delta_1/2\nu = 0.39$ ($\Delta_1 \approx 0.49$) which is much closer to the Golner and Riedel[22] estimate of $\Delta_1/2\nu = 0.44$. The exponent estimates obtained by Rehr

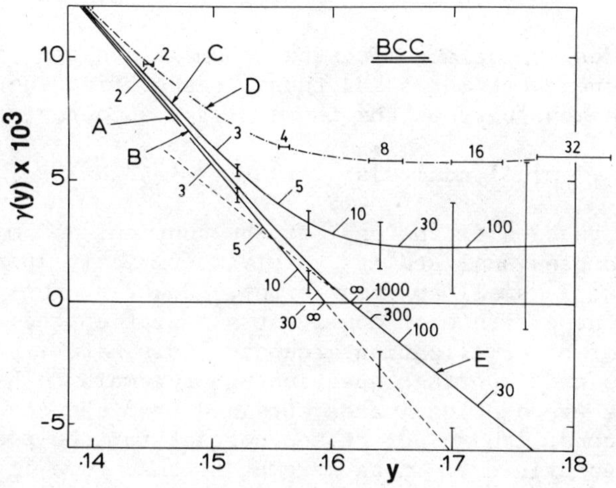

Fig. 9. Curves A - D are estimates of the Callan Symanzik analog $\gamma(y)$ defined by 3.7 and labelled by ξ/a ratios. Curve E is the equivalent continuum model $\gamma(y)$ (cf. Ref. 2) and is labelled by the dimensionless bare expansion parameter defined by the left hand side of 4.6. A - Dlog Padé estimate from Ref. 2. B - Second order differential approximant as discussed in Ref. 3. This particular curve is biased with the continuum model y^*. C - Temperature plane approximant from Ref. 2. D - Estimated $\gamma(y)$ from the tabulated data of Ref. 27. Note that this last curve is based on 2 fewer series terms than curves A to C.

are fairly insensitive to changes in the assumed y^* and one can obtain reasonable approximants biased with $y^* = 0.1616$ which is the continuum ϕ^4 model estimate for this quantity.[4,23] One such biased approximant has been integrated to obtain $\gamma(y)$ which is shown in Fig. 9. It is clearly a very reasonable interpolation between the simpler estimates that ignore confluent corrections and as such we also expect the amplitude D obtained from this approximant to be reasonable. We find $C_1 = 0.12$ which combined with the Fisher and Burford estimate $B = 0.285$[24] leads to

$$3A_{\Delta_1}(\xi/a) + 2A_{\Delta_1}(\chi) - A_{\Delta_1}(\chi'') = -0.33 \quad . \tag{3.8}$$

From the preceding ratio analysis (cf. eqn. 2.12) we estimate the first two contributions on the left hand side of 3.8 as -0.65 which suggests $A_{\Delta_1}(\chi'')$ must be large. However I believe factors of 2 in these amplitudes can be accommodated by small changes in critical point and exponent estimates and thus I rather find the order of magnitude agreement between the two analyses encouraging.

IV Continuum ϕ^4 model analysis

The perturbation expansion[25] of the continuum ϕ^4 model in three dimensions has been analysed[4,23] to yield critical exponent estimates with exceptionally small apparent error. However, these analyses have ignored the presence of confluent singular corrections which, on the basis of renormalization group analysis, are expected to be present. Thus two important questions that remain to be answered are first, are the exponent estimates obtained from the continuum model biased and second, can one detect and correct for the presence of confluent singularities? In an attempt to shed some light on these questions I will discuss an analysis, as described by Le Guillou and Zinn-Justin[4], of the continuum ϕ^4 model in one dimension. I find the known confluent singularities are undetectable from series of practicably attainable length and conclude that unless one can independently develop bounds on the amplitudes of the confluent corrections in the d = 3 model, one must assume the quoted uncertainties in the exponent estimates[4,23] are unrealistically small.

The continuum ϕ^4 model is defined by the dimensionless hamiltonian

$$H = \int_\Omega d^d r \left\{ \frac{1}{2} (\nabla\phi)^2 + \frac{1}{2} m_o^2 \phi^2 + \frac{\lambda}{4} \phi^4 + h\phi \right\} \qquad (4.1)$$

for a system of volume $\Omega = L^d$. Averages of any function $A(\phi(\vec{r}_1), \phi(\vec{r}_2), \ldots)$ are defined by

$$\langle A \rangle = \int \mathcal{D}\phi A e^{-H} \Big/ \int \mathcal{D}\phi e^{-H} \qquad (4.2)$$

where the functional integral $\int \mathcal{D}\phi$ is to be interpreted as the zero lattice spacing limit of a product of ordinary, appropriately normalized, integrals over fields defined at lattice sites i, $\Pi_i \left(N^{-1} \int d\phi_i \right)$. At the same time in eqn. 4.1, the gradient is to be replaced by an appropriate difference operator and the integral by a sum. Some care must be exercised in performing the zero lattice spacing limit as discussed below. Of course, one need not take this limit; such models in which the lattice spacing a remains as an extra parameter have recently been discussed by Bender, et al[26] and Baker and Kincaid.[27]

The averages of polynomial functions A in 4.2 can be evaluated by expanding the $\lambda\phi^4$ part of the exponential in a power series since the remaining quadratic part of the exponential can then be diagonalized and the functional integral replaced by a product of ordinary integrals over gaussians of normal mode coordinates. The remaining technical effort consists of first, cataloging the results of the gaussian integrations, i.e. representing their contributions as a set of topologically distinct graphs, and second, evaluating each graph which represents a multidimensional sum over functions of normal mode labels, i.e. integrals over functions of momenta $\vec{k}_i$ in the thermodynamic $\Omega \to \infty$ limit. It is this last step which has limited the calculation[25] of the d = 3 model. To proceed to even one higher order will require 9-dimensional numerical quadrature unless advances are made in the analytic evaluation[28] of certain integrals.

An additional complication in the perturbation expansion of 4.2 originates with the definition of the model as a continuum limit. If the limit of zero lattice spacing is naively taken at the beginning of the calculation, divergences can appear in certain calculated quantities. For example, in the continuum and thermodynamic limits, the inverse second moment of the 2-point correlations, $m^2 = \xi^{-2}$, is given to first order in λ by

$$m^2 = m_o^2 + 3\lambda \int \frac{d^d k}{(2\pi)^d} \frac{1}{m_o^2 + k^2} \tag{4.3}$$

which is undefined for $d \geq 2$. Keeping the lattice spacing finite eliminates the divergence but at the expense of making higher order calculations impractical. The expedient, called mass renormalization, which is conventionally adopted is to simply define m^2 as a finite parameter of the theory and formally eliminate m_o^2 order by order in λ from all physically observable correlation functions. I will assume in the following that this procedure has been adopted although for $d = 1$ replacing m_o^2 by m^2 as the independent variable is only convenient, not essential. For dimension $d \geq 4$ additional divergences force one to eliminate λ as a parameter in favor of a renormalized coupling constant u; although I will introduce such a parameter in the discussion below it is essential to realize that for $d < 4$, u is a <u>calculable</u> function order by order in λ.

The form of the perturbation expansion in the continuum model can be deduced by dimensional analysis. From the definition of the renormalized mass as the inverse correlation length and the definition 4.1 of the dimensionless hamiltonian we conclude

$$\left[\xi^2 = m^{-2}\right] = \left[L^2\right], \quad \left[\phi^2\right] = \left[L^{2-d}\right],$$

$$\left[\lambda\right] = \left[L^{d-4}\right], \quad \left[h^2\right] = \left[L^{-2-d}\right].$$

Thus, for example, the susceptibility which is an extensive variable must have the form

$$\chi = \left\langle \left(\int d^d r\, \phi(\vec{r}) \right)^2 \right\rangle = \Omega\, m^{-2}\, Z_3\left(\lambda/m^{4-d}\right) \tag{4.4}$$

where the so called wavefunction renormalization function Z_3 is dimensionless and is expected to vanish as $\left(m/\lambda^{1/(4-d)}\right)^{\eta}$ as the critical point, m = 0, is approached. The dimensionless renormalized coupling constant, an intensive variable defined by

$$u \equiv -N_u\, m^d\, \Omega\, \chi^{-2} \left\langle \left(\int d^d r\, \phi(\vec{r}) \right)^4 \right\rangle_c = u\left(\lambda/m^{4-d}\right) \tag{4.5}$$

where N_u is a convenient numerical normalization constant of order unity, will approach a constant u^* as $m \to 0$ if hyperscaling is satisfied. In general, correlation functions can be written as prefactors times dimensionless series in $\lambda\xi^{4-d}$ which one can either view as coupling constant expansions for fixed ξ, or more appropriately for critical phenomena, as expansions in powers of ξ^{4-d} for fixed λ. In a sense $\lambda^{-1/(4-d)}$ in the continuum model plays the same role as a in the lattice Ising models in defining a microscopic length scale against which measurements of the correlation length ξ are compared. Note that if the continuum limit in 4.1 and 4.2 is not taken, the resulting model depends on two microscopic lengths and will appear Ising like or continuum like depending on whether $\lambda^{-1(4-d)}$ is much less or much greater than the lattice spacing a.[27]

Just as in the discussion in section III, the variable $\lambda\xi^{4-d}$ is not particularly useful as an expansion parameter because of the infinite range $0 \leq \lambda\xi^{4-d} \leq \infty$. Whether or not u is a useful independent variable depends on whether hyperscaling is satisfied. If hyperscaling fails, $u \to 0$ as $\xi \to \infty$ and hence functions of u will be multivalued and not susceptible to any simple analysis. If hyperscaling is satisfied as we assume below, u is almost certainly a monotonic increasing function of ξ in the interval $0 \leq u \leq u^*$. The perturbation expansion for u can be reverted and, with the choice $N_u = 3\Gamma(2 - d/2)\, 2^{-d-1}\, \pi^{-d/2}$ in eqn. 4.5, normalized to the form

$$6N_u\, \lambda\xi^{4-d} = u + u^2 + \ldots \tag{4.6}$$

For d = 1 the expected behavior near $u = u^*$ is worked out in detail later while for d = 2 or 3 the discussion leading to 3.5 can be repeated here with $m_o^2 - m_{oc}^2$ playing the role of $T - T_c$. Near the critical point

$$\lambda\xi^{4-d} \approx \hat{D}(u^* - u)^{-(4-d)\nu/\Delta_1} \Big\{1 + \hat{D}_1(u^* - u) + \hat{D}_2(u^* - u)^{(1-\Delta_1)/\Delta_1} + \hat{D}_3(u^* - u)^{(\Delta_2-\Delta_1)/\Delta_1} + \ldots\Big\} \quad (4.7)$$

and hence the inverse logarithmic derivative

$$\beta(u) \equiv \left[-\frac{d}{du}\ln\left(\lambda\xi^{4-d}\right)\right]^{-1} = -u + u^2 - \ldots$$

$$\approx \frac{\Delta_1}{(4-d)\nu}(u - u^*) + \hat{\hat{D}}_1(u^* - u)^2 + \hat{\hat{D}}_2(u^* - u)^{1/\Delta_1} + \hat{\hat{D}}_3(u^* - u)^{\Delta_2/\Delta_1} + \ldots \quad (4.8)$$

has a zero with positive slope at $u = u^*$ but with confluent branch point singularities. The Golner and Riedel[22] estimates are $1/\Delta_1 = 1.82$, $\Delta_2/\Delta_1 = 2.13$ and thus because both exponents are so close to 2 it is conceivable that one can legitimately combine the $\hat{\hat{D}}_2$ and $\hat{\hat{D}}_3$ terms into an effective $\hat{\hat{D}}_1(u^* - u)^2$ in 4.8. However, if this is the case I do not understand why the confluent terms in the analog $\gamma(y)$ function discussed in section III can apparently not be ignored.

Although the Callan-Symanzik function $\beta(u)$ is a close analog of $\gamma(y)$ the actual analysis must be entirely different because the power series expansion of $\beta(u)$ is asymptotic. The large order behavior of $\beta(u)$ has been worked out in detail by Brézin, et al[29] and Brézin and Parisi[30] who give

$$\beta(u) = \sum_k \beta_k u^k ,$$

$$\beta_k \approx (-A)^k\, \Gamma\left(k + \frac{d}{2} + 3\right) C , \quad k \to \infty \quad (4.9)$$

where

$$A = 0.14777422 , \; C = 0.03996 , \; d = 3;$$

$$A = 0.238659 , \; C = 0.04886 , \; d = 2;$$

$$A = 1/3 , \; d = 1 . \quad (4.10)$$

A relevant, scale independent, measure of the divergence of the $\beta(u)$ series is the product Au^* which is 3/4 for d = 1 and 0.209 for d = 3. Thus the d = 3 series should be much more amenable to analysis and it is not surprising that we find the d = 1 test series requires many more terms for comparable accuracy.

The function $\beta(u)$ can in principle be obtained from the generalized Borel transform function

$$B(t) \equiv \sum \beta_k \, t^k/\Gamma\left(k + \frac{d}{2} + \ell + 1\right) \tag{4.11}$$

via the integral

$$\beta(u) = \int_0^\infty \frac{dt}{u} e^{-t/u} \left(\frac{t}{u}\right)^{\ell+d/2} B(t) \quad . \tag{4.12}$$

The transform function B(t) is convergent for $|t| < A^{-1}$ but eqn. 4.12 requires its analytic continuation to the entire positive real axis. This is a non-trivial numerical problem in that as u approaches u^*, it is precisely the $t \to \infty$ regime that becomes important. In particular, if the confluent corrections to $\beta(u)$ indicated in 4.8 are present, then some higher derivative of β must diverge as $u \to u^*$. The simplest functional form for B(t) consistent with such divergence is $B(t) \sim t^{-p} e^{t/u^*}$, $t \to \infty$, where $p > \ell + \frac{d}{2} + 1$ No published analysis of the continuum ϕ^4 model has attempted to build in such a limiting asymptotic form and in fact a Padé analysis of B(t) as used by Baker, et al[23] is in principle incorrect. Padé approximants will attempt to simulate an exponential divergence by poles on an arc intersecting the positive real axis in direct contradiction to the known analytic behavior.[31]

A simple analysis that does not violate analyticity requirement is the conformal mapping method introduced by Le Guillou and Zinn-Justin.[4] The transformation

$$t = 4\,A^{-1}\,\omega(1-\omega)^{-2} \tag{4.13}$$

maps the t plane cut from $-A^{-1}$ to $-\infty$ onto the unit circle in the ω plane. If B(t) is analytic except for this cut then the power series expansion of the transformed function $\hat{B}(\omega)$ is convergent on the required interval $0 \leq \omega < 1$ and we can obtain $\beta(u)$ from the integral

$$\beta(u) = \int_0^\infty \frac{dt}{u} e^{-t/u} \left(\frac{t}{u}\right)^{\ell+d/2} \hat{B}(\omega(t)) \tag{4.14}$$

in which we use only the power series representation of $\hat{B}(\omega)$.

Whether this postulated analytic behavior for B(t) is correct is not known but at least the series do not show any obvious structure to indicate otherwise. The sequence of approximants generated from 4.14 for $d = 3$, $\ell = 5/2$ are shown in Fig. 10. Approximate geometric convergence is observed even for u well beyond the limit $u = u^*$; there is no indication of any divergence in the integral representation of $\beta(u)$. Thus either confluent singularities are absent or are not observable from short series. In an attempt to decide between these two possibilities we describe below a similar analysis for the $d = 1$ model.

The continuum ϕ^4 model for $d = 1$ is particularly attractive both because results can be derived exactly numerically and extended series can be efficiently generated without the technical complications of a graphical expansion. These simplifications are based on the observation that if the fields at the endpoints of the interval $0 \le x \le L$ are fixed at ϕ_0 and ϕ_L then the partition function defined by

$$Z\left(\phi_L, \phi_0; L\right) \equiv \int \mathcal{D}\,\phi(x)\, e^{-H} \tag{4.15}$$

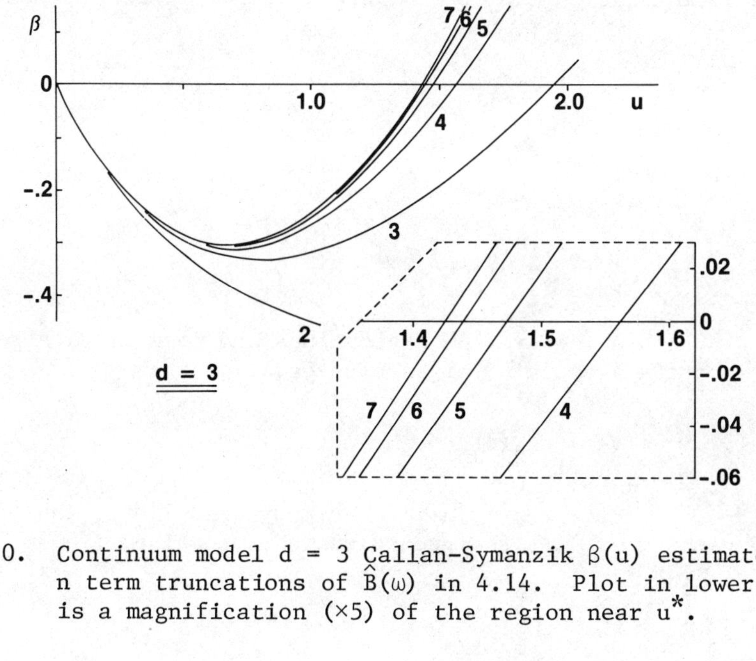

Fig. 10. Continuum model $d = 3$ Callan-Symanzik $\beta(u)$ estimated from n term truncations of $\hat{B}(\omega)$ in 4.14. Plot in lower right is a magnification (×5) of the region near u^*.

is similar to the path integral of quantum mechanics[32] and one can write a partial differential equation for Z that is similar to the Schroedinger equation. To derive this differential equation we simply follow the Feynman and Hibbs[32] prescription of adding one extra site to the chain before taking the continuum limit to obtain the recursion formula

$$Z\left(\phi_{L+a}, \phi_o; L+a\right) = N^{-1} \int d\,\phi_L\, e^{-\frac{1}{2a}\left(\phi_{L+a} - \phi_L\right)^2}$$

$$\cdot\, e^{-aV\left((\phi_{L+a} + \phi_L)/2\right)} \cdot Z\left(\phi_L, \phi_o; L\right)$$

$$\approx \left(1 - a\, V\left(\phi_{L+a}\right)\right) N^{-1} \int d\,\phi_L\, e^{-\frac{1}{2a}\left(\phi_{L+a} - \phi_L\right)^2}$$

$$\cdot \left(1 + \frac{1}{2}\left(\phi_{L+a} - \phi_L\right)^2 \frac{\partial^2}{\partial\phi^2_{L+a}}\right) Z\left(\phi_{L+a}, \phi_o; L\right) \tag{4.16}$$

where

$$V(\phi) = \frac{1}{2} m_o^2\, \phi^2 + \frac{\lambda}{4}\, \phi^4 + h\phi \quad . \tag{4.17}$$

With the normalization choice $N = \sqrt{2\pi a}$ we find, in the limit $a \to 0$,

$$-\frac{\partial}{\partial L} Z\left(\phi, \phi_o; L\right) = \left(-\frac{1}{2}\frac{\partial^2}{\partial\phi^2} + V(\phi)\right) Z\left(\phi, \phi_o; L\right) \tag{4.18}$$

and hence

$$Z\left(\phi_L, \phi_o; L\right) = \sum_i A_i\left(\phi_L\right) A_i\left(\phi_o\right) e^{-LE_i} \tag{4.19}$$

where the E_i and $A_i(\phi)$ are the eigenvalues and normalized eigenfunctions of the associated eigenvalue equation

$$\left(-\frac{1}{2}\frac{\partial^2}{\partial\phi^2} + V(\phi)\right) A_i(\phi) = E_i\, A_i(\phi) \quad . \tag{4.20}$$

From 4.19 we conclude that, in the thermodynamic limit, the free energy density f is just the ground state eigenvalue, i.e.

$$f \equiv \lim_{L\to\infty} \frac{1}{L}\left(-\ln Z\right) = E_o \tag{4.21}$$

and is independent of the boundary conditions imposed at $x = 0,L$.

Correlation functions in the $d = 1$ model are equally easy to obtain. For example, the 2-point average

$$<\phi_x \phi_y> = \int d\phi_x \, d\phi_y \, Z(\phi_L, \phi_x; L - x) \, \phi_x \, Z(\phi_x, \phi_y; x - y) \, \phi_y$$

$$\cdot \, Z(\phi_y, \phi_o; y) \Big/ Z(\phi_L, \phi_o, L) \quad , \quad x > y \quad , \qquad (4.22)$$

in the limit $L \to \infty$ reduces to

$$<\phi_x \phi_y> = \sum_i \left(\int d\phi \, A_i(\phi) \, \phi \, A_o(\phi) \right)^2 e^{-|x-y|(E_i - E_o)} \, . \qquad (4.23)$$

In zero field h the ground state wavefunction is even and for $|x - y|$ large the sum in 4.23 is then dominated by the first excited state. Thus we identify the true inverse correlation length $m = \xi^{-1}$ as

$$m = E_1 - E_o \, . \qquad (4.24)$$

The susceptibility is simply

$$\chi = <\left(\int dx \, \phi_x \right)^2> = - L \left. \frac{\partial^2 E_o}{\partial h^2} \right|_{h=0} \qquad (4.25)$$

and corresponding formulae apply for the higher field derivatives. In particular, the renormalized coupling constant defined by 4.5 is

$$u = \frac{3}{8} m \frac{\partial^4 E_o}{\partial h^4} \Big/ \left(\frac{\partial^2 E_o}{\partial h^2} \right)^2 \qquad (4.26)$$

Bender and Wu[33] have described an efficient numerical recursion procedure for obtaining the zero field E_o as a series in λ and a straightforward extension applies to calculations of E_1, χ, and χ''. The structure of these series follows from the dimensional arguments given earlier; for example $m = m_o \, \rho_1(\lambda/m_o^3)$ and $u = \frac{m}{m_o} \rho_2(\lambda/m_o^3)$ where the ρ_i are dimensionless. The ρ_i series can be reverted to yield λ/m^3 as a series in u. Further straightforward manipulations yield $\beta(u)$ and $\hat{B}(\omega)$ although I have found that in the last step of obtaining $\hat{B}(\omega)$ cancellations result in a great loss of significance so that even with 112 bit accuracy at all intermediate stages it is not possible to generate more than about 40 terms.

The behavior of $\beta(u)$ for u near u^* can be determined from an analysis of the anharmonic oscillator problem in the deep double well regime. The wavefunctions in zero field in the limit $m_o^2 \to -\infty$ are pairs of harmonic oscillator wavefunctions centered at $\phi = \pm |m_o|/\sqrt{\lambda}$ and coupled symmetrically or antisymmetrically through the barrier. Explicitly, the A_i in 4.20 are given by

$$A_{2i}(\phi) \approx \frac{1}{\sqrt{2}}\left\{\psi_i\left(\phi - \frac{|m_o|}{\sqrt{\lambda}}\right) + \psi_i\left(\phi + \frac{|m_o|}{\sqrt{\lambda}}\right)\right\},$$

$$A_{2i+1}(\phi) \approx \frac{1}{\sqrt{2}}\left\{\psi_i\left(\phi - \frac{|m_o|}{\sqrt{\lambda}}\right) - \psi_i\left(\phi + \frac{|m_o|}{\sqrt{\lambda}}\right)\right\} \qquad (4.27)$$

where the ψ_i are wavefunctions for a harmonic oscillator of frequency $\sqrt{2}\,|m_o|$. The corresponding energies E_i in 4.20 are separated by $E_{i+2} - E_i \approx \sqrt{2}\,|m_o|$ while the mass gap $m = E_1 - E_0$ is given in lowest order W.K.B. approximation by

$$m \approx \frac{\sqrt{2}\,|m_o|}{\pi} \exp\left(-\frac{|m_o|^2}{\sqrt{2\lambda}} \int \sqrt{\left(1 - \frac{\lambda\,\phi^2}{|m_o|^2}\right)^2 - \frac{2\sqrt{2}\,\lambda}{|m_o|^3}}\, d\phi\right)$$

$$\approx \frac{8|m_o|}{\pi}\sqrt{\frac{|m_o|^3 e}{\sqrt{2}\,\lambda}}\exp\left(\frac{-2\sqrt{2}\,|m_o|^3}{3\lambda}\right), \quad m_o^2 \to -\infty, \qquad (4.28)$$

which simplifies further to the approximately lograithmic dependence $|m_o| \sim |\ell n\, m|^{1/3}$. The second and fourth order shifts in E_o in the presence of the field perturbation $h\phi$ are

$$\Delta_2 = \sum_i \frac{\phi_{oi}^2}{E_o - E_i} \approx -\frac{\phi_{01}^2}{m} - \frac{\phi_{03}^2}{\sqrt{2}\,|m_o|},$$

$$\Delta_4 = \sum_{ijk} \frac{\phi_{oi}\,\phi_{ij}\,\phi_{jk}\,\phi_{ko}}{(E_o - E_i)(E_o - E_j)(E_o - E_k)} - \Delta_2 \sum_i \frac{\phi_{oi}^2}{(E_o - E_i)^2}$$

$$\approx \frac{\phi_{01}^2}{m^2}\left(-\frac{\phi_{12}^2}{\sqrt{2}\,|m_o|} - \Delta_2\right) \qquad (4.29)$$

from which we deduce the ratio

$$\frac{\Delta_4}{\Delta_2^2} \approx \frac{1}{m}\left(1 - \frac{m}{\sqrt{2}\,|m_o|}\,\frac{\phi_{03}^2 + \phi_{12}^2}{\phi_{01}^2}\right) . \tag{4.30}$$

Note that the truncation of the sums in 4.29 to a finite number of terms follows from the harmonic oscillator form 4.27 for the wave-functions. Furthermore, the remaining matrix elements in 4.30 can be simply worked out as

$$\phi_{01}^2 \approx |m_o|^2/\lambda \ , \quad \phi_{03}^2 \approx \phi_{12}^2 \approx 1/(2\sqrt{2}\,|m_o|) \ . \tag{4.31}$$

The expression 4.26 for u is

$$u = \frac{9\,m\,\Delta_4}{4\,\Delta_2^2} \approx \frac{9}{4} - \frac{9\,m\,\lambda}{8\,|m_o|^4} \tag{4.32}$$

so that $u^* = 9/4$ and

$$\frac{\lambda}{m^3} \approx \left(\frac{9}{8}\right)^3 \left(\frac{\lambda}{|m_o|^3}\right)^4 (u^* - u)^{-3} \tag{4.33}$$

The Callan-Symanzik function $\beta(u)$ as defined by 4.8 is then

$$\beta(u) = -u + u^2 - \ldots \approx \frac{1}{3}(u - u^*)\left(1 - 4\,\frac{\partial\,\ell n\,|m_o|}{\partial\,\ell n\,m}\right) \tag{4.34}$$

and thus because of the approximate logarithmic dependence $|m_o| \sim |\ell n\, m|^{1/3}$ the simple zero of $\beta(u)$ is confluent with a singularity of approximate form $(u^* - u)/\ell n(u^* - u)$. Such a logarithmic confluence implies all derivatives of β beyond the first diverge at $u = u^*$ and the approach to the limiting slope $\beta'(u^*) = 1/3$ is exceedingly slow. This latter effect is well illustrated by the sequence of approximants 4.14 for $d = 1$, $\ell = 5/2$ shown in Fig. 11 and in greater detail for u near u^* in Fig. 12. The overall convergence in Fig. 11 is impressive although the number of terms required is greater here than for $d = 3$ as expected (cf. eqn. 4.10 and subsequent comments). On the other hand the convergence in the immediate vicinity of u^* is very poor with, for example, the estimated slope from 35 terms differing from the asymptotic slope by factor 2!

Although one does not expect such extreme behavior for $d = 3$ note that it is knowledge of the slope of the β-function rather than just u^* that is crucial for the accurate determination of critical

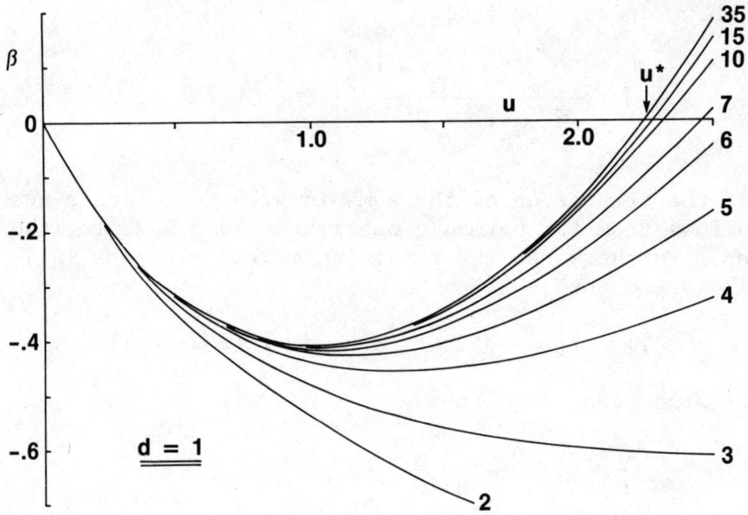

Fig. 11. Continuum model d = 1 Callan-Symanzik β(u) from eqn. 4.14 with $\hat{B}(\omega)$ truncated to various n.

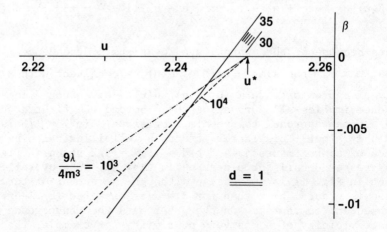

Fig. 12. Expanded view of Fig. 11 near u^*. Solid curves are n = 30 to 35 term approximants to 4.14. Dash curve is the asymptotic estimate determined from 4.34 combined with 4.28 and 4.33. Dot-dash is a straight line through $u = u^*$, $\beta(u^*) = 0$, with the correct limiting slope $\beta'(u^*) = 1/3$.

exponents. For example, if Z_3 as defined by 4.4 vanishes as $\left(m/\lambda^{1/(4-d)}\right)^{\eta}$ then η is determined from the series

$$\eta = \lim_{u \to u^*} \left\{ \frac{d \ln Z_3}{d \ln m} = \frac{d \ln Z_3}{du} \frac{du}{d \ln m} \right.$$

$$\left. = (4-d)\, \beta(u) \frac{d \ln Z_3}{du} \right\} \qquad (4.35)$$

where all derivatives are calculated with fixed λ. The pole of $\frac{d \ln Z_3}{du}$ at $u = u^*$ is cancelled by the corresponding zero of β and thus either explicitly or implicitly one is assuming a smooth extrapolation of the slope of β. Whether this is justified depends specifically on the exponents and amplitudes of the expected confluent terms in $\beta(u)$ given in 4.8; what I believe the $d = 1$ example has shown is that such confluent terms cannot be detected from series alone.

Acknowledgements

The financial support of the Natural Sciences and Engineering Research Council Canada through grant A9348 and the University of Guelph Computer Services Council through their special computing program is gratefully acknowledged. I also thank Bruce Sharpe for some of the programming for the series generation and Ralph Roskies for helpful comments.

References

1. F. J. Wegner, Corrections to scaling laws, Phys. Rev. B. 5:4529 (1972).

2. B. G. Nickel and B. Sharpe, On hyperscaling in the Ising model in three dimensions, J. Phys. A:Math. Gen. 12:1819 (1979).

3. J. J. Rehr, Confluent singularities and hyperscaling in the spin ½ Ising model, J. Phys. A:Math. Gen. 12:L179 (1979).

4. J. C. LeGuillou and J. Zinn-Justin, Critical exponents from field theory, Phys. Rev. B 21:3976 (1980) and Critical exponents for the n-vector model in three dimensions from field theory, Phys. Rev. Lett. 39:95 (1977).

5. D. S. Gaunt and A. J. Guttmann, Asymptotic analysis of coefficients, in: "Phase transitions and critical phenomena" vol III, C. Domb and M. S. Green, eds., Academic Press, London (1974).

6. J. Oitmaa and J. Ho-Ting-Hun, The critical exponent γ for the three-dimensional Ising model, J. Phys. A:Math. Gen. 12: L281 (1979).

7. A. J. Guttmann and G. S. Joyce, On a new method of series analysis in lattice statistics, J. Phys. A:Gen. Phys. 5: L81 (1972).

8. M. Wortis, Linked cluster expansion, in: "Phase transitions and critical phenomena" vol III, C. Domb and M. S. Green, eds., Academic Press, London (1974).

9. B. R. Heap, The enumeration of homeomorphically irreducible star graphs, J. Math. Phys. 7:1582 (1966).

10. D. S. Gaunt and M. F. Sykes, The critical exponent γ for the three-dimensional Ising model, J. Phys. A:Math. Gen. 12:L25 (1979).

11. T. T. Wu, B. M. McCoy, C. A. Tracy, and E. Barouch, Spin-spin correlation functions for the two-dimensional Ising model: Exact theory in the scaling region, Phys. Rev. B 13:316 (1976)

12. M. F. Sykes, D. S. Gaunt, P. D. Roberts, and J. A. Wyles, High temperature series for the susceptibility of the Ising model. I. Two dimensional lattices, J. Phys. A:Gen. Phys. 5:624 (1972).

13. S. McKenzie, High-temperature reduced susceptibility of the Ising model, J. Phys. A:Math. Gen. 8:L102 (1975).

14. S. McKenzie, The exponent γ for the spin ½ ising model on the face-centered cubic lattice, J. Phys. A:Math. Gen. 12:L185 (1979).

15. A. Aharony and G. Ahlers, Universal ratios among correction-to-scaling amplitudes and effective critical exponents, Phys. Rev. Lett. 44:782 (1980).

16. M. Chang and A. Houghton, Universal ratios among correction-to-scaling amplitudes on the coexistence curve, Phys. Rev. Lett. 44:785 (1980).

17. M. Ferer, Amplitude universality and confluent corrections to scaling for Ising models, Phys. Rev. B 16:419 (1977).

18. D. M. Saul, M. Wortis, and D. Jasnow, Confluent singularities and the correction-to-scaling exponent for the $d = 3$ fcc Ising model, Phys. Rev. B 11:2571 (1975).

19. M. Abramowitz and I. A. Stegun, "Handbook of Mathematical Functions", Dover, New York (1965).

20. R. Schrader, New rigorous inequality for critical exponents in the Ising model, Phys. Rev. B 14:172 (1976).

21. J. W. Essam and D. L. Hunter, Critical behavior of the Ising model above and below the critical temperature, J. Phys. C 1:392 (1968).

22. G. R. Golner and E. K. Reidel, Scaling-field approach to the isotropic n-vector model in three dimensions, Phys. Lett. 58A:11 (1976).

23. G. A. Baker, Jr., B. G. Nickel, and D. I. Meiron, Critical indices from perturbation analysis of the Callan-Symanzik equation, Phys. Rev. B 17:1365 (1978).

24. M. E. Fisher and R. J. Burford, Theory of critical-point scattering and correlations. I. The Ising model, Phys. Rev. 156:583 (1967).

25. B. G. Nickel, D. I. Meiron, and G. A. Baker, Jr., Compilation of 2-pt. and 4-pt. graphs for continuous spin models, University of Guelph report (1977).

26. C. M. Bender, F. Cooper, G. S. Guralnik, and D. H. Sharp, Strong-coupling expansion in quantum field theory, Phys. Rev. D 19:1865 (1979).

27. G. A. Baker, Jr. and J. M. Kincaid, The continuous-spin Ising model, $g_0:\phi^4:_d$ field theory, and the renormalization group, Los Alamos report LA-UR-80-869 (1980).

28. B. G. Nickel, Evaluation of simple Feynman graphs, J. Math. Phys. 19:542 (1978).

29. E. Brezin, J. C. LeGuillou, and J. Zinn-Justin, Perturbation theory at large order. I. The ϕ^{2n} interaction, Phys. Rev. D 15:1544 (1977).

30. E. Brezin and G. Parisi, Critical exponents and large-order behavior of perturbation theory, J. Stat. Phys. 19:269 (1978).

31. E. B. Saff and R. S. Varga, On the zeros and poles of Padé approximants to e^z. II, in: "Padé and rational approximation", E. B. Saff and R. S. Varga, eds., Academic Press, London (1977).

32. R. P. Feynman and A. R. Hibbs, "Quantum mechanics and path integrals", McGraw-Hill, New York (1965).

33. C. M. Bender and T. T. Wu, Anharmonic oscillator, Phys. Rev. 184:1231 (1969).

DIFFERENTIAL APPROXIMANTS AND CONFLUENT SINGULARITY ANALYSIS*

John J. Rehr

Dept. of Physics, University of Washington

Seattle, WA 98195 USA

Developments in the understanding of critical phenomena have led to an increasingly accurate and detailed picture of critical behavior. One such development is the appreciation that critical point singularities are generally accompanied by weaker, confluent singular structure.[1] For example, thermal properties, such as the magnetic susceptibility, near a critical point should behave as

$$\psi(z)=A_0(z)+A_1(z)|z-z_c|^{-\gamma_1}+A_2(z)|z-z_c|^{-\gamma_2}+\ldots,\quad z\to z_c, \quad (1)$$

where z is an appropriate temperature variable and $A_i(z)$ are regular functions at z_c.

This paper reviews some recent developments[2-4] in the area of series analysis[5], which permit an analysis of high temperature series consistent with Eq. (1). In particular we discuss (I) the basic theory of differential approximants; (II) a generalization for confluent singularity analysis; and (III) an application based on the field theoretic method of analysis developed by Nickel and Sharpe.[6] Possible implications of these developments in resolving the problem of hyperscaling are also discussed.

I Differential Approximants

Differential approximants are perhaps best introduced by considering the Dlog Padé approximant to a series, $\psi(z)=\Sigma_n c_n z^n$, for which the leading N+1 coefficients are known. These approximants

*Supported in part by the NSF through grant DMR-79-07238.

are based on a fit of dlog $\psi(z)/dz$ to the ratio of two polynomials, as in

$$\psi'(z)/\psi(z)=\left(\sum_{j=0}^{L}P_j z^j\right) \Big/ \left(\sum_{j=0}^{M}Q_j z^j\right), \tag{2}$$

where $Q_0=1$. Substituting the series expansion for $\psi(z)$ into Eq. (2) leads to a set of linear equations for the coefficients P_j,Q_j; hence the approximants are uniquely defined by L, M and the coefficients $c_0,c_1,\ldots,c_N$, where $N=L+M+1$. The critical point is identified as a pole of Eq. (2), and the associated critical exponent is the residue at the pole.

Differential approximants arise by viewing Eq. (2) as a member of the class of K-th order, linear differential equations with polynomial coefficients[2],

$$\left(\sum_{i=0}^{K}Q_i(z)D^i\right)\psi(z) = P(z), \tag{3}$$

where $Q_i(z)=\Sigma_{j=0}^{M_i}Q_{i,j}z^j$, $P(z)=\Sigma_{j=0}^{M_i}P_j z^j$ and the differential operator $D=d/dz$ or zd/dz. A solution to Eq. (3), $\psi_{[\vec{M}/L]}$, which is specified by the integer array $[M_0,M_1,\ldots,M_K;L]$, is termed a differential approximant. From this point of view, critical properties follow from the classic theory of ordinary differential equations.[8] For example, the singular points of $\psi(z)$ are the zeros of $Q_K(z)$, and the critical exponents are solutions of the indicial equation at these zeros. Eq. (3) also implies a linear recurrence relation among the coefficients c_i; for $D=zd/dz$ this relation is

$$\sum_{i=0}^{K}\sum_{j=0}^{M_i}Q_{i,j}(n-j)^i c_{n-j}=P_n. \tag{4}$$

Eq. (4) provides a set of linear equations relating the coefficients $Q_{i,j}$,P_j and c_n. Thus these approximants are no more difficult to obtain than are Dlog Padé approximants, though the possible singular structure a solution can possess is much richer. As the number of series coefficients is usually limited, however, it is impractical to go to very high order. Thus the generalizations of the Dlog Padé which have received most attention are 2nd order, homogeneous or $[M_0,M_1,M_2]$ approximants[2-4,9] and 1st order, inhomogeneous or $[M_0,M_1;L]$ approximants[10,11], the latter of which are also called integral approximants.[10] Provided the roots of $Q_K(z)$ are distinct, which is usually the case in practice, these approximants allow for a power law singularity and an analytic background at each singular point, $\psi(z)=A_0(z)+A_1(z)|z-z_c|^{-\gamma_1}$.

Various generalizations of these approximants are possible, in addition to those discussed below. For example, one can develop differential equations to fit functions with essential singularities, as in asymptotic perturbation expansions or at Kosterlitz-Thouless singularities. A generalization to partial differential approximants has been developed by Fisher and co-workers.[12]

II Confluent Approximants

As we have noted, quantities of interest in critical phenomena generally have confluent singularities as in Eq. (1). The leading pair of singularities can be represented by an $[M_0,M_1,M_2]$ approximant in which two of the zeros of $Q_2(z)$ are degenerate at z_c; also $Q_1(z_c)$ must vanish. The critical exponents are then solutions of a quadratic indicial equation; for D=zd/dz the form is

$$\gamma(\gamma+1)(z_c^2/2)Q_2''(z_c)-\gamma z_c Q_1'(z_c)+Q_0(z_c)=0. \tag{5}$$

Since Eq. (5) involves only Q_2'' and Q_1', the confluence at z_c does not have to be precise. However, it is often desirable to construct biased approximants, in which this confluence is built in exactly. This is done by forcing $Q_2'(z_c)=Q_2(z_c)=Q_1(z_c)=0$, which gives three linear constraints on the coefficients $Q_{i,j}$. The critical point z_c is then regarded as a parameter which can be varied to obtain a best fit. By forcing Eq. (5), one can also obtain approximants biased with the exact critical indices. Approximants without any such built-in parameters are referred to as unbiased.

Tests of this method[3] were carried out on functions formed from hypergeometric functions, which have a confluent structure as in Eq. (1). These tests showed that such confluent approximants usually give better estimates of the critical parameters than do Dlog Padé approximants. For example, estimates of z_c with unbiased approximants were usually better in accuracy by an order of magnitude. Similarly the accuracies of estimates of the dominant exponent γ_1 were roughly comparable for 10-term series, better by an order of magnitude at 20 terms, and by two orders of magnitude at 30 terms. Note that a significant number of terms are needed before substantial improvement is noticeable. The tests also showed that estimates of the confluent exponent γ_2 were improved by at least half an order of magnitude when the approximants were biased with the exact value of z_c, and that on variation of z_c, the scatter among the various approximants was minimal at the exact value.

The method was then used to analyze 12-term, spin-∞, FCC Ising model series[13] for χ the susceptibility and M_2, the second moment of the spin-spin correlation function. This analysis was carried out with biased $[M_0,M_1,M_2]$ approximants, in which both z_c and γ_1 were fixed, with γ_2 determined by minimizing the scatter. Central values for the Wegner correction-to-scaling exponent, $\Delta_1=\gamma_1-\gamma_2$, obtained in this way, ranged between 0.47 and 0.55. These results agree with those from other methods of confluent singularity analysis, including the Baker-Hunter method[14,15], the "method of 4-fits"[13], and a generalized ratio method[16], all of which are intrinsically non-linear. The differential approximants have the advantage that only linear equations are involved in their construction and they also give a global representation of the function which agrees exactly with the given series through the order used. Though a more definitive analysis is precluded by the shortness of these series, the test-series suggest that a significant improvement should be possible with more terms. An analysis of the extended 21-term series presented by Nickel[20] at this Institute is currently in progress.

III Confluent Singularities and Hyperscaling

An application[4] of differential approximants to Ising model series based on the method of Nickel and Sharpe (NS)[6,20] is now described. This study was designed to attack two puzzles concerning the spin-½ Ising model in 3-dimensions, as observed in a conventional high temperature series analysis: i) the apparent absence of the confluent singularities found in the spin-s Ising model and expected by universality; and ii) the apparent failure[17] of hyperscaling, the scaling law $d\nu=2-\alpha$. The NS method is similar to the field theoretic approach.[18,19] The validity of hyperscaling implies that the dimensionless, renormalized coupling constant $u\sim\xi^{-d}(\partial^2\chi/\partial H^2)/\chi^2$ remains finite as the correlation length ξ diverges; i.e.,

$$u=u^*+C_1\xi^{-\omega_1}+C_2\xi^{-\omega_2}+\ldots,\quad \xi\to\infty, \tag{6}$$

where $\omega_i=\Delta_i/\nu$ are correction-to-scaling indices; if hyperscaling fails, $u\sim\xi^{-d^*}$, where d^* is the "anomalous dimension of the vacuum." Turned around, Eq. (6) is equivalent to $x=\xi^2$ diverging at a finite value of $y=u^{-3/2}$, where x and y are the variables in the NS approach. Thus the NS approach focuses directly on confluent corrections. The resulting expression for x(y) has the form of Eq. (1) with $\gamma_1=2/\omega_1$, $\gamma_2=(2-\omega_2+\omega_1)/\omega_1$, etc. As in the field theoretic method, NS construct Padé approximants to $\gamma(y)=(\partial\ln x/\partial y)^{-1}$, the analog of the Callen-Symanzik β-function. This is equivalent to a Dlog Padé analysis of x(y).

Their analysis of the BCC γ(y) series gave $y^*=0.1594$, in support of hyperscaling and consistent with the value 0.1615 obtained for the continuum model.[18,19] However, puzzle i) takes on a new twist; their value $\omega_1\cong1.18$ differed substantially from 0.79, the value both for the continuum model and the spin-s models. This is not inconsistent with earlier series analyses[13,15,16], which suggested that the leading confluent singularity vanishes for spin-½. However, an alternate, more satisfying possibility is that the leading correction is simply weak. Thus the dominant behavior and the approach to criticality is controlled by higher order terms until $\xi>>1$. Since such behavior is inconsistent with a Dlog Padé analysis, a confluent singularity analysis was performed on the series for x(y) using biased $[M_0,M_1,M_2]$ approximants. These exhibited a point of minimal scatter at $y^*\cong 0.1602$, and gave $\omega_1\cong 0.79$, remarkably close to the expected value, and $\omega_2\cong 1.4$. As a check on the possibility that this agreement might be accidental, a numerical integration of the differential approximants was carried out. The amplitude of the leading singularity was found to be quite small, so it is not surprising that the Dlog Padé did not "see" it. Translated to thermal variables, i.e., writing $y=y^*(1+at^{\Delta_1})$, the amplitude a =-.14. This is several times larger than estimates for the confluent term in χ but not unreasonably so. Although longer series are needed to verify these results, we feel this study provides strong evidence in the spin-½ Ising model for the existence of confluent singularities consistent with the (n=1,d=3) universality class. It is not clear to us why similar confluent singular structure should not be present in the continuum model as well. If so this suggests yet another application of the differential approximants.

Acknowledgements: Portions of this work were carried out in collaboration with G.S. Joyce and A.J. Guttmann. The author thanks M.E. Fisher and B.G. Nickel for criticism and advice.

REFERENCES

1. F. Wegner, Phys. Rev. B5, 4529(1972)
2. A.J. Guttmann and G.S. Joyce, J. Phys. A5, L81(1972)
3. J.J. Rehr, G.S. Joyce, and A.J. Guttmann, J. Phys. A13,1587(1980)
4. J.J. Rehr, J. Phys. A12,L179(1979)
5. D.S. Gaunt and A.J. Guttmann, Phase Transitions and Critical Phenomena Vol. 3, C. Domb and M.S. Green ed(NY,Academic,1974)
6. B.G. Nickel and B. Sharpe, J. Phys. A12,1819(1979)
7. G.A. Baker, Jr., Essentials of Pade Approximants(NY,Academic,1974)
8. E.L. Ince, Ordinary Differential Equations(London, Longmans,1927)
9. A.J. Guttmann, J. Phys. A8,1081(1975)
10. D.L. Hunter and G.A. Baker, Jr., Phys. Rev. B19,3808(1979)
11. M.E. Fisher and H.Au-Yang, J. Phys. A12,1677 (1979)
12. M.E. Fisher and R.M. Kerr, Phys. Rev. Lett. 39, 667(1977)
13. D.M. Saul, M. Wortis, and D. Jasnow, Phys. Rev. B11,2571(1975)
14. G.A. Baker, Jr. and D.L. Hunter, Phys. Rev. B7,3377(1973)
15. W.J. Camp, D.M. Saul, J.P. VanDyke, and M. Wortis, Phys. Rev. B 14,3990(1976)
16. W.J. Camp and J.P. VanDyck, Phys. Rev. B11,2579(1975)
17. G.A. Baker, Jr., Phys. Rev. B15,1552(1977)
18. G.A. Baker, Jr., B.G. Nickel, and D.I. Meiron, Phys. Rev. B17, 1365(1978)
19. J.C. LeGuillou and J. Zinn-Justin, Phys. Rev. Lett. 39, 95(1977)
20. B.G. Nickel, 1980 Cargese Lecture Notes (this volume)

CRITICAL BEHAVIOUR FROM THE FIELD THEORETICAL RENORMALIZATION GROUP TECHNIQUES

Edouard Brézin

Service de Physique Théorique
CEN-Saclay
BP n°2
91190 Gif-sur-Yvette (France)

Many review articles and books deal at length with this subject[1]. Any reader who feels interested by these topics should not content himself with these few pages which are nothing but a brief and incomplete survey of some of the ideas involved in the renormalization group approach.

INTRODUCTION

The renormalization group was introduced as a technical device in quantum electrodynamics[2] : the conventional definition of the renormalized electric charge breaks down for massless particles (because of infra-red divergences). The proper definition of the charge introduces then an arbitrariness and the physics is independent of these choices. This led to the first formulation of the RG which found a few applications but the meaning of the whole approach remained unclear until it was realized that the real underlying concept was asymptotic scale invariance[3]. A theory like massless QED in (3+1) dimensions has no dimensional parameter, and thus is classically scale invariant ; however the quantum theory cannot be defined without the introduction of an additional length scale, and scale invariance is broken. However a scale transformation, can be absorbed into a redefinition of the charge and of the field strengths, and in the asymptotic regime at distances very large (or very small) compared to the additional length scale, asymptotic scale invariance can be restored whenever the RG flow of coupling constants leads to stable fixed points. K. Wilson realized that these concepts could be applied to many physical situations, such as critical phenomena, and that there were many alternative ways, without reference to conventional field theory calculations, to study the effect of a scale transformation on the physical parameters. This led to considerable developments and the influence of these ideas on condensed matter physics in general is growing every day. The aim of these few lectures is to introduce the usual field theory techniques. There are several reasons to spend some time studying this approach. The RG flow equations induced by a dilatation on the parameters of the theory are simpler. It is a theory in which a lot can be calculated, critical exponents, amplitude ratios, universal correlation functions and equation of state, etc... The ε-expansion[4] and three-dimensional calculations[5] can be brought into an accurate numerical scheme[6].

LIMITATIONS OF LANDAU THEORY

Let us introduce a coarse grained order parameter (magnetization) obtained by averaging the spins over a region small compared to the size of the system, but large compared to the microscopic length scale a (such as the lattice spacing)

$$\vec{\phi}(x) = \frac{1}{|D|} \sum_{i \in D} \vec{S}_i \quad .$$

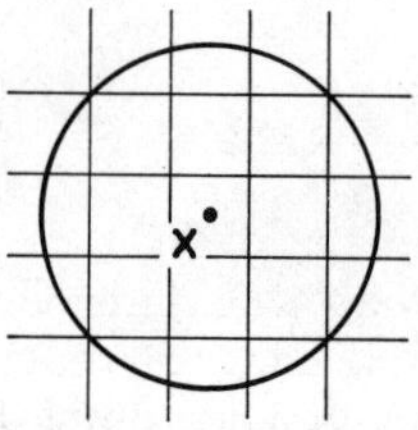

The Landau Hamiltonian $H(\phi)$ can be obtained with the following assumptions :

a) The Hamiltonian is invariant under the symmetry group of the high temperature phase. (For simplicity we will limit ourselves here to a rotation group).

b) In the critical region (T near T_c , small external fields, large correlation length) $\phi(x)$ is small and slowly varying. Therefore we can write (in zero external field)

$$\beta H = H = \int d^d x \left\{ \frac{1}{2} r_o \vec{\phi}^2 + \frac{1}{4!} u_o (\vec{\phi}^2)^2 + \frac{1}{6!} v_o (\vec{\phi}^2)^3 + \ldots + \frac{1}{2} (\nabla\vec{\phi})^2 + \ldots \right\} \tag{1}$$

the dots include terms like $\vec{\phi}^2(\nabla\vec{\phi})^2$, $(\phi_\alpha\vec{\nabla}\phi_\alpha)^2$, etc...

In addition one should remember that by definition of $\phi(x)$, it cannot vary when x is moved over one lattice cell. It is more convenient to express this in momentum space : $\phi(x)$ has no Fourier component larger that $a^{-1} = \Lambda$.

The partition function in an external field $H(x)$ is obtained as a sum over all field configurations compatible with this cut-off

$$Z(\vec{H}) = \int D\vec{\phi} \quad \exp\left\{- \mathcal{H} + \int d^d x \; \vec{H}(x).\vec{\phi}(x)\right\} \tag{2}$$

Landau Theory

It consists in the approximation that only the contribution from the most likely field configuration should be retained. Let $\vec{\phi}^*(x)$ be the solution which maximizes $\{-\mathcal{H} + \int dx \; \vec{H}.\vec{\phi}\}$:

$$\frac{\delta \mathcal{H}}{\delta \phi^*_\alpha(x)} = H_\alpha(x) \qquad (\alpha = 1,\ldots,n) \tag{3}$$

Then the approximation is simply

$$Z \simeq Z_0 = \exp\left\{ - \mathcal{H}(\phi^*) + \int dx \; \vec{H}.\vec{\phi}^*\right\} \tag{4}$$

It becomes slightly more familiar if we take the free energy $F(H) = -\text{Log}\, Z$, and Legendre transforms it to $\Gamma(M)$ by the usual procedure :

$$M\ (x) = - \frac{\delta F}{\delta H_\alpha(x)} \qquad \text{is solved for } H \tag{5}$$

and the thermodynamical potential $\Gamma(M)$ is thus given as

$$\Gamma(M) = F + \int dx \; \vec{H}(x).\vec{M}(x) \quad . \tag{6}$$

In Landau approximation, using the stationarity Eq.(3) we simply obtain

$$\vec{M}(x) = \vec{\phi}^*(x)$$

and thus

$$\Gamma_0(M) = \mathcal{H}(M) \quad . \tag{7}$$

This leads to the standard consequences of mean field theory[7].

Fluctuations

Landau theory assumes that in the sum over configurations,

all the stochastic fluctuations of the order parameter around its most likely value can be neglected. Therefore it is valid, if and only if the statistical fluctuations of ϕ around ϕ^* are not important. In order to examine this point, we shall go one step further and calculate the first effect of these fluctuations :

$$\vec{\phi} = \vec{\phi}^* + \vec{\psi} \tag{8}$$

$$H(\phi) = H(\phi^*) + \frac{1}{2}\int dx\, dy \left.\frac{\delta^2 H}{\delta\phi(x)\,\delta\phi(y)}\right|_{\phi^*} \psi(x)\,\psi(y) + O(\psi^3) \tag{9}$$

(There is no term linear in ψ , by definition of ϕ^*). The substitution of (9) into the partition function leads to a Gaussian integral. Noting that

$$\int \prod_{i=1}^{N} d\,\psi_i \;\exp\left\{-\frac{1}{2}\;\psi_i A_{ij}\psi_j\right\} = (2\pi)^{N/2}\;(\det A)^{-1/2}$$

and that $\log\det A \equiv \operatorname{tr}\log A$, we obtain a first correction to Landau result :

$$F_1 = F_o + \frac{1}{2}\operatorname{tr}\log A$$

$$A(x,y) = \left.\frac{\delta^2 H}{\delta\phi(x)\;\delta\phi(y)}\right|_{\phi(x)=\phi^*(x)}$$

and it is a straightforward (but worth doing) exercise to verify that this leads to

$$\Gamma_1(M) = \Gamma_o(M) + \frac{1}{2}\;\operatorname{tr}\log A_{\alpha\beta}(M) \tag{10}$$

$$A_{\alpha\beta}(x,y) = \left.\frac{\delta^2 H}{\delta\phi_\alpha(x)\;\delta\phi_\beta(y)}\right|_{\phi(x)=M(x)} \tag{11}$$

There are many consequences of these formulae. Let us for simplicity study the simple example of the zero field susceptibility above T_c . In a translationally invariant situation the formulae (10),(11) reduce to

$$A_{\alpha\beta}(x,y) = -\vec{\nabla}^2\;\delta_{\alpha\beta}\;\delta(x-y) + \left(r_o + \frac{u_o}{2}\,\vec{M}^2\right)\frac{M_\alpha M_\beta}{M^2}$$

$$+ \left(r_o + \frac{u_o}{6}\,\vec{M}^2\right)\left(\delta_{\alpha\beta} - \frac{M_\alpha M_\beta}{M^2}\right) + O(M^4)$$

and then

$$\gamma(M) = \frac{\Gamma(M)}{\text{volume}} = \frac{1}{2}\,r_o M^2 + \frac{1}{2}\int\frac{d^d p}{(2\pi)^d}\;\log\left(p^2 + r_o + \frac{u_o}{2}\,M^2\right)$$

$$+ \left(\frac{n-1}{2}\right)\int\frac{d^d p}{(2\pi)^d}\;\log\left(p^2 + r_o + \frac{u_o}{6}\,M^2\right) + O(M^4) \quad .$$

The uniform field H is $\frac{\partial\gamma}{\partial M}$, and thus the susceptibility :

$$\chi^{-1} = \frac{\partial H}{\partial M} = \frac{\partial^2\gamma}{\partial M^2} .$$

In zero field above T_c this leads to :

$$\chi^{-1} = \left.\frac{\partial^2\gamma}{\partial M^2}\right|_{M=0} = r_o + \frac{n+2}{6} u_o \int \frac{d^dp}{(2\pi)^d} \frac{1}{p^2+r_o} . \quad (12)$$

The critical temperature T_c is defined by an infinite susceptibility :

$$0 = r_{oc} + \frac{n+2}{6} u_o \int \frac{d^dp}{(2\pi)^d} \frac{1}{p^2+r_{oc}} . \quad (13)$$

(In the integral, we can replace r_{oc} by its lowest order value zero, since we are doing a first order calculation).

Subtraction of (13) from (12) gives

$$\chi^{-1} = r_o - r_{oc} - \left(\frac{n+2}{6}\right) u_o (r_o - r_{oc}) \int \frac{1}{p^2(p^2+r_o)} \frac{d^dp}{(2\pi)^d} \quad (14)$$

All the momentum integrals are of course cut-off at Λ. We can now answer the question of the validity of Landau theory : when r_o approaches r_{oc}, the correction term does not modify the leading result (a simple pole for χ) provided $\lim_{r_o \to r_{oc}} \int^{\Lambda} \frac{d^dp}{p^2(p^2+r_o)} < \infty$, and at this order, this means simply that $\int^{\Lambda} \frac{d^dp}{p^4} < \infty$, which is only true above four dimensions. Below four dimensions (up to four included) the fluctuations of the statistical distribution of the order parameter are infinite and the mean field picture is not valid. The RG will be essential to deal with these strong fluctuations.

DILATATIONS

We begin by elementary dimensional analysis. Deciding by convention that a wave number has dimension 1, and hence a length has dimension -1, from the dimensionless character of H we deduce

$$[(\nabla\phi)^2] = [r_o\phi^2] = [u_o\phi^4] = [v_o\phi^6] = [H(x)\,\phi(x)] = d$$

and therefore

$$[\Lambda] = 1$$

$$[\phi(x)] = [M] = \frac{1}{2}(d-2) \quad (15)$$

$$
\begin{aligned}
[r_o] &= 2 \\
[H] &= \frac{1}{2}(d+2) \\
[u_o] &= (4-d) \\
[v_o] &= 2(4-d)-2 \qquad \text{etc ...} \\
[\chi^{-1}] &= 2
\end{aligned}
\tag{15}
$$

It will be convenient therefore to use the reduced temperature

$$r_o - r_{oc} = t$$

and the dimensionless parameters

$$
\begin{aligned}
\frac{u_o}{\Lambda^{4-d}} &= u_1 \\
\frac{v_o}{\Lambda^{2(4-d)-2}} &= v_1 \qquad \text{etc...}
\end{aligned}
\tag{16}
$$

and thus $\chi = \chi(t,u_1,v_1,\ldots,\Lambda)$.

From dimensional considerations alone we do not go very far :

$$(t,u_1,v_1,\ldots,\lambda\Lambda) = \lambda^{-2} \quad \chi(\frac{t}{\lambda^2}, u_1,v_1,\ldots,\Lambda) \quad . \tag{17}$$

In the l.h.s. the problem of strong fluctuations arises when $t/(\lambda\Lambda)^2$ is much smaller than one. In the r.h.s. when $(t/\lambda^2)/\Lambda^2$ is much smaller than one. It is of course the same problem.

The RG equations are of a different nature. They express that a change of the length scale $\Lambda \to \lambda\Lambda$ can be absorbed into an adequate modification of the parameters $t \to t(\lambda)$, $u_1 \to u_1(\lambda)$, and a rescaling $\zeta^{-1/2}(\lambda)$ of the order parameter. (We will assume momentarily that coupling constants like v_1 are not present and discuss later their real effect). These properties, as well, as the explicit construction of these scale transformations, will be explained below. If we accept this, we find

$$\chi(t,u_1,\Lambda) = \zeta(\lambda) \; \chi\Big(t(\lambda) \, , u_1(\lambda) \, , \lambda\Lambda\Big) \tag{18}$$

and with dimensional analysis

$$\chi(t,u_1,\Lambda) = \zeta(\lambda) \; \lambda^{-2} \; \chi\left(\frac{t(\lambda)}{\lambda^2} \, , \, u_1(\lambda) \, , \Lambda\right) \tag{19}$$

We have now the possibility with this free parameter λ of escaping from the strong fluctuation regime. In the l.h.s. we are dealing with $t/\Lambda^2 \ll 1$. In the r.h.s. we can use the freedom of the dilatation in order to prevent $(t(\lambda)/\lambda^2)/\Lambda^2$ to become small. We thus make a dilatation by a factor λ_o^{-1} such that

$$\frac{t(\lambda_o)}{\lambda_o^2} = t_o \quad \text{and} \quad \frac{t_o}{\Lambda^2} \text{ is a fixed arbitrary number.} \tag{20}$$

Equation (20) defines λ_o as a function of the physical temperature t. The meaning of λ_o is made clearer if we apply the same equations to the correlation length $\xi(t,u_1,\Lambda)$: it is a renormalization group invariant (i.e. it is not affected by the rescaling of the order parameter). Therefore

$$\begin{aligned} \xi(t,u_1,\Lambda) &= \xi\left(t(\lambda)\,,\, u_1(\lambda)\,,\, \lambda\Lambda\right) \\ &= \lambda^{-1}\,\xi\left(\frac{t(\lambda)}{\lambda^2}\,,\, u_1(\lambda)\,,\, \Lambda\right) \\ &= \lambda_o^{-1}\,\xi\left(t_o\,,\, u_1(\lambda)\,,\, \Lambda\right) \end{aligned} \tag{21}$$

and if the dependence upon λ_o of $\xi(t_o\,,\, u_1(\lambda_o))$ is weak, we see that λ_o is proportional to ξ^{-1} : we are using the dilatation up to the correlation length itself. The flow equations under rescaling cannot be obtained without approximations. But before we describe them, we can examine what exactly we are looking for ; the equations that we will obtain will have the following form :

defining $\dot{X} \equiv \lambda \frac{dX}{d\lambda}$, we will find

$$\begin{aligned} \dot{u}_1 &= \beta(u_1) & u(1) &= u \\ \dot{t} &= t\,\kappa(u_1) & t(1) &= t \\ \dot{\zeta} &= \zeta\,\eta(u_1) & \zeta(1) &= 1 \end{aligned} \tag{22}$$

For t/Λ^2 small, ξ is large, and thus λ_o is approaching zero. The first equation shows that $u_1(\lambda)$ approaches in the small λ limit, a zero of the β-function with positive slope (an infrared stable fixed point, in the usual terminology) :

$$\begin{aligned} \beta(u_1^*) &= 0 \\ \omega &= \beta'(u_1^*) > 0 \end{aligned} \tag{23}$$

$$u_1(\lambda) - u_1^* \underset{\lambda_o \to 0}{\sim} \lambda^{\omega} \quad . \tag{24}$$

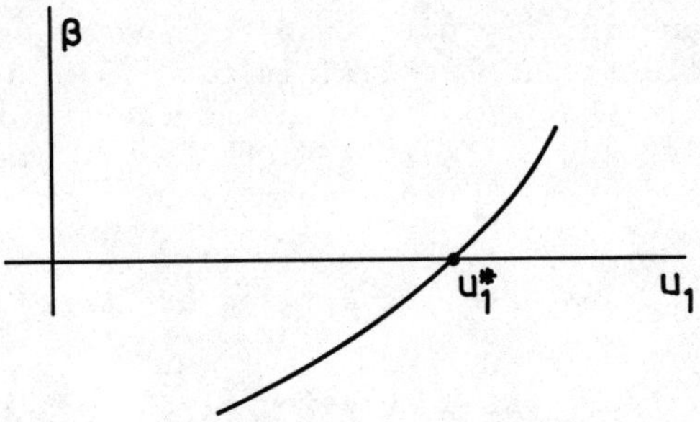

Then, integration of the last equations (22) leads to

$$\frac{t(\lambda)}{t} \underset{\lambda\to 0}{\sim} \lambda^{\kappa} \qquad \text{with} \qquad \kappa \equiv \kappa(u_1) \tag{25}$$

$$\zeta(\lambda) \underset{\lambda\to 0}{\sim} \lambda^{\eta} \qquad \text{and} \qquad \eta \equiv \eta(u_1) \quad . \tag{26}$$

The equation for λ_o is thus

$$t_o \frac{\lambda_o^2}{t} \sim \lambda_o \quad , \quad \text{i.e.} \quad \lambda_o \sim \left(\frac{t}{t_o}\right)^{1/(2-\kappa)}$$

but since $\xi \sim \lambda_o^{-1}$ this means that

$$\xi \underset{t\to 0}{\simeq} \xi_o \, t^{-\nu}$$

with

$$\frac{1}{\nu} = 2-\kappa \quad . \tag{27}$$

Finally equation (19) becomes

$$\chi(t,u_1,\Lambda) \underset{t/\Lambda^2\to 0}{=} \lambda_o^{\eta-2} \, \chi(t_o,u_1^*,\Lambda)$$

or

$$\chi(t,u_1,\kappa) \underset{t/\Lambda^2\to 0}{=} \chi_o \, t^{-\gamma}$$

with

$$\gamma = \nu(2-\eta) \tag{28}$$

Thus scaling and universality of the critical exponents (here independence with respect to u_1) follow from this RG flow.

Remark : η in these formulae has been defined as $\eta(u_1^*)$ and it is not clear at this stage that it coincides with the usual exponent. In order to convince oneself that it is so, one writes the RG equation for the order-parameter correlation function

$$G(p) = \int e^{i\vec{p}\,\vec{x}} \langle\vec{\phi}(x) \cdot \vec{\phi}(0)\rangle \, d^d x$$

$$G(p,t,u_1,\Lambda) = \zeta(\lambda)\, G\Big(p\,,\, t(\lambda)\,,\, u_1(\lambda)\,,\, \lambda\Lambda\Big)$$

$$= \zeta(\lambda)\, \lambda^{-2}\, G\left(\frac{p}{\lambda}\,,\, \frac{t(\lambda)}{\lambda^2}\,,\, u_1(\lambda)\,,\, \Lambda\right) \quad .$$

Now, going again to the scale $\lambda_o = \xi^{-1}$,

$$G(p,t,u_1,\Lambda) = \xi^{2-\eta}\, G(p\xi) = p^{-(2-\eta)}\, F(p\xi)$$

which shows that η agrees with its conventional definition.

Irrelevant Variables

Up to now we have not discussed the role of operators of higher dimensions, ϕ^6 , $\phi^2(\nabla\phi)^2$ etc... . Let us for instance look at the operator $v_o\phi^6 = \frac{1}{\Lambda^{2-2(4-d)}} v_1\phi^6$. In four dimensions this operator is weighted by a coupling constant of order $1/\Lambda^2$. This is not sufficient to eliminate this operator, even in the critical region where Λ is larger than all the parameters of same dimension ($t^{1/2}$, $H^{2/(d/2)}$). Indeed if we start with a small v_1 , and includes its effect perturbatively, the operator ϕ^6 provides Λ^2 divergent integrals. For instance in the 4-point correlation function, in addition to the usual diagrams

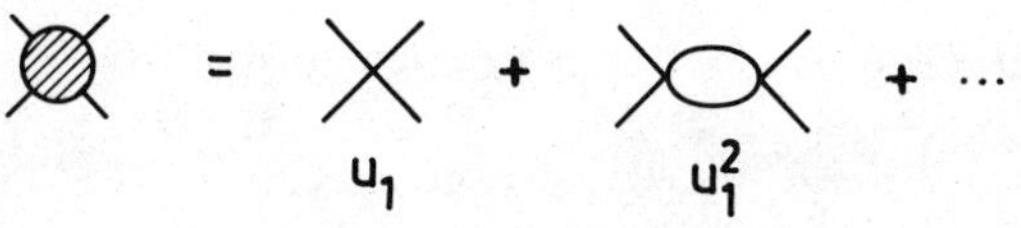

we now have to consider diagrams such as 

v_1 (A)

or 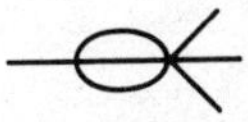 , etc...

$u_1\, v_1$ (B)

Their rigorous analysis is cumbersome[8] but its result is easy to understand. A diagram of type A is a (constant) $\times\,\Lambda^2$. However it is multiplied by $1/\Lambda^2$ by definition of v_1 ; it can thus be absorbed into a redefinition of u_1 . The diagram (B) has a (constant) $\times\Lambda^2$ plus momentum dependent contributions multiplied by $\log\Lambda$. If we divide by Λ^2 , the only piece which remains is the constant. The general result is that the perturbative insertion of ϕ^6 with a weight $1/\Lambda^2$ in the large Λ limit, is to modify the coefficients of all the operators of lower dimensions, ϕ^4 , ϕ^2 , $(\nabla\phi)^2$. However since we will treat the most general Hamiltonian with operators of dimension smaller or equal to four, the critical behaviour will not be different if we insert these higher "irrelevant" operators. This is true in four dimensions, and remains true within the (4-d) expansion.

The Origin of the RG equations

Why does a dilatation can be absorbed into a redefinition of the parameters, and why is related to renormalization problems ? In principle we should go into the theory for instance perturbatively, perform a change in the length scale and try to adjust t , u_o , the order parameter normalization in order to go back to the old length scale ; for instance we calculate the expansion for the zero-momentum four-point function, near four dimensions ($\varepsilon = 4-d$ is thus a second expansion parameter)

$\Gamma^{(4)} =$  $= u_o - \frac{3}{2}\left(\frac{n+8}{9}\right) u_o^2$ 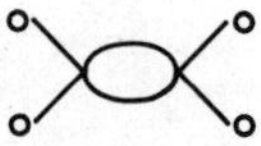 $+ O(u_o^3)$

$$= \Lambda^{\varepsilon}\left[u_1 - \frac{n+8}{6} u_1^2 \int^{\Lambda} \frac{d^d p}{(2\pi)^d (p^2+\Gamma_o)^2} + O(u_1^3)\right]$$

$$= u_1 - \frac{n+8}{6} u_1^2 \frac{2\pi^2}{(2\pi)^4} \log\Lambda + \varepsilon u_1 \log\Lambda + O(u_1^3, \varepsilon^2 u_1, \varepsilon u_1^2) \ .$$

If we multiply by $\frac{2\pi^2}{(2\pi)^4}$ we redefine $u_1 \frac{2\pi^2}{(2\pi)^4}$ as the new coupling constant, noted again u_1 , we obtain

$$\Gamma^{(4)} = u_1 - \frac{n+8}{6} u_1^2 \log\Lambda + \varepsilon u_1 \log\Lambda + O(u_1^3, \varepsilon u_1^2, \varepsilon^2 u_1) \ .$$

We now look for $u_1(\lambda)$ such that

$$\Gamma^{(4)}(\lambda\Lambda, u_1(\lambda)) = \Gamma^{(4)}(\Lambda, u_1)$$

and it leads immediately to

$$u_1(\lambda) = u_1 - \varepsilon u_1 \log\lambda + \frac{n+8}{6} u_1^2 \log\lambda + O(u_1^3, \varepsilon u_1^2, \varepsilon^2 u_1)$$

and therefore to

$$\dot{u}_1 = \beta(u_1) = -\varepsilon u_1 + \frac{n+8}{6} u_1^2 + O(u_1^3, \varepsilon u_1^2, \varepsilon^2 u_1) \tag{29}$$

Similarly if we return to the expansion (14) for the susceptibility

$$\chi^{-1} = t + \left(\frac{n+2}{6}\right) t\, u_1 \log\Lambda + O(u_1^2, \varepsilon u_1)$$

and express that at this order

$$\chi^{-1}(u_1(\lambda), \lambda\Lambda) = \chi^{-1}(u_1, \Lambda)$$

$(\zeta(u_1) = 1 + O(u_1^2))$, we obtain

$$\frac{t(\lambda)}{t} = 1 + \frac{n+2}{6} u_1 \log\lambda + O(u_1^2, \varepsilon u_1)$$

and therefore

$$\kappa(u_1) = \frac{\dot{t}}{t} = +\frac{n+2}{6} u_1 + O(u_1^2, \varepsilon u_1) \tag{30}$$

This procedure is not very convenient, but it could be carried further without difficulty. The reason of its success can indeed be related to renormalization. In four dimensions (and within a double series expansion in powers of u_o and $(4-d)$) the limit $\Lambda \to \infty$ of the correlation functions exists provided they are expressed in terms of renormalized parameters t_R, u_R and that a change of scale of the order parameter is performed, [9] . For the N-point (one-particle irreducible) correlation functions the statement is

$$\Gamma^{(N)}(u_o, t, \Lambda) = Z^{N/2}(\Lambda)\, \Gamma_R^{(N)}(u_R, t_R) + O(1/\Lambda^2) \quad . \tag{31}$$

Of course u_o and t are functions of u_R, t_R and of .

We now define $u_o(\lambda) = u_o(u_R, t_R, \lambda\Lambda)$

$$t(\lambda) = t(u_R, t_R, \Lambda)$$

From Eq.(31) we obtain :

$$\frac{\Gamma^{(N)}\left(u_o(\lambda), t(\lambda), \lambda\Lambda\right)}{\Gamma^{(N)}(u_o, t, \Lambda)} = \left[\frac{Z(\lambda\Lambda)}{Z(\Lambda)}\right]^{N/2} = \zeta^{N/2}(\lambda) \tag{32}$$

which is the equation that we assumed previously. We can therefore forget the r.h.s. of (31), but it is its invariance under a re-scaling of the length Λ^{-1} which leads directly to the RG equations.

THE RENORMALIZED THEORY AND THE ε-EXPANSION

This section will review some useful technical devices. Sometimes these techniques do not only simplify the calculations, they also make possible to solve some problems that would be too difficult otherwise. We will briefly describe the following simplifications :

i) the diagrams involved in the calculation of RG flows are simpler in the renormalized theory (no cut-off at all).

ii) all the ε-expansion can be generated from the 4-dimensional theory without any additional calculation if one uses an appropriate subtraction scheme.

iii) all the critical exponents can be calculated at T_c , and the diagrams are much simpler for $T = T_C$.

The Renormalized Theories and the Minimal Subtraction Scheme

We start again from

$$H = \int d^d x \left[\frac{1}{2} (\nabla\vec{\phi})^2 + \frac{1}{2} r_o \vec{\phi}^2 + \frac{u_o}{4!} (\vec{\phi}^2)^2\right]$$

and rescale the order parameter, by a factor that will be fixed later

$$\vec{\phi} = \sqrt{Z}\ \vec{\phi}_R \quad . \tag{33}$$

The correlation functions of ϕ_R will be expressed in terms of parameter u_R and t_R , which will be defined later, and are such that these correlation functions are finite when $\Lambda \to \infty$ and $\varepsilon \to 0$. Z itself will be a power series in power of u_o (or rather u_R) determined by the same requirements. The effect of a scale transformation on these renormalized functions may be obtained by the same method that we used for the physical ones. We write for instance

$$u_R = u_R\left(u_o , t , \frac{\mu}{\Lambda}\right)$$

in which μ is the length scale in the renormalized theory :

$$u_R(\lambda) = u_R\left(u_o , t , (\lambda \frac{\mu}{\Lambda})\right)$$

$$\dot{u}_R = \beta(u_R) = \mu \frac{\partial}{\partial\mu}\bigg|_{u_o, t} u_R(u_o , t , \frac{\mu}{\Lambda})$$

and similarly for the other functions.

There are several equivalent ways of working out this renormalization program. One of the simplest is the minimal subtraction scheme[10] : u_o and Z are determined as powers series in u_R in $(4-\varepsilon)$

dimensions with coefficients which have only poles when $\varepsilon \to 0$ (without any finite part) :

$$u_o = \Lambda^{\varepsilon} u_1 = \mu^{\varepsilon} u_R Z_u \tag{34}$$

with μ being an arbitrary parameter which makes u_R dimensionless :

$$Z_u = 1 + u_R \frac{a_1}{\varepsilon} + u_R^2 \left(\frac{a_{21}}{\varepsilon^2} + \frac{a_{22}}{\varepsilon}\right) + u_R^n \left(\frac{a_{n1}}{\varepsilon^n} + \ldots + \frac{a_{nn}}{\varepsilon}\right) + \ldots \tag{35}$$

$$Z = 1 + 0.u_R + u_R^2\left(\frac{b_{21}}{\varepsilon}\right) + \ldots + u_R^n \left(\frac{b_{n1}}{\varepsilon^{n-1}} + \ldots + \frac{b_{n,n-1}}{\varepsilon}\right) + \ldots \tag{36}$$

These equations, plus the finiteness requirement of the correlation functions, determine u_R and the coefficients $a_{n,p}$ and $b_{n,p}$ uniquely and by calculations which are simpler than the corresponding cut-off diagrams. For instance we consider the one-loop contributions to the 4-point function (at this order Z=1)

$$\Gamma_R^{(4)}(\underset{\sim}{p}) = \mu^{\varepsilon} u_R Z_u Z^2 - \mu^{2\varepsilon} u_R^2 Z_u^2 Z^4 \frac{n+8}{18} \int \frac{d^d q}{(2\pi)^d} [q^2 (p_1+p_2+q)^2]^{-1}$$

$$+ 2 \text{ crossed terms} + O(u_R^3)$$

$$= \mu^{\varepsilon} \; u_R + u_R^2 \frac{a_1}{\varepsilon} - \frac{n+8}{6} u_R^2 \frac{2\pi^2}{(2\pi)^4} \frac{1}{\varepsilon} + \text{finite part} \quad .$$

Therefore

$$a_1 = \frac{n+8}{6} \quad \left(\text{if } \frac{2\pi^2}{(2\pi)^4} \text{ is included again in } u_R\right) \quad . \tag{37}$$

Let us work out the β function in this scheme. Differentiating (34) with respect to μ with fixed u_o we obtain :

$$0 = \varepsilon + \beta(u_R) \frac{1}{u_R} + \beta(u_R) \frac{d}{du_R} \log Z_u$$

i.e.

$$\beta(u_R) = - \frac{\varepsilon u_R}{1 + u_R \frac{d}{du_R} \log Z_u} \quad . \tag{38}$$

We can write (35) as

$$Z_u = 1 + \frac{1}{\varepsilon} A_1(u_R) + \frac{1}{\varepsilon^2} A_2(u_R) + \ldots \tag{39}$$

$$A_1 = a_1 u_R + a_{22} u_R^2 + \ldots + a_{nn} u_R^n + \ldots$$

$$A_2 = a_{21} u_R^2 + \ldots \ldots\ldots + a_{n\,n-1} u_R^n + \ldots$$

$$\log Z_u = 1 + \frac{1}{\varepsilon} A_1 + \frac{1}{\varepsilon^2}(A_2 - \frac{1}{2} A_1^2) + \dots$$

$$\left(1 + u_R \frac{d}{du_R} \log Z_u\right)^{-1} = 1 - \frac{1}{\varepsilon} u_R \frac{dA_1}{du_R} + \frac{1}{\varepsilon^2}\left[u_R^2\left(\frac{dA_1}{du_R}\right)^2 - u_R \frac{d}{du_R}(A_2 - \frac{1}{2} A_1^2)\right] + \dots$$

Putting this into (38) and remembering that $\beta(u_R)$ is finite when $\varepsilon \to 0$ we obtain

$$\beta(u_R) = -\varepsilon u_R + u_R^2 \frac{dA_1}{du_R}$$

$$0 = u_R^2\left(\frac{dA_1}{du_R}\right)^2 - u_R \frac{d}{du_R}(A_2 - \frac{1}{2} A_1^2)$$

etc...

$$u_R^2 \frac{dA_1}{du_R} = a_1 u_R^2 + 2a_{22} u_R^3 + \dots + n\, a_{nn} u_R^{n+1} + \dots$$

is the four-dimensional β function. We thus have a connection valid to all orders in ε between the d-dimensional and the 4-dimensional β-functions :

$$\beta_d(u_R) = -(4-d)u_R + \beta_4(u_R) \quad . \tag{40}$$

Similarly we have (see the last equation (22))

$$\eta(u_R) = \mu \left.\frac{\partial}{\partial\mu}\right|_{u_o, t} \log Z \tag{41}$$

$$= \beta(u_R) \frac{d}{du_R} \log Z \quad . \tag{42}$$

If we write

$$Z = 1 + \frac{B_1(u_R)}{\varepsilon} + \frac{B_2(u_R)}{\varepsilon^2}$$

$$\log Z = \frac{B_1(u_R)}{\varepsilon} + \frac{1}{\varepsilon^2}\left(B_2(u_R) - \frac{1}{2} B_1^2(u_R)\right) + \dots$$

$$\frac{d}{du_R} \log Z = \frac{1}{\varepsilon} B_1'(u_R) + \frac{1}{\varepsilon^2}(B_2 - \frac{1}{2} B_1^2)' + \dots$$

we obtain

$$\eta_d(u_R) = \left[-\varepsilon\, u_R + \beta_4(u_R)\right]\left[\frac{1}{\varepsilon} B_1'(u_R) + \frac{1}{\varepsilon^2}(B_2 - \frac{1}{2} B_1^2)' + \dots\right]$$

The limit of η_d when $d \to 4$ exists (from the renormalizability of the 4-dimensional theory). This gives

$$\eta_d(u_R) = - u_R B_1'(u_R) \equiv \eta_4(u_R) \tag{43}$$

plus the identities

$$\beta_4 B_1' - u_R(B_2 - \frac{1}{2} B_1^2)' = 0 \qquad \text{etc} \ldots$$

Therefore the calculation of the critical exponents in the ε-expansion can be done from the knwoledge of the four-dimensional functions $\beta_4(u_R)$, $\eta_4(u_R)$, $\kappa_4(u_R)$.

The fixed point u_R^* is the solution of the equation

$$\varepsilon u_R^* = \beta_4(u_R) = a_1 u_R^{*2} + 2 a_{22} u_R^{*3} + \ldots$$

$$u_R^* = \frac{\varepsilon}{a_1} - 2 \frac{a_{22}}{a_1^3} \varepsilon^2 + \ldots \tag{44}$$

and for instance the critical exponent η is equal to

$$\eta \equiv \eta_4(u_R) = -2 b_{21} u_R^{*2} - 3 b_{32} u_R^{*3} + \ldots$$

$$= - \frac{2 b_{21}}{a_1^2} \varepsilon^2 + \left(\frac{8 b_{21} a_{22}}{a_1^4} - \frac{3 b_{32}}{a_1^3} \right) \varepsilon^3 + O(\varepsilon^4) \ . \tag{45}$$

Therefore the critical exponents are determined entirely by the coefficients of the poles in ε of the subtraction constant. This is to be contrasted with ordinary schemes in which

$$\beta_d(u_R) = -\varepsilon u_R + \beta_4(u_R) + \varepsilon(d_1 u_R^2 + d_2 u_R^3 + \ldots) + \varepsilon^2(e_1 u_R^2 + e_2 u_R^3 + \ldots) + \ldots$$

$$\eta_d(u_R) = \eta_4(u_R) + \varepsilon(f_1 u_R^2 + f_2 u_R^3 + \ldots) + \varepsilon^2(g_1 u_R^2 + \ldots) + \ldots \ .$$

There a calculation of η to order ε^n requires a calculation of β_4 to order u_4^{n+1} but also of all the extra-terms up to $\varepsilon^{n-1} u_R^2$, of η_4 up to order (n-2) and of the extra-terms up to $\varepsilon^{n-2} u_R^2$

Calculation of the Critical Exponents at T_c

The exponent η is related to the 2-point correlation function at T_c by :

$$G^{(2)}(p,T_c)^{-1} \equiv \Gamma^{(2)}(p,T_c) \underset{p \ll \Lambda}{\sim} p^{2-\eta} \tag{46}$$

but other exponents like ν and γ are related to the temperature behaviour of the correlation functions and they involve a priori calculations away from T_c which are more difficult because the propagator is not a simple $1/p^2$. However one can relate ν to the correlation functions at T_c involving ϕ^2. For instance we consider

$$\Gamma^{(2,1)}(p_1,p_2\,;T_c) = \int \langle \vec{\phi}(x_1).\vec{\phi}(x_2)\ (\vec{\phi}^2(x_3))\rangle_{\text{amputated}} \times \exp\ ip_1(x_1-x_3) + ip_2(x_2-x_3) \quad ;$$

in terms of diagrams

$\Gamma^{(2,1)} =$

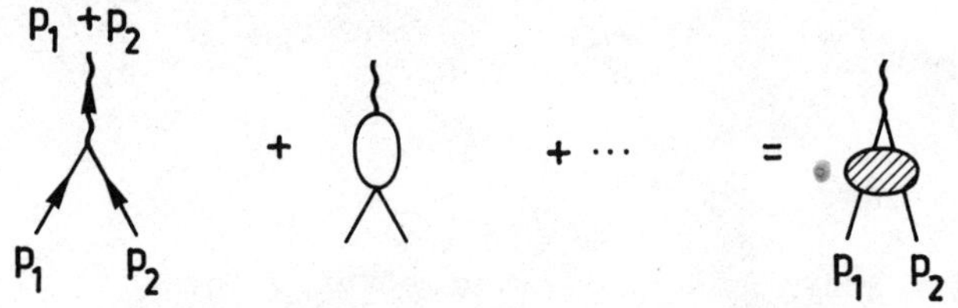

From the RG equations one verifies that

$$\Gamma^{(2,1)}(\lambda p_1\,,\lambda p_2\,;T_c) \underset{\lambda\to 0}{\sim} \lambda^{2-\frac{1}{\nu}-\eta} \tag{47}$$

Therefore this makes the calculation of ν possible from the knowledge of the $\vec{\phi}^2$ renormalization constant

$$\Gamma^{(N,M)} = \langle \phi_1 \dots \phi_N\ (\phi^2)_1 \dots (\phi^2)_M \rangle_{\text{amputated}}$$

$$= Z^N\,\hat{Z}^M\,\Gamma_R^{(N,M)} \tag{48}$$

$$\kappa(u_R) = \beta(u_R)\,\frac{d}{du_R}\log\hat{Z}\,(u_R) \tag{49}$$

and calculating the $1/\varepsilon$-poles of the series for $\hat{Z}$ we obtain the ε-expansion for ν from

$$\kappa(u_R^*) = 2 - \frac{1}{\nu} \quad .$$

<u>The ε-expansion</u>

It has carried up to order ε^4 for γ (or ν) and to order ε^5 for η [11]. The results for $\varepsilon=1$ have the following features : they give a succession of approximations for the exponent which are (for example if n=1) :

$\gamma^{(0)} = 1 \qquad \gamma^{(1)} = 1.167 \qquad \gamma^{(2)} = 1.244 \qquad \gamma^{(3)} = 1.195 \qquad \gamma^{(4)} = 1.375$

to be compared with experimental values around 1.24 . These results indicate a divergence of the ε-expansion at $\varepsilon=1$, and it is sometimes concluded that as a consequence the ε-expansion should be limited to second-order with an accuracy in principle limited. If the divergence is certainly present, it is not certain that the accuracy of the ε-expansion is limited. In fact the large order behaviour of this expansion is known[12]. If we write

$$\gamma = \sum_0^\infty \gamma_k \varepsilon^k$$

then

$$\gamma_k \underset{k\to\infty}{\sim} k! \left(-\frac{3}{n+8}\right)^k k^{3+\frac{1}{2}n} \quad . \qquad (50)$$

This shows that the ε-expansion is divergent for any ε . However from (50) one can make transformations, like Padé-Borel, on the ε-series which lead to seemingly convergent algorithms. The accuracy with the available terms is of the order of 1% and it seems clear that if more terms were available, this would lead to a good precision. However it turns out that direct three-dimensional calculations[5], which are not a priori better since they also involve similarly divergent series which are to be summed for a parameter of order one (the fixed point), have been pushed to larger orders[13] and the results are more accurate[6] than those of the ε-expansion.

REFERENCES

1. K.G. Wilson and J. Kogut, Phys.Rep. 12C, 75 (1974),
S.K. Ma (1976) Modern Theory of Critical Phenomena (Reading, Massachussets, Benjamin Advanced Book Program).
Phase transitions and Critical Phenomena Vol.6 (Eds. C.Domb and M.S.Green) London, Academic Press (1976),
D.J. Amit (1978) Field Theory, The Renormalization Group and critical phenomena (New-York, Mc Graw-Hill).

2. E.C.Stueckelberg and A.Petermann, Helv.Phys.Acta 26, 499 (1953),
M.Gell-Mann and F.Low, Phys.Rev. 95, 1300 (1954).

3. C.G.Callan, Phys.Rev. D2, 1541 (1970),
K.Symanzik, Commun.Math.Phys. 18, 227 (1970),
K.Wilson, Phys.Rev. D2, 1473 (1970).

4. K.G.Wilson and M.E.Fisher, Phys.Rev.Letters 28, 240 (1972),
K.G.Wilson, Phys.Rev.Letters 28, 548 (1972),
E.Brézin, J.C.Le Guillou, J.Zinn-Justin and B.G.Nickel, Phys. Letters 44A, 227 (1973).

5. G.Parisi, Lectures given at Cargèse Summer Institute (1973) (unpublished),
B.G.Nickel, J.Math.Phys. 19, 542 (1978).

6. J.C.Le Guillou and J.Zinn-Justin, Phys.Rev.Letters 39, 95 (1977), G.A.Baker, B.G.Nickel, M.S.Green and P.I.Meiron, Phys.Rev. Letters, 36, 1351 (1976).

7. See, for instance, L.D.Landau and E.M.Lifshitz, Statistical Physics (London ; Pergamon Press) 1959.

8. E.Brézin, J.C.Le Guillou and J.Zinn-Justin, in Phase transitions and Critical Phenomena, Vol.6 (op.cit.).

9. See ref.[8] and for a modern introduction to the theory of renormalization, C.Itzykson and J.B.Zuber, "Quantum Field Theory", Mc Graw-Hill, New-York (1980).

10. G. 't Hooft, Nucl.Phys. B61, 455 (1973).

11. A.A.Vladimirov, D.I.Kazakov and O.V.Tarasov, Sov.Phys. JETP, 50, 521 (1979).

12. E.Brézin, J.C.Le Guillou and J.Zinn-Justin, Phys.Rev. D15, 1544 and 1588 (1977).

13. Ref.5, see also E.Brézin and G.Parisi, J.Stat.Phys.19,269(1978).

CALCULATION OF CRITICAL EXPONENTS FROM FIELD THEORY

J. Zinn-Justin

Service de Physique Théorique
CEN-Saclay, Orme des Merisiers, BP N°2
91190 Gif-sur-Yvette, France

I - INTRODUCTION

In the main part of this lecture, we shall discuss the calculation of the critical exponents of the n-vector model in 3 dimensions, using field theoretical techniques [1,2]. This calculation is based on a perturbative expansion at fixed dimension of the various renormalization group (R.G.) functions. Since in 3 dimensions the fixed point S^4 coupling constant is not small, the perturbative expansion has to be resummed. We shall present a summation technique using a Borel transformation and a conformal mapping on the variable, argument of the Borel transform. Such a conformal mapping has been made possible by the existence of methods to evaluate the large order behavior of perturbation theory [3].

In addition we shall also present the application of the same summation technique to the n=1 (Ising like) S^4 model in two dimensions, and the Wilson-Fisher ε-expansion [4,5,6] .

Finally we shall compare the n=1, d=3 results for the exponents, with results obtained by analyzing the longer B.C.C. high temperature (H.T.) series obtained by Nickel [7], which seem to have eliminated the discrepancies between the R.G. predictions and lattice model calculations [7,8].

II - RENORMALIZATION GROUP CALCULATIONS AT FIXED DIMENSIONS [9,10]

In his lectures Brézin [11] has shown how starting from a Landau-Ginsburg-Wilson hamiltonian in d dimensions :

$$A(\vec{S}) \equiv \beta H(\vec{S}) = \int d^d x \left[\frac{1}{2} (\partial\mu\vec{S})^2 + \frac{1}{2} r_o \vec{S}^2 + \frac{1}{4!} g_o (\vec{S}^2)^2\right] \tag{1}$$

in which $\vec{S}(x)$ in a n component vector field, and using R.G. arguments, it is possible to derive the critical behavior of correlation functions. He has then described a method to calculate critical quantities using the dimensional $\varepsilon=4-d$ expansion.

At fixed dimension $d < 4$, it is not possible to construct directly in perturbation the critical theory ($T=T_c$, $\xi^{-1}=0$) because it is infrared divergent. But it is still possible to write R.G. equations (called Callan-Symanzik equations) taking the correlation length $\xi=m^{-1}$ as one of the parameter of the theory.

When the correlation length becomes large compared to the inverse cut-off Λ representative in this theory of the original lattice spacing, and if one is interested only in the critical region in which the typical distances $|x|$ satisfy :

$$\xi >> |x| >> \Lambda^{-1}$$

one can take order by order in perturbation theory the large Λ limit, except for one quantity :

the expansion parameter g_o which has a dimension. It can be written :

$$g_o = \Lambda^{4-d} u_o$$

in which u_o is dimensionless.

Thus in the critical region g_o is infinite in the scale fixed by the correlation length. It is therefore convenient to express the theory in terms of a different quantity g, called the renormalized coupling constant, which is formally proportional to g_o for g_o small, but has a finite large g_o limit g^*. This infrared fixed point value g^* is the zero of a R.G. function $W(g)$ calculable in perturbation theory. The critical exponents are then obtained by calculating two other R.G. functions $\gamma(g)$ and $\eta(g)$ for $g = g^*$:

$$\gamma = \gamma(g^*)$$
$$\eta = \eta(g^*)$$

III - RENORMALIZATION GROUP FUNCTIONS

The renormalized coupling constant g, and the R.G. renormalization function $W(g)$ are defined by :

$$W(g) = \frac{(d-4)}{\frac{\partial \ell n\, g_o}{\partial g}} \qquad m^{4-d}\, g = \Gamma^{(4)}_{ren.}(0,0,0,0) \tag{2}$$

In three dimensions :

$$W(g) = -g + g^2 - W_3 g^3 + \ldots + (-1)^K W_K\, g^K + \ldots \tag{3}$$

The infra-red fixed point is defined by :

$$\begin{cases} W(g^*) = 0 \\ W'(g^*) = \omega > 0 \end{cases} \tag{4}$$

At second order in g, W(g) has a zero, but unfortunately, the coefficients W_K at order K increase too fast, and no reliable prediction about an existence of a zero can be made directly from the perturbative expansion. If such a zero has been found by some other method, the critical exponents are then given by :

$$\begin{aligned} \gamma &= \gamma(g^*) \\ \eta &= \eta(g^*) \end{aligned} \tag{5}$$

In three dimensions Nickel has calculated W(g) up to order g^7, and $\gamma(g)$ and $\eta(g)$ up to order 6.

Notice that in the R.G. framework, only two exponents are independent. The series expansion for all other exponents can be obtained using scaling laws.

IV - LARGE ORDER BEHAVIOR AND ASYMPTOTIC SERIES [13]

Since there is some numerical evidence that the perturbative expansion cannot be used directly, a natural question arises : What is the nature of the perturbative expansion in the S^4 field theory. Classical arguments based on the number of Feynman diagrams which increases like K ! at order K, and also on the instability of the ground state in the anharmonic oscillator for any negative value of the coupling constant suggest a behavior in K ! of the perturbation series at order K.

This assumption was used by Baker et al [2] to try to sum the various series for the R.G. functions.

In recent years a method first introduced by Lipatov [3], has been developed which allows one to estimate the large order behavior of the perturbative expansion for many field theories, in particular the $(\vec{S}^2)^2$ field theory in dimension d : d < 4.

For the $(\vec{S}^2)^2_d$ theory the result is that for any quantity A(g):

$$A(g) = \sum_{K=0}^{\infty} A_K \, g^K$$

the large order behavior is of the form :

$$A_K \sim C \, K^b \, [-a(d)]^K \, K! \, (1+0(\frac{1}{K})) \tag{6}$$

with $a(d) > 0$.

In this expression a depends only on the classical action, while b,c depend on the particular function one is considering.

Consequently, as expected, the perturbative expansion is divergent for all values of the coupling constant g. It is only an asymptotic series. Let us remind that an asymptotic series is defined in the following way :

$$\left| A(g) - \sum_{n=0}^{K} A_n g^n \right| < D_{K+1} |g|^{K+1} \tag{7}$$

for $|g| < \varepsilon \quad , \quad |\text{Arg } g| < \frac{\pi}{2} \alpha$.

In what follows we shall assume that :

$$D_K = M \, D^K \, K! \tag{8}$$

Such a series does not define in general uniquely A(g). Indeed, at g fixed, the smallest error $\varepsilon(g)$ we can obtain from the series is given by :

$$\varepsilon(g) = \min_{\{K\}} D_K \, |g|^K \sim \exp - \frac{D}{|g|} \tag{9}$$

As a result the determination of A(g) is ambiguous, since one can add to a function A(g) satisfying the bound (7), any function analytic in the domain defined in (7) and bounded in this domain by $\varepsilon(g)$.

But if α is larger than 1, a classical theorem on analytic functions (Phragmen-Lindelöf) tells us that no such function exists [14].

The idea is that functions of this type are of the form $\exp - \frac{\gamma}{g}$, which decrease only in the half plane :

$$\text{Re}(g) > 0$$

In the latter case A(g) is uniquely defined by the asymptotic expansion.

For the $S^4_{2,3}$ field theory, such a result has been proven rigorously [15].

V - THE PROBLEM OF RESUMMATION : THE BOREL TRANSFORMATION [1]

Once we know that the A(g) is uniquely defined by its asymptotic expansion, we still face the problem of reconstructing it from its asymptotic expansion. Many methods are available [16]. We shall expose here one of the most extensively used, based on the Borel transformation.

We define the Borel transform B(g) of the function A(g) by :

$$B(g) = \sum_{K=0}^{\infty} B_K \, g^K \tag{10}$$

with

$$B_K = \frac{A_K}{K!} \quad . \tag{11}$$

Then the condition (7) shows that :

$$|A_K| < M \, D^K \, K! \quad \Rightarrow \quad |B_K| < M \, D^K \quad .$$

Therefore B(g) is analytic in a circle of radius at least equal to D^{-1}.

In addition it can be shown that if α is larger than 1, B(g) is analytic in the angle $\pi \, \alpha_B$:

$$\alpha_B = \alpha - 1 \tag{12}$$

Now in the sense of power series, we have the formal identity:

$$A(g) = \int_0^{\infty} e^{-t} \, B(gt) \, dt \tag{13}$$

If α is larger than 1, A(g) is called Borel summable because the integral representation (13) exists and gives the uniquely defined function A(g).

The integral representation (13) reduces the problem of the reconstruction of the function A(g), to the analytic continuation of the function B(g) outside of its circle of convergence, in a neighborhood of the real positive axis.

In the absence of any information about the analytic structure of B(g), it has been proposed [2] to use the Padé approximation for this purpose.

But the large order behavior analysis gives us some information about the singularities of B(g). For the S_d^4 theory :

$$A_K \sim (-a)^K K! K^b \qquad a > 0$$

Therefore :

$$B_K \sim (-a)^K K^b \quad .$$

As a result we know the location and the nature of the singularity of B(g) closest to the origin :

$$B(g) \sim (1+ag)^{-1-b} \quad , \quad \text{for} \quad g \to -\frac{1}{a} \quad . \tag{14}$$

Further considerations suggest that all finite distance singularities of B(g) lie on the real negative axis in the case of the S_d^4 field theory ($d < 4$). If we assume that this property holds we can map the cut-plane into a circle of radius 1, leaving the origin invariant :

$$g = \frac{4}{a} \frac{u}{(1-u)^2} \quad \Leftrightarrow \quad u = \frac{\sqrt{1+ag}-1}{\sqrt{1+ag}+1} \quad . \tag{15}$$

We get then a new expansion for A(g) of the form :

$$A(g) = \sum_{K=0}^{\infty} U_K I_K(g) \tag{16}$$

with :

$$I_K(g) = \int_0^{\infty} e^{-t} [u(gt)]^K dt \quad . \tag{17}$$

Although the new expansion of B(g) in terms of u(g) is convergent in the whole cut plane, we must still discuss the convergence properties of the expansion (16).

If on the other hand, the analyticity hypothesis is incorrect, we have produced a new asymptotic series, just less badly divergent than the initial one. For g is small enough such a series can still provide accurate estimates.

A last obvious remark : If we started with K terms of the expansion of A(g), after all these transformations, we still have K terms in the new expansion of equation (16).

VI - THE NATURAL DOMAIN OF CONVERGENCE [1]

We can say nothing about the coefficients U_K except that the series :

$$F(U) = \sum_{0}^{\infty} U_K U^K$$

converges in a circle of radius 1.

But we can estimate the behavior for K large of the integrals $I_K(g)$, calculating them by steepest descent. It is easy to show that :

$$I_K(g) \sim \exp -3(K^2/ag)^{1/3} \qquad \text{for} \qquad K \to +\infty \tag{18}$$

We see now that various situations can arise :

(i) The U_K are bounded by :

$$|U_K| < M_\varepsilon \quad \exp \varepsilon K^{2/3} \qquad \forall \varepsilon > 0 \tag{19}$$

The new expansion (16) converges in the domain :

$$\mathrm{Re}[g^{-1/3}] > 0$$

$$\text{or} \quad |\mathrm{Arg}\ g| < \frac{3\pi}{2}$$

The zero dimensional S^4 theory (a simple integral) presents an example of such a situation.

(ii) The generic case : the coefficients U_K behave like :

$$U_K \sim \exp C\ K^{2/3} \qquad K \to \infty \tag{20}$$

The expansion converges in the domain

$$\mathrm{Re}\ [(ag)^{-1/3}] > \frac{C}{3} \tag{21}$$

For any value of C, this condition still implies some analyticity in a part of the second sheet of the function, for small values of $|g|$.

(iii) The U_K grow faster, for example like $\exp K^{4/5}$.

The expansion remains a kind of asymptotic expansion. In this case it means that there exist too strong singularities on the cut, which we have mapped onto the circle. We have then to map a smaller part of the domain of analyticity into the circle.

A last remark concerning the Borel transform : We have used a slightly generalized form of the Borel transformation [6,1] :

$$A(g) = \int_0^\infty \frac{e^{-t}\, t^\beta}{(1-u(gt))^\alpha} \sum_{K=0}^\infty U_K(\alpha,\beta)[u(gt)]^K \tag{22}$$

The advantage of this form is that it allows us, by varying α and β, to decrease the singularity of the function B(g) in the u plane at $u = \pm 1$. For example we take in general :

$$\beta \geq b + 1.5$$

If in addition A(g) behaves for g large like :

$$A(g) \sim |g|^\lambda \qquad |g| \to \infty$$

the best choice of α is

$$\alpha = 2\lambda .$$

Of course, asymptotically the result should become independent of α and β.

VII - THE PSEUDO ε-EXPANSION

To check the consistency and reliability of our result, we have summed as if they were independent series :

$$\gamma(g),\ \gamma^{-1}(g),\ \nu(g),\ \nu^{-1}(g),\ \beta^{-1}(g),\ \eta(g)$$

and $\dfrac{1}{\nu(g)} - 2 + \eta(g) = \eta^{(2)}(g)$.

In the direct method, we have first calculated g^* by solving:

$$W(g^*) = 0$$

and then summed the other series for $g = g^*$. In the language of H. T. series these are biased exponents, in the sense that a slightly erroneous evaluation of g^*, reflects itself in a shift in the value of all critical exponents. Fortunately, in contrast to the situation in H.T. series, the values of the exponent are not too sensitive to the value of g^*. Nevertheless we have also tried to construct unbiased approximants in the following way. Let us introduce the function $W(g,\varepsilon)$:

$$W(g,\varepsilon) = -\varepsilon\, g + \sum_2^\infty (-1)^K W_K\, g^K \tag{23}$$

so that :

$$W(g) = W(g,1) \tag{24}$$

We can now calculate g* and the critical exponents as power series in ε [17]. Of course this new ε expansion is unrelated to the Wilson-Fisher [4] ε expansion, ε being here just an artificially introduced parameter.

We can then sum this pseudo-ε expansion by the methods exposed in previous sections and set $\varepsilon = 1$. The consistency of these results with the other ones is a check of our methods. [1]

VIII - NUMERICAL APPLICATION TO EXACTLY SOLVABLE CASES

The simple integral

Let us consider the integral

$$Z(g) = \int_{-\infty}^{+\infty} dx \exp-(x^2 + gx^4)$$

The coefficient of g^K counts the number of vacuum Feynman diagrams at order K in the S^4 theory.

It is easy to estimate the large order behavior :

$$Z_K \sim (-4)^K \frac{K!}{K} .$$

One can prove that the transformed expansion (16) converges as $\exp-K^{2/3}$. In addition for special values of α and β, a finite number of terms give the exact answer.

The ground state energy of the anharmonic oscillator

After the trivial S_0^4 theory, it is natural to study the S_1^4 theory, on the particular example of the ground state energy of the anharmonic oscillator.

$$E(g) = \sum_{K=0}^{\infty} E_K g^K$$

One can show that [18,3] :

$$E_K \sim (-3)^K K! K^{-1/2} . \tag{25}$$

In addition we know that :

$$E(g) \sim g^{1/3} \qquad \text{for} \qquad |g| \to \infty$$

We should verify that $\alpha = 2/3$ is the best choice for the parameter α [19,1] .

Figures (1,2,3) show some results [1] for a value of g(g=0.5) for which the speed of convergence is similar to the apparent speed of convergence observed for the critical exponents.

IX - RESULTS FOR THE CRITICAL EXPONENTS [1]

We have calculated all exponents for the n-vector model for n=0,1,2 and 3 in three dimensions.

The series are known up to order 6 for $\frac{W(g)}{g}$, $\gamma(g)$ and $\eta(g)$.

In addition we have calculated the exponents for Ising-like systems in two dimensions. Only 4 terms are known in this case, and unfortunately in a natural scale the coupling constant is twice larger than in three dimensions, so that the comparison with the exact results of the two dimensional Ising model is not so instructive as one could hope.

Finally we have summed by the same method the Wilson-Fisher ε expansion which has been extended [6] up to order 4.

The apparent accuracy of the expansion at fixed dimensions, and of the ε-expansion are roughly similar at the same order and the results are consistent.

Figures (4-8) show some examples of the series presented in reference [1].

Our results are given in Tables 1,2,3.

It has been argued [20] that the point $g = g^*$ could be a singularity of the R.G. functions. No indications of a singular behavior has been found in our calculations. Still, since the existence of such a singularity implies a power law convergence, we have plotted the estimate of g^* at order K versus 1/K on figure 9. One sees that the assumption of a power law convergence would not affect our conclusions.

On figure 10 we present for n=1 a detailed comparison between R.G., H.T. and experimental results [21]. To the standard H.T. series results we have added the new estimates obtained from the longer B.C.C. series generated by Nickel, as analyzed by Nickel[7] and Zinn-Justin [8]. It is striking that the old discrepancies between R.G. results and H.T. series results have now almost completely disappeared and that the agreement between experiments and theoretical calculations is excellent.

Figures 11,12,13 show the same comparison for n = 2,3,0 systems respectively.

For completeness we have added on figures 14 and 15 the estimates for the B.C.C. exponents as taken from reference [8].

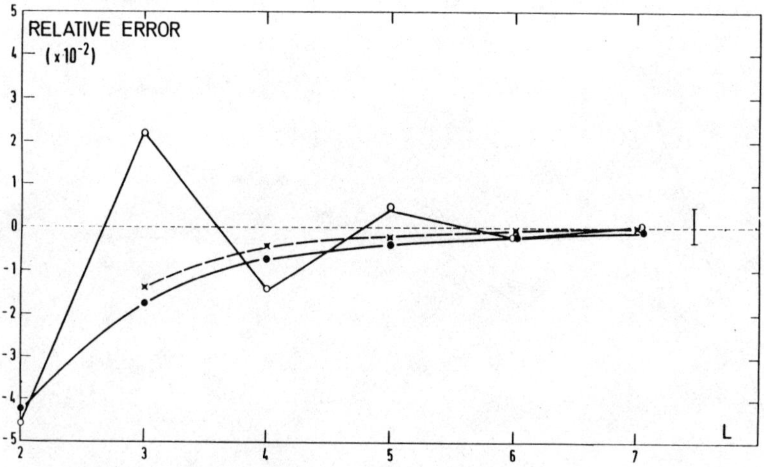

Fig. 1 - Anharmonic oscillator $(g = 0.5)$: Typical results for the relative error $(E/E_{exact}-1)$(times 10^{-2}) as a function of the order L of the perturbation series, with $\alpha = 0$ and different values of β, α and β being the parameters used in formula (22). The vertical bar ⊢⊣ on the right represents the estimated error bar at order $L = 6$.

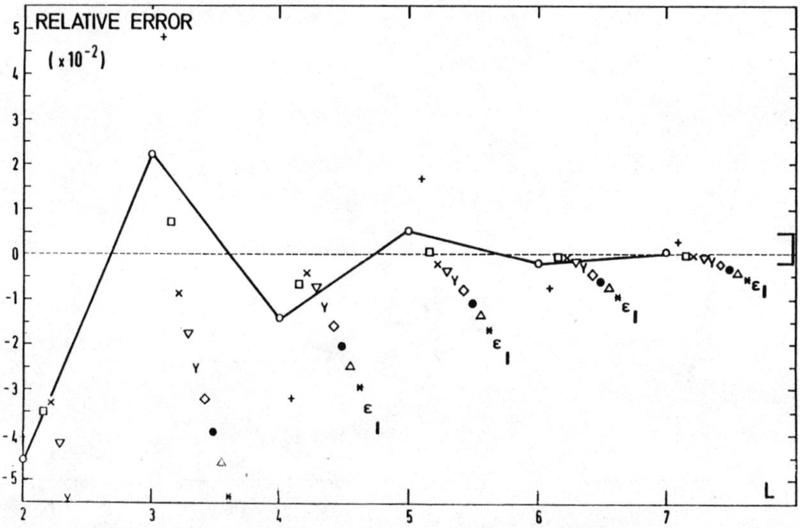

Fig. 2 - Anharmonic oscillator $(g = 0.5)$: Same as Fig. 1. At each order L, points from left to right correspond to increasing values of the parameter β, starting from $\beta = -1$ (for the point +) and increasing by steps of 1.

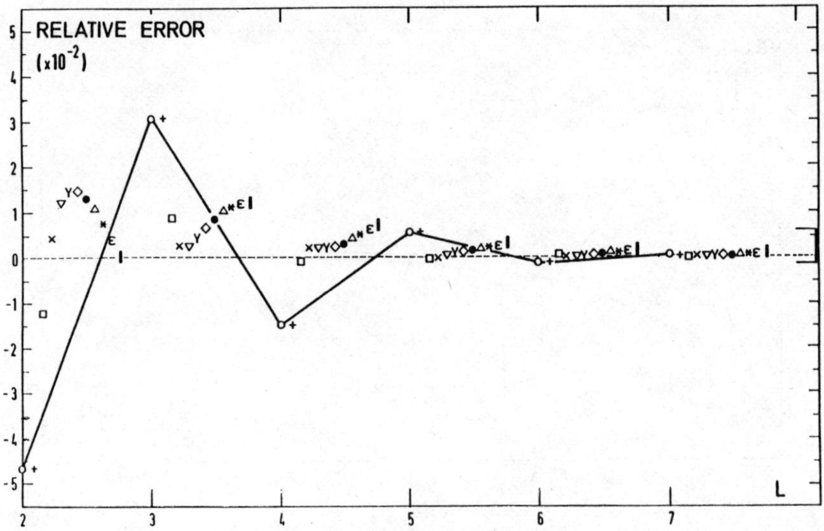

Fig.3 - Anharmonic oscillator ($g = 0.5$) : Same as Fig.2, but with the parameter α equal to 1 (β starts from 0).

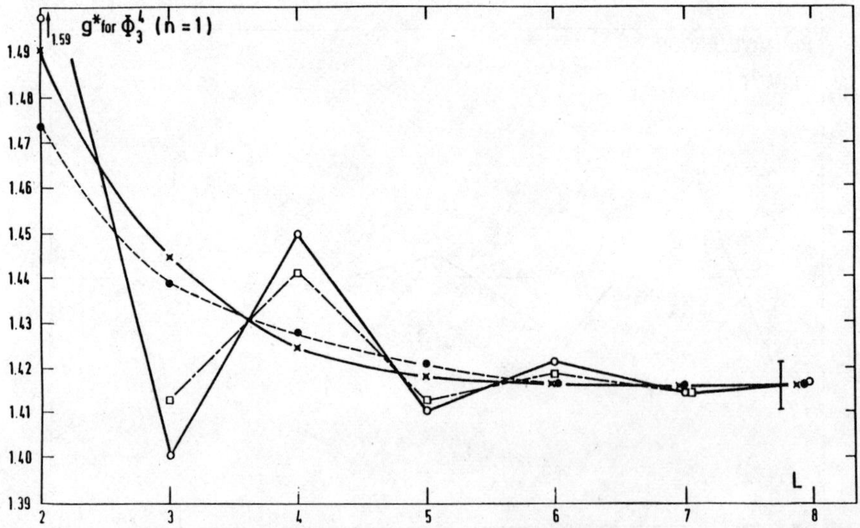

Fig.4 - Field Theory, with d=3 and n=1 : Some typical convergences for the value of g*, zero of W(g), as a function of the order L of the perturbation series of W(g)/g. The two last orders (L=7 and 8) correspond to a fit of the known terms (up to L=6) of the series (see Ref.1). The vertical bar ⊢⊣ on the right represents the final estimate arg* (see Ref.1).

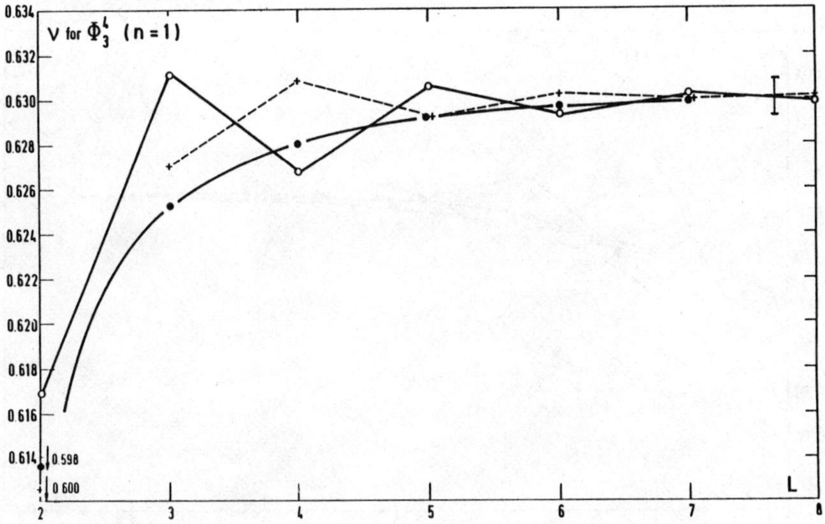

Fig.5 - Field Theory, with d=3 and n=1 : Some typical convergences for the critical exponents ν, as a function of the order L of the perturbation series of ν(g*), here with g* = 1.416.

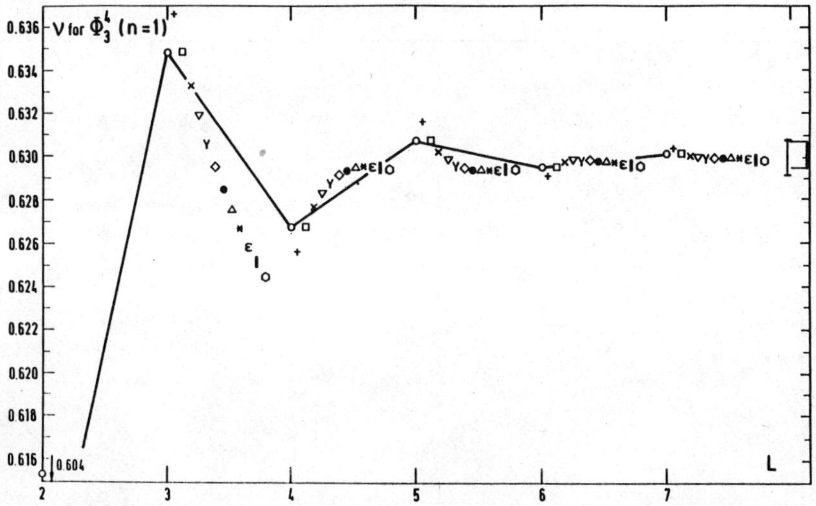

Fig.6 - Field Theory, with d=3 and n=1 : Critical exponents ν=ν(g*) with g* = 1.416, as a function of the order L of the perturbation series. One has here α = 4, and, at each order L, points from left to right correspond to increasing values of β (starting from β = 4 for point +, and increasing by steps of 1), α and β being the parameters used in formula (22).

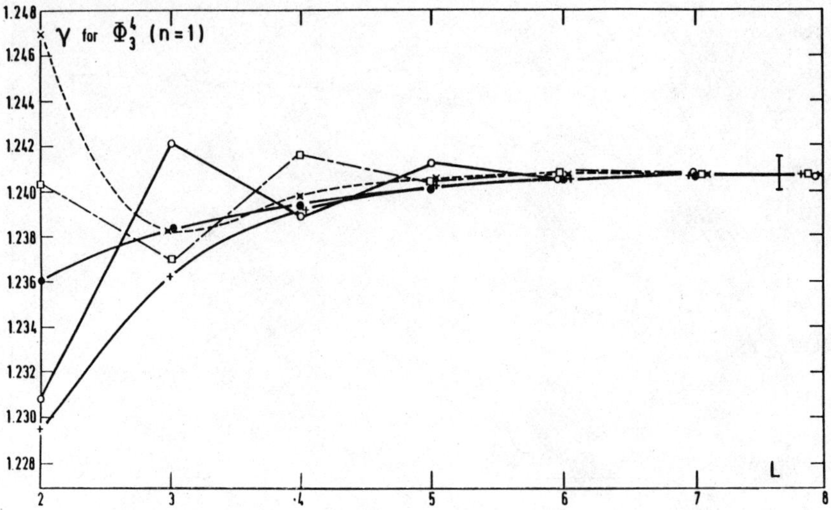

Fig.7 - Field Theory, with d=3 and n=1 : Same as Fig.5, but for the critical exponent γ.

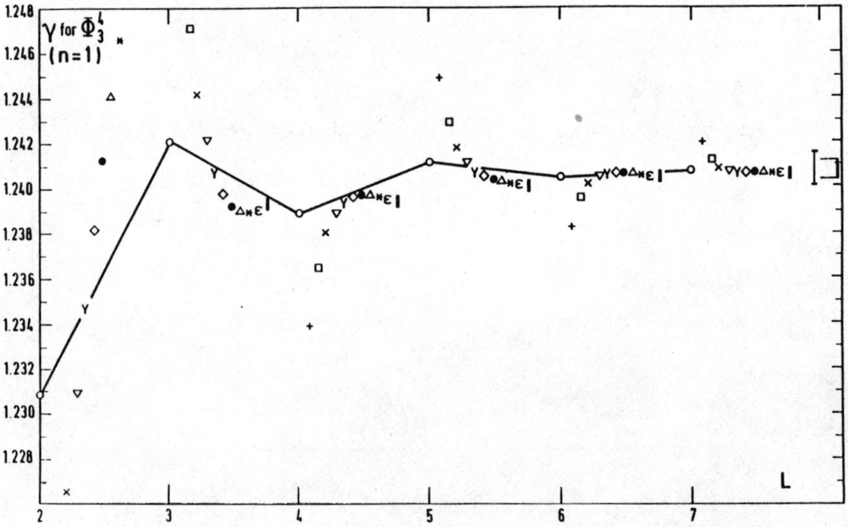

Fig.8 - Field Theory, with d=3 and n=1 : Same as Fig.6, but for the critical exponent γ, for the parameter $\alpha = 2$ and with the parameter β starting from 1.

Table 1. Estimates of critical exponents in the ϕ^4 field theory in d = 2 (Reference [1])

g^*	$= 1.85 \pm 0.10$
ω	$= 1.3 \pm 0.2$
γ	$= 1.79 \pm 0.09$
ν	$= 0.97 \pm 0.08$
η	$= 0.13 \pm 0.07$
$\eta^{(2)}$	$= -0.85 \pm 0.07$

Table 2. Estimates of critical exponents in the three dimensional n-vector model $(\underset{\sim}{\phi}^2)^2_3$ (Reference [1])

	n = 0	n = 1	n = 2	n = 3
g^*	1.421±0.008	1.416±0.005	1.406±0.004	1.391±0.004
ω	0.80 ± 0.04	0.79 ± 0.03	0.78 ± 0.025	0.78 ± 0.02
γ	1.1615±0.0020	1.241±0.0020	1.316±0.0025	1.386±0.0040
ν	0.588 ± 0.0015	0.630±0.0015	0.669±0.0020	0.705±0.0030
η	0.027 ± 0.004	0.031±0.004	0.033±0.004	0.033±0.004
β	0.302 ± 0.0015	0.325±0.0015	0.3455±0.0020	0.3645±0.0025
$\eta^{(2)}$	-0.2745±0.0035	-0.3825±0.0030	-0.474±0.0025	-0.549±0.0035
α	0.236 ± 0.0045	0.110±0.0045	-0.007±0.006	-0.115±0.009
$\Delta_1, \omega\nu$	0.470 ± 0.025	0.498±0.020	0.522±0.018	0.550±0.016

Table 3. Values of critical exponents obtained in the framework of the ε expansion, (Reference [1])

d=3 ; n=0					
	$\omega = 0.80$	$\gamma = 1.163$	$\nu = 0.589$	$\eta = 0.034$	$\beta = 0.305$
d=3 ; n=1					
	$\omega = 0.80$	$\gamma = 1.242$	$\nu = 0.632$	$\eta = 0.04$	$\beta = 0.330$
d=3 ; n=2					
	$\omega = 0.79$	$\gamma = 1.324$	$\nu = 0.676$	$\eta = 0.05$	$\beta = 0.357$
d=3 ; n=3					
	$\omega = 0.78$	$\gamma = 1.395$	$\nu = 0.713$	$\eta = 0.05$	$\beta = 0.379$
d=2 ; n=0					
	$\omega = 1.53$	$\gamma = 1.43$	$\nu = 0.77$	$\eta = 0.30$	$\beta = 0.58$
d=2 ; n=1					
	$\omega = 1.40$	$\gamma = 1.82$	$\nu = 1.03$		
d=1 ; n=0					
	$\omega = 2.44$	$\gamma = 2$	$\nu = 1.3$		

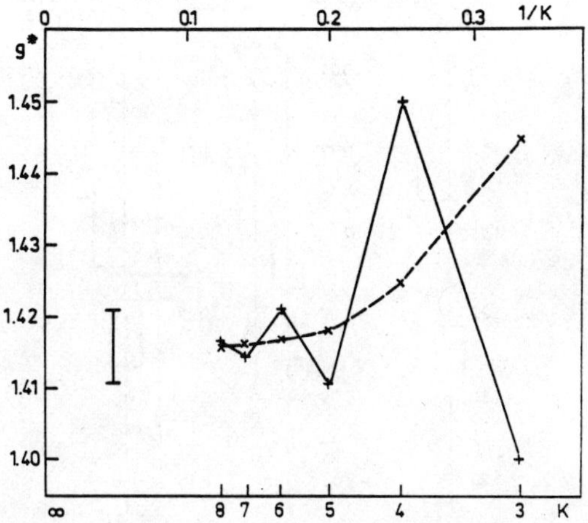

Fig.9 - Field Theory, with d=3 and n=1 : Examples of results for the zero g* of W(g)/g at order K versus 1/K. The two last orders (K=7 and 8) correspond to a fit of the known terms (up to K=6) of the series using the large order behavior (see Ref.1). The vertical bar on the left represents the estimate of g* from the full analysis of Ref.1.

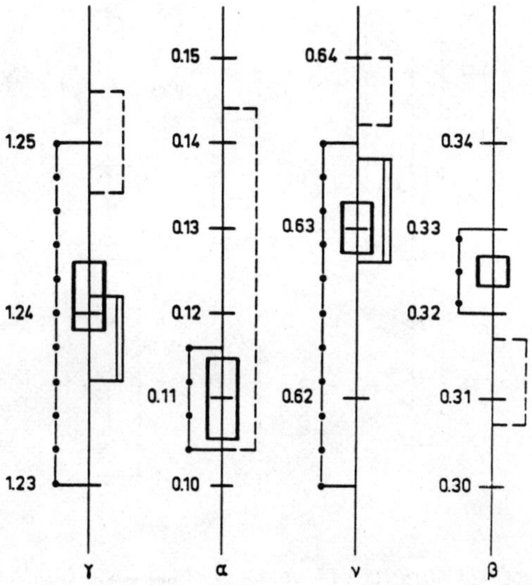

Fig.10 - Critical exponents for n=1 : ——— Renormalization group (Ref.1). ●—●—● Experiments (Ref.21). - - - - Series, before Nickel (Ref.22). ═══ Series, after Nickel (Refs 7 and 8).

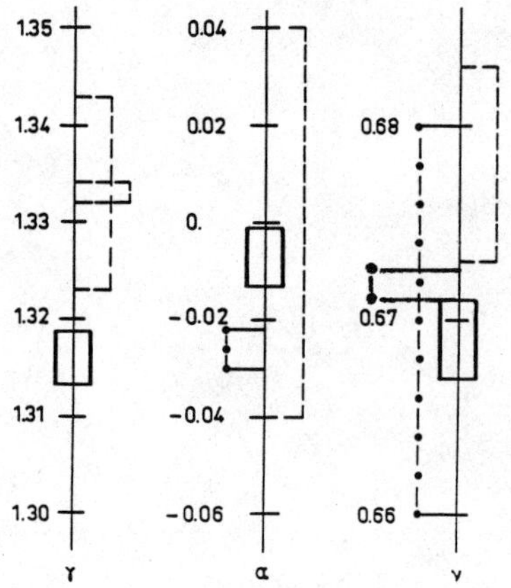

Fig.11 - Critical exponents for n=2 : ▬▬ *Renormalization group (Ref.1)*. ●—●—● *Experiments*(*): D.S.Greywall and G.Ahlers, Phys.Rev. A7,2145 (1973);K.H.Mueller, F.Pobell and G.Ahlers, Phys.Rev.Lett. 34,513 (1975). ---- *Series* : J.Rogiers, M.Ferer and E.R.Scaggs, Phys.Rev. B19,1644 (1979); J.Rogiers, D.D.Betts and T.Lookman, Can. J.Phys. 56,420 (1978).(*) and G. Ahlers : 1980 Cargèse lecture notes.

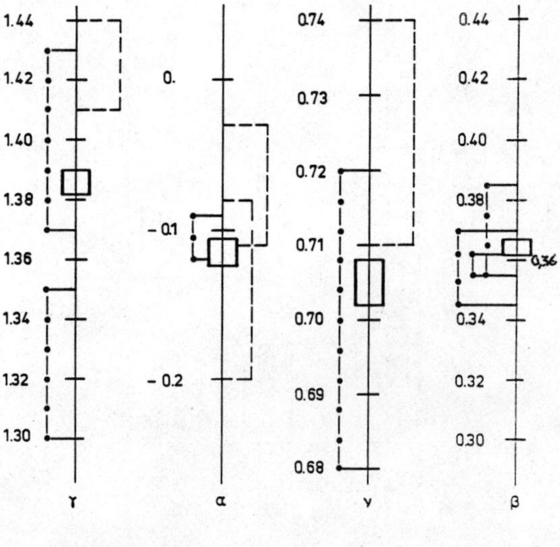

Fig.12 - Critical exponents for n=3 : ▬▬ *Renormalization group (Ref.1)*.●—●—● *Experiments* : J.Als-Nielsen in "Phase transitions and Critical Phenomena" edited by C.Domb and M.S.Green (Academic, NY,1976) Vol.5A;J.D.Cohen and T.R.Carver, Phys.Rev. B15,5350 (1977). ---- *Series* : W.J.Camp and J.P.Van Dyke, J.Phys. A9,731 (1976); D.S.Ritchie and M.E.Fisher, Phys.Rev.B5,2668 (1972);M.Ferer, M.A. Moore and M.Wortis, Phys.Rev.B4,3954 (1971).

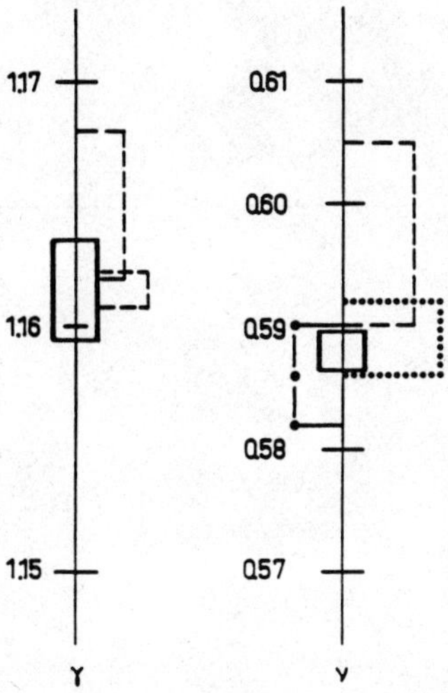

Fig. 13 - Critical exponents for n=0 : ——— *Renormalization group (Ref. 1).* ●—●—● *Experiments* : J.P.Cotton, J.Phys.Lett.41,L231(1980). – – – – *Series* : S.Mc Kenzie, J.Phys. A12,267 (1979); M.G.Watts, J.Phys. A8,61 (1975) and A7,489 (1974);J.L.Martin and M.G.Watts, J. Phys.A4,456(1971). •••••••• *Monte-Carlo Renormalization Group* : A. Baumgärtner, J.Phys.A,13,L39(1980). For this method, see also Ref.23.

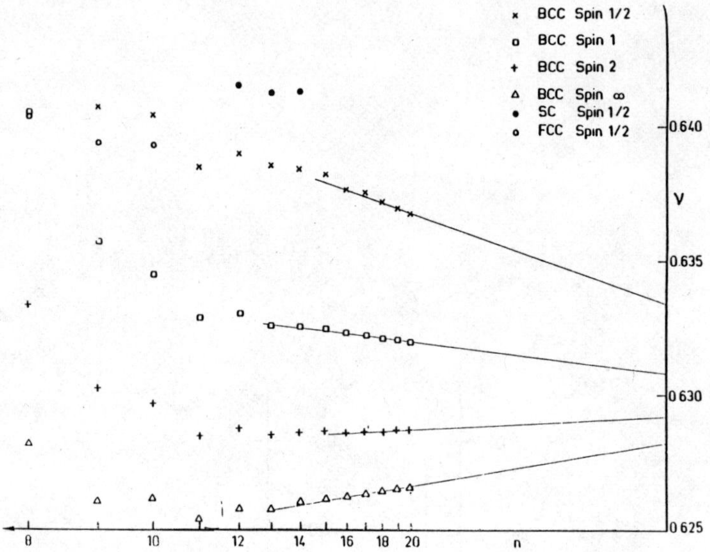

Fig. 14

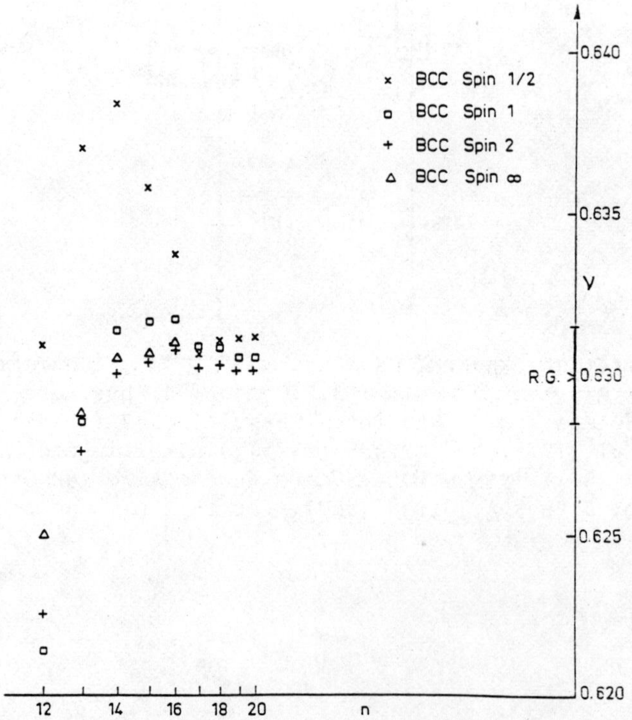

Fig. 15

Fig. 14 and 15 - The figures 14 and 15 show a plot of estimates of ν as a function of $1/n$ (n is the order) obtained by a modified ratio method (ref.8). The convergence of the estimates of figure 15 is improved by assuming a confluent singularity with $\omega\nu = 0.5$.

A last comment is in order : the same renormalization group method allows in principle to calculate not only critical exponents, but all other universal quantities. In particular Bervillier, Godrèche and Bagnuls have obtained various universal ratios of amplitude [24].

REFERENCES

1. J.C.Le Guillou and J.Zinn-Justin, Phys.Rev.Lett. 39, 95 (1977) and Phys.Rev. B21, 3976 (1980)

2. G.A.Baker, B.G.Nickel and D.I.Meiron, Phys.Rev. B17, 1365 (1978)
G.A.Baker, B.G.Nickel, M.S.Green and D.I.Meiron, Phys.Rev.Lett. 36, 1351 (1976)

3. L.N.Lipatov, JETP 72, 411 (1977)
E.Brézin, J.C.Le Guillou and J.Zinn-Justin, Phys.Rev. D15, 1544, 1558 (1977)

4. K.G.Wilson and M.E.Fisher, Phys.Rev.Lett. 28, 240 (1972)

5. E.Brézin, J.C.Le Guillou, B.G.Nickel and J.Zinn-Justin, Phys.Lett. A44, 227 (1973)

6. A.A.Wladimirov, D.I.Kazakov and O.V.Tarasov, Sov.Phys. JETP, 50, 521 (1979)

7. B.G.Nickel, Proceedings of the 1980 Cargèse Summer Institute on "Phase Transitions"

8. J.Zinn-Justin, Saclay preprint DPh-T/80-129, submitted to the "Journal de Physique"

9. E.Brézin, J.C.Le Guillou and J.Zinn-Justin in "Phase Transitions and Critical Phenomena" edited by C.Domb and M.S.Green (Academic, New-York, 1976), Vol.VI p.177

10. G.Parisi, in Proceedings of the 1973 Cargèse Summer Institute and J.Stat.Phys. 23, 49 (1980)

11. E.Brézin, Proceedings of the 1980 Cargèse Summer Institute on "Phase transitions"

12. B.G.Nickel, unpublished

13. For a review see : J.Zinn-Justin, Freie Universität Berlin preprint FUB/HEP 8/80, Physics Report in print

14. E.Borel, Leçons sur les séries divergentes (2ème ed., Gauthier-Villars, Paris, 1928)
G.Hardy, Divergent series (Oxford Univ.Press, Oxford 1949)

15. J.P.Eckmann, J.Magnen and R.Sénéor, Comm.Math.Phys. 39, 251 (1975) ; J.S.Feldman and K.Osterwalder, Ann.Phys. (N.Y.) 97, 80 (1976) ; J.Magnen and R.Sénéor, Commun.Math.Phys. 56, 237 (1977)

16. Another method for example is exposed in : R.Seznec and J.Zinn-Justin, J.Math.Phys. 20, 1398 (1979)

17. B.G.Nickel, private communication

18. C.Bender and T.T.Wu, Phys.Rev. 184, 1231 (1969) ; Phys.Rev. Lett. 27, 461 (1971) ; Phys.Rev. D7, 1620 (1973)

19. D.I.Kazakov, D.V.Shirkov and O.V.Tarasov, Teor.Math.Phys. 38, 9 (1979)

20. G.Parisi, private communication and reference 7.

21. D.Beysens, Proceedings of the 1980 Cargèse Summer Institute on "Phase Transitions".

22. D.S.Gaunt, Proceedings of the 1980 Cargèse Summer Institute on "Phase Transitions".

23. R.H.Swendsen, Proceedings of the 1980 Cargèse Summer Institute on "Phase Transitions".

24. C. Bervillier, Phys. Rev. B14, 4964 (1976) ; C. Bervillier and C. Godrèche, Phys. Rev. B21, 5427 (1980) ; C. Bagnuls and C. Bervillier, preprint Saclay DPh-T/PSRM/1801/80.

THEORY OF POLYMERS IN SOLUTION

Jacques des Cloizeaux

Département de Physique Théorique, CEN-Saclay

BP n°2, 91190 Gif-sur-Yvette France

INTRODUCTORY REMARKS

We are interested here, in the properties of long polymers in good solvents, because we believe that they have a universal behaviour. An infinitely long polymer chain can be considered as scale invariant and therefore as a critical object. The existence of critical indices and scaling laws for long polymers is a consequence of this fact. Moreover, a correspondence has been found between the theory of polymers in solution and the limit of the $(\varphi^2)^2$ Lagrangian theory, when the number n of components of the field goes to zero.

Thus, the renormalization methods apply to polymer physics. However as these methods will be extensively studied during this session, we shall not concentrate on them, but we shall describe more specific characteristics of polymer theory.

1. MODELS AND FIELD THEORY

In a solution, the segments of a polymer chain interact in a complex way. The excluded volumes produce a repulsion, the van der Waals forces produce an attraction. For temperatures higher than the Flory temperature θ , the repulsive interactions dominate and the solvent is a good solvent ; for temperatures below θ , the attractive interactions dominate, the chains collapse and a demixtion occurs. Here, we shall study only solutions in good solvents. For this purpose, we shall represent the complex physical structure by models.

1.1. *Self avoiding walks*

The simplest model consists in representing polymers by self avoiding walks on a lattice.

We consider $\mathbb{N}$ polymers with N links. We submit them to various constraints C. A partition function Z(C), which is a pure number, corresponds to each situation and all the physical quantities can be expressed in terms of these partition functions. This model can be used to define the critical indices. We consider a long chain, with a fixed origin, made of N links. For large N, the partition function is

$$Z_N \propto N^{\gamma-1} \mu^N$$

where μ is a lattice dependent number.

Let $\vec{r}_o$ be the vector defining the origin of the chain and $\vec{r}_N$ the vector defining the extremity. For large N, $R^2 = \langle(\vec{r}_N - \vec{r}_o)^2\rangle_N \propto N^{2\nu}$.

These equalities define the main indices γ and ν ; other ones could be defined in a similar way.

The model has the advantage that everything is well defined and finite. However, it is lattice dependent and not adapted to perturbative calculations.

1.2. *The two parameter model*

The best model is probably the continuous model. To introduce this model, let us start from a chain with independent links on a lattice, in a space of dimension d. The probability distribution of the end to end vector is asymptotically gaussian. Thus, in a continuous space of dimension d, a chain without interaction can be seen as a chain made of independent links with gaussian probabilities. Accordingly, the weight associated with a given configuration can be written :

$$P = \exp\left[-\frac{1}{2\ell^2}\sum_1^N (\vec{u}_j)^2\right]$$

where $\vec{u}_j$ is the vector defining a link.

With this model, we find

$$\vec{r}_N - \vec{r}_o = \sum_1^N \vec{u}_j$$

$$R^2 = \langle(\vec{r}_N - \vec{r}_o)^2\rangle = d\, N\, \ell^2$$

Now, let us pass to the continuous limit. We represent the polymer by a curve; a point on the curve is given by $\vec{r}(s)$ $(0 < s < S)$. With each curve, we associate the weight

$$P_o = \exp\left[-\frac{1}{2}\int_0^S \left(\frac{d\vec{r}(s)}{ds}\right)^2\right] \quad (1.1)$$

We note immediately that s has the dimension of area. We define S as the area of the Brownian chain. The definition is consistent with the fact that the Hausdorff dimension of a Brownian chain is two, not one. Thus a Brownian chain is a curve which has the properties of a surface. Its area must be considered as proportional to the number of links.

Averages are now found by using functional integrals. Let us, for instance, calculate

$$Z_o(\vec{k}) = \langle \exp(i\vec{k}.[\vec{r}(S) - \vec{r}(o)])\rangle_o \quad (1.2)$$

We have

$$Z_o(\vec{k}) = \frac{\int d\{r\}\exp\left(i\vec{k}.[\vec{r}(S) - \vec{r}(o)] - \frac{1}{2}\int_0^S ds\left(\frac{d\vec{r}(s)}{ds}\right)^2\right)}{\int d\{r\}\exp\left(-\frac{1}{2}\int_0^S ds\left(\frac{d\vec{r}(s)}{ds}\right)^2\right)}$$

$$= \frac{\int d\{r\}\exp\left(-\frac{1}{2}\int_0^S ds\left[\left(\frac{d\vec{r}(s)}{ds} - ik\right)^2 - k^2\right]\right)}{\int d\{r\}\exp\left(-\frac{1}{2}\int_0^S ds\left(\frac{d\vec{r}(s)}{ds}\right)^2\right)}$$

The gaussian integrals compensate each other and it remains

$$Z_o(\vec{k}) = e^{-Sk^2/2} \quad (1.3)$$

Hence, we get by expanding (1.2) and (1.3) with respect to $\vec{k}$

$$R^2 = \langle[\vec{r}(S) - \vec{r}(o)]^2\rangle_o = dS \quad (1.4)$$

Thus, S defines the size of the chain.

Let us now introduce interactions. The weight associated with an isolated chain can be written

$$P = \exp\left(-\frac{1}{2}\int_0^S \left(\frac{dr(s)}{ds}\right)^2 - \frac{b}{2}\int_0^S ds' \int_0^S ds''\ \delta(\vec{r}(s') - \vec{r}(s''))\right) \tag{1.5}$$

As the number of configurations is infinite, it is only possible to define a renormalized partition function.

$$^{+}Z(S) = \left\langle \exp\left(-\frac{b}{2}\int_0^S ds' \int_0^S ds''\ \delta(\vec{r}(s') - \vec{r}(s''))\right)\right\rangle_0$$

(first renormalization)

The symbol + is introduced to show that a short range cut-off has to be introduced ; otherwise, the mean value is infinite. For instance, it is convenient to assume that in (1.5), we have $|s' - s''| > s_0$ where s_0 is the cut-off.

However, all the terms which become infinite when $s_0 \to 0$ can be extracted and eliminated by a renormalization of the form*

$$^{+}Z(S) = \exp\left\{S\ c(s_0, b)\right\}\ Z(S) \tag{1.6}$$

(second renormalization)

Now, the model depends only on two parameters S and b. From S and b, a dimensionless parameter can be constructed

$$z = b\ S^{2-d/2}\ (2\pi)^{-d/2}$$

and we see in this way[1] that Z(S) is just a function of z.

The value z = 0 corresponds to the "Brownian" chain which is the asymptotic limit of a random walk and the value z = ∞ corresponds to the "Kuhnian" chain which is the asymptotic limit of the chain with excluded volume.

In the same way, the weight associated with $\mathbb{N}$ chains with areas $S_1, \ldots, S_{\mathbb{N}}$ is

*it corresponds in field theory to an ultraviolet renormalization i.e. to a mass subtraction.

$$P = \exp\left\{-\frac{1}{2}\sum_{i=1}^{\mathbb{N}}\int_0^{S_i} ds\left(\frac{dr_i(s)}{ds}\right)^2 - \frac{b}{2}\sum_{i=1}^{\mathbb{N}}\sum_{j=1}^{\mathbb{N}}\int_0^{S_i} ds'\int_0^{S_j} ds''\,\delta(\vec{r}_i(s') - \vec{r}_j(s''))\right\} \tag{1.7}$$

Again, we may define a partition function $^{+}Z_G(S_1,\ldots,S_{\mathbb{N}})$. The cumulants are the connected partition functions $^{+}Z(S_1,\ldots,S_{\mathbb{N}})$ which are very important from a physical point of view.

By renormalization, we find $Z(S_1,\ldots,S_{\mathbb{N}})$ which does not depend any more on the cut-off.

In the same way, we may define partition functions with constraints. For instance, we may assume that, for all j, the origin of the polymer j corresponds to $\vec{r}_{2j-1}$ and the extremity to $\vec{r}_{2j}$.

The corresponding partition function can be written

$$Z(\vec{r}_1,\vec{r}_2,\ldots,\vec{r}_{2\mathbb{N}}\,;\,S_1,\ldots,S_{\mathbb{N}})$$

and its Fourier transform (with $\vec{k}_1 + \ldots + \vec{k}_{2\mathbb{N}} = 0$) is

$$\bar{Z}(\vec{k}_1,\ldots,\vec{k}_{2\mathbb{N}}\,;\,S_1,\ldots,S_{\mathbb{N}})$$

$$= \frac{1}{V}\int d^dr_1\ldots\int d^dr_{2\mathbb{N}}\; e^{i(\vec{k}_1\vec{r}_1+\ldots+\vec{k}_{\mathbb{N}}\vec{r}_{\mathbb{N}})}\; Z(\vec{r}_1,\ldots,\vec{r}_{\mathbb{N}}\,;\,S_1,\ldots,S_{\mathbb{N}})$$

For instance, we have

$$Z(S_1,\ldots,S_{\mathbb{N}}) \equiv \bar{Z}(\vec{0},\ldots,\vec{0}\,;\,S_1,\ldots,S_{\mathbb{N}})$$

1.3. *Diagrams*

$\bar{Z}(\vec{k}_1,\ldots,\vec{k}\,;\,S_1,\ldots,S_{\mathbb{N}})$ can be expanded in terms of diagrams. Each diagram is made of $\mathbb{N}$ polymer lines ; points on these lines are connected by interaction lines (see Fig. 1) ; these points are called interaction points. Each diagram is connected. The interaction points cut each polymer line into polymer segments. Each polymer segment and each interaction line carries a wave vector. The external vectors $\vec{k}_1,\ldots,\vec{k}_{2\mathbb{N}}$ are carried by the external segments of the polymer lines (see Fig.1). At each interaction point, the flow of wave vectors is conserved (see Fig.2).

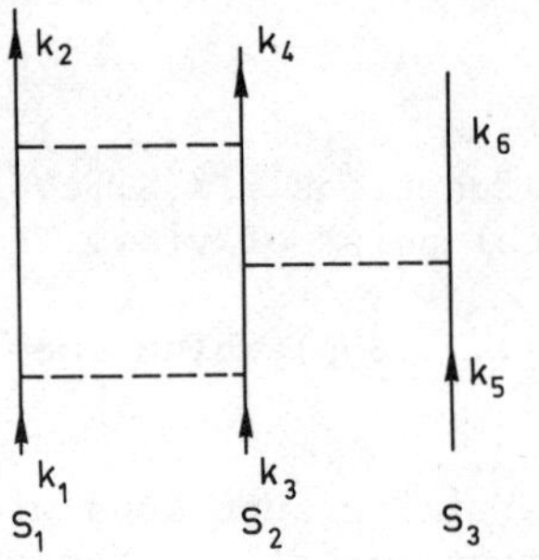

Fig.1. A diagram. Polymer lines are solid lines. Interaction lines are dashed lines.

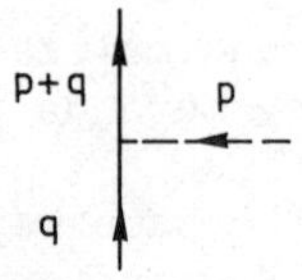

Fig.2. Conservation of wave vector flow at an interaction point.

Each polymer segment has an area s ; the sum of the areas of the polymer segments of the polymer line j is S_j.

The contribution of a diagram is an integral of a product of factors. The factors are the following :

1) With each interaction line is associated a factor -b.

2) With each segment carrying a wave vector k and an area s is associated a factor $\exp(-sk^2/2)$.

To calculate the contribution of a diagram, we may integrate first with respect to the independent internal wave vectors and each integration will be of the form $\frac{1}{(2\pi)^d}\int d^dq\ldots.$

Then, we integrate with respect to the independent surfaces of the segments. To calculate ${}^{+}Z(\ldots)$, we give to the surface of each segment a lower bound s_o in order to avoid divergences.

More precisely, if the polymer j of surface S_j is made of p segments of areas $s_1,\ldots,s_p$, we perform the integration

$$\int_{s_o}^{\infty} ds_1 \ldots \int_{s_o}^{\infty} ds_p \quad \delta(S_j - s_1 \ldots - s_p)\ (\ldots)$$

To calculate $\bar{Z}(\ldots)$, we do the same but we forget the divergences when $s_o \to 0$ by writing $\int_o^S ds\, s^n = \frac{S^{n+1}}{n+1}$ for any n (n > -1 or n < -1).

1.4. The generalized de Gennes' transformation

Let us now consider the Laplace transform

$$G(\vec{k}_1,\ldots,\vec{k}_{2N}\ ;\ a_1,\ldots,a_{\mathbb{N}})$$

$$\times\int_o^\infty d\,S_1\ldots\int_o^\infty d\,S_{\mathbb{N}}\ \exp\{-(a_1S_1+\ldots+a_{\mathbb{N}}S_{\mathbb{N}})\}\bar{Z}(\vec{k}_1,\ldots\vec{k}_{2N};a_1,\ldots,a_{\mathbb{N}}) \tag{1.8}$$

This expression can be easily expressed in terms of diagrams. Consider, in a diagram contributing to $\bar{Z}(\ldots)$, a polymer line made of p polymer segments. A polymer segment, labelled by q $(q = 1,\ldots,p)$ carries an area s_q and a wave vector $\vec{k}_q$.

Thus, when calculating $G(\ldots)$ we have to perform the integration

$$\int e^{-a_jS_j}\,dS_j\int_o^\infty ds_1\ldots\int_o^\infty ds_p\ \delta(S_j-s_1-\ldots-s_p)\exp\{-\tfrac{1}{2}(s_1k_1^2+\ldots+s_pk_p^2)\}$$

$$=\left(\int_o^\infty ds_1\ \exp\{-a_js_1\}\ \exp\{-s_1k_1^2/2\}\right)\ldots\left(\int_o^\infty ds_p\ \exp\{-a_js_p\}\exp\{-s_pk_p^2/2\}\right)$$

$$=\frac{1}{a_j+k_1^2/2}\ \ldots\ \frac{1}{a_j+k_p^2/2}\ .$$

The diagrams which give $G(k_1,\ldots,k_{2N}\ ;\ a_1,\ldots,a_{\mathbb{N}})$ and $\bar{Z}(k_1,\ldots,k_{2\mathbb{N}}\ ;\ S_1,\ldots,S_{\mathbb{N}})$ have the same structure. However, at present, to each polymer line j, corresponds a coefficient a_j. Consider now a polymer segment of this polymer line, if it carries a wave vector $\vec{k}$, the corresponding factor is $1/(a_j+k^2/2)$.

Let us now consider M fields $\vec{\varphi}_m$, $(m = 1,\ldots,M)$ and let us assume that each field has n components. Thus, a component of $\vec{\varphi}_m$ is $\varphi_{m,\alpha}$ $(\alpha = 1,\ldots,n)$.

A field theory can be associated with the Lagrangian

$$L = \frac{1}{2} \sum_{m,\alpha} \left[\frac{1}{2}(\vec{\nabla}\varphi_{m\alpha})^2 + a_m \varphi^2_{m,\alpha}\right] + \frac{b}{8}(\sum_m \varphi^2_{m,\alpha})^2$$

and Green's functions can be defined.

Consider for instance for $\mathbb{N} \leq M$

$$<[\varphi_{11}(k_1)\varphi_{11}(k_2)] \ldots [\varphi_{\mathbb{N},1}(k_{2\mathbb{N}-1})\varphi_{\mathbb{N},1}(k_{2\mathbb{N}})]>$$

$$= (2\pi)^d \, \delta(k_1+\ldots+k_{\mathbb{N}}) \; G(k_1,\ldots,k_{2\mathbb{N}};\{a\}).$$

It is easy to see that the structure of the diagrams which contribute to these Green's functions is the same as the structure of the polymer diagram but for one fact. These diagrams may contain closed solid lines (see Fig.3). However, each closed solid line corresponds to one component of a field and it is necessary to sum over all components of all fields. Therefore, each closed solid line produces a factor proportional to n, and the contribution of each diagram is a power of n. An analytic continuation is easily done for values of n which are not integers. In the limit n = 0, the diagrams with closed loops give no contributions, and we may write

$$G(\vec{k}_1,\ldots,\vec{k}_{\mathbb{N}},\{a\}) = G(\vec{k}_1,\ldots,\vec{k}_{\mathbb{N}}; a_1,\ldots,a_{\mathbb{N}})$$

This result expresses in a general form the correspondence found by P.G. de Gennes[2] in 1972.

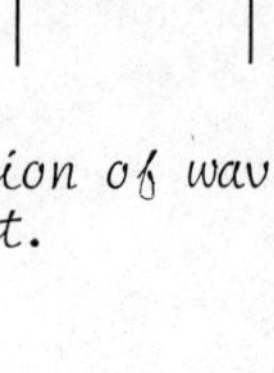

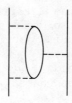

Fig.3. Conservation of wave vector flow at an interaction point.

This correspondence has been derived in different forms (Emery, Sarma, Hilhorst). It can also be obtained by analytical methods (Edwards, Freeds) since it is related to the replica trick used in other domains. A rather nice analytical derivation has been recently given by B. Duplantier[3].

1.5. *Renormalization*

Long polymers correspond to the vicinity of the critical point. Hence, the results which are obtained, by renormalization, in field theory can be translated into the language of polymer theory. Therefore, the transformation found by de Gennes has played a crucial historical role. However, the method is indirect and is not well adapted to the study of polymer solutions or to the exploration of the behaviour of internal parts of chains, as a large number of fictitious fields has always to be introduced.

However, it is possible to apply directly methods of renormalization in polymer theory, and it can be done with the same efficiency. However, this new approach[4] has not been yet fully developed.

2. THEORETICAL AND EXPERIMENTAL RESULTS

2.1. *Critical indices*

The best values of the current critical indices seem to be those which have been obtained by J.C. Le Guillou and J. Zinn-Justin[5] directly for $d = 3$, by analyzing series calculated on a computer by B.G. Nickel.

Their values are

$$\nu = 0.5880 \pm 0.0010$$

$$\gamma = 1.1615 \pm 0.0011$$

$$\omega = 0.790 \pm 0.015$$

Until now γ and ω have not been measured with any precision.

On the other hand, for years, it has been thought that the value $\nu = 0.6$ was exact and old measurements are not reliable. Recently, several groups of Japanese physicists[6,7,8] have made careful light scattering experiments. They use solutions of polyethylene in benzene.

The intensity of scattered light is proportional to the form factor

$$S_1(\vec{q}) = \left\langle \sum_{i=1}^{N} \sum_{j=1}^{N} \exp\{i\vec{q}(\vec{r}_i - \vec{r}_j)\} \right\rangle \qquad (2.1)$$

In this way, the size of the radius of gyration R_G is obtained

$$R_G^2 = \frac{1}{2N^2} < \sum_{i=1}^{N} \sum_{j=1}^{N} (\vec{r}_i - \vec{r}_j)^2 >$$

and we know that for $N >> 1$, $R_G \propto N^\nu$ like the end to end distance R.

However, all the polymers in the solutions have not exactly the same number of links and this leads to uncertainties. The results have been carefully reanalyzed a few months ago by J.P. Cotton [9]. By making proper corrections in two different ways, he finds

$$\nu = 0.586 \pm 0.004 \quad \text{or} \quad \nu = 0.588 \pm 0.002$$

in good agreement with the theory.

The index ν could be obtained also from measurements of the second virial coefficient.

2.2. Scaling laws

In the asymptotic limit of long chains, scaling laws can be written. Some of these laws have been discovered with the help of the de Gennes' transformation, by brutal application of results obtained by renormalization in field theory [10].

However, these scaling laws can be obtained also by simple arguments and this approach has been used systematically by P.G. de Gennes in a recent and interesting book [11].

The scale is given by X

$$R^2 = <(\vec{r}_N - \vec{r}_0)^2> = dX^2$$

(where $X \propto N^\nu$)

Here X defines the size of an isolated chain.

The probability distribution $P(\vec{r})$ of the end to end vector is

$$P(r) = \frac{1}{X^d} f(r/X) \qquad (2.2)$$

where f(x) is, at given d, a universal function which is normalized by the conditions

$$\int d^d x \; f(x) = 1$$

$$\int d^d x \; x^2 \; f(x) = d$$

Thus, for a Brownian chain

$$f_B(x) = (2\pi)^{-d/2} \; e^{-x^2/2}$$

For a Kuhnian chain the shape of f(x) is rather well known[12,13].

For small x

$$f(x) \propto x^{\gamma - 1/\nu}$$

For large x

$$f(x) \propto x^{\sigma} \; \exp\{- A \; x^{1/(1-\nu)}\}$$

where

$$\sigma = \frac{1 - \gamma + \nu d - d/2}{1 - \nu}$$

and A is a constant.

Unfortunately, it is difficult to determine experimentally this function but it seems that, with computer experiments on a lattice, it can be done more easily.

Correlations between points belonging to the interior of a chain can be studied. The ratio R_G^2/R^2 is for a critical chain a universal number. For a Brownian chain, $R_G^2/R^2 = 1/6 \simeq 0.1667$. For a Kuhnian chain, T. Witten and L. Schäfer[14] have calculated this ratio to first order in 2 = 4-d and for d = 3 they obtained

$$R_G^2/R^2 \approx \frac{1}{6}\,(1 - \frac{\varepsilon}{96}) = 0.165$$

which does not agree exactly with the result (ν = 0.155) of the computer experiments of C. Domb and F.T. Hioe[15].

Probability distributions for the vector joining two internal points of an infinite chain, or one internal point and one extremity of a semi-infinite chain have been studied[16]. These functions also obey scaling laws similar to Eq. (2.2) but their properties at short distances depend on new critical indices[16].

For the osmotic pressure Π of a monodisperse polymer solution, a scaling law can also be written. Let $\mathbb{C}$ be the number of polymers per unit volume. When a polymer is isolated, it occupies a volume proportional to X^d. The volume fraction occupied by the polymers would be $\mathbb{C}\, X^d$ if they were not overlapping.

Thus, we have

$$\Pi\beta = \mathbb{C}\, F\, (\mathbb{C}\, X^d) \tag{2.2}$$

For small x, F(x) can be expanded in terms of x. We may write

$$F(x) = 1 + \frac{1}{2}(2\pi)^d\, g\, x + \ldots$$

where g has been recently calculated[17] to second order in ε

$$g = \frac{\varepsilon}{8} + \frac{\varepsilon^2}{16}\left(\frac{25}{16} + \ln 2\right)$$

When x is large, the chains overlap very much, this is the semi dilute regime. We believe that, in this limit, the osmotic pressure will depend only on the number of monomers per unit volume i.e. $\mathbb{C}\, N$ (where N is the number of links). As X is proportional to N^{ν}, this means that in Eq. (2.2), $\Pi\beta$ must be a function of $C = \mathbb{C}\, X^{1/\nu}$ Thus for large x, we must find

$$F(x) \simeq F\, x^{\frac{1}{\nu d - 1}}$$

$$\Pi\beta \simeq F\, (\mathbb{C}\, X^{1/\nu})^{\frac{\nu d}{\nu d - 1}} \propto C^{9/4} \qquad (\nu \simeq \tfrac{3}{5},\ d = 3)$$

This relation seems to be verified experimentally (see Fig.4, extracted from reference 18).

For $\mathbb{C} > \mathbb{C}^*$, the chains overlap and $\mathbb{C}^X$ can be defined by

$$\mathbb{C}^*\, X^d \simeq C^*\, X^{d-1/\nu} \simeq 1.$$

For $C > C^*$, the chains are quasi-brownian and we may consider them as made of quasi monomers of size ξ where ξ is a correlation length which can be defined by

$$C\, \xi^{d-1/\nu} \simeq 1 \qquad (C > C^*)$$

which gives $\xi \propto C^{-3/4} \qquad (\nu \simeq \frac{1}{5},\ d = 3)$

in agreement with the experimental results given on Fig. 5.

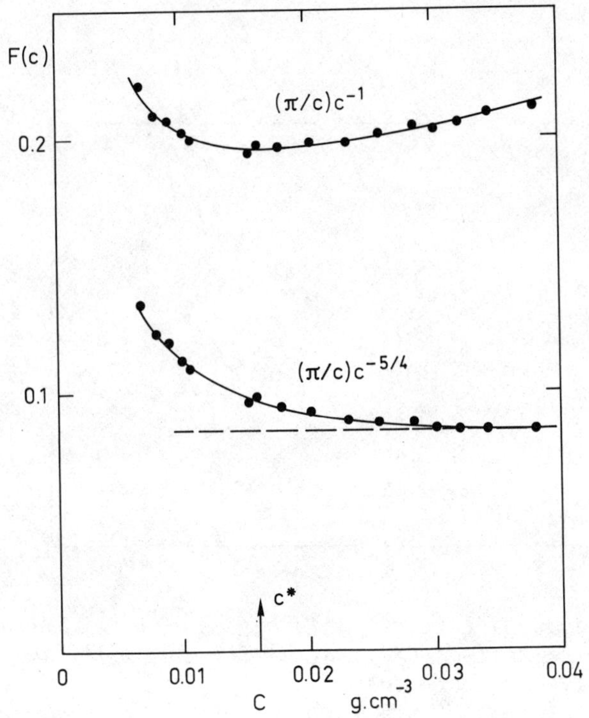

Fig. 4. Osmotic pressure versus concentration in the semi dilute regime. The same data are are plotted on both curves.

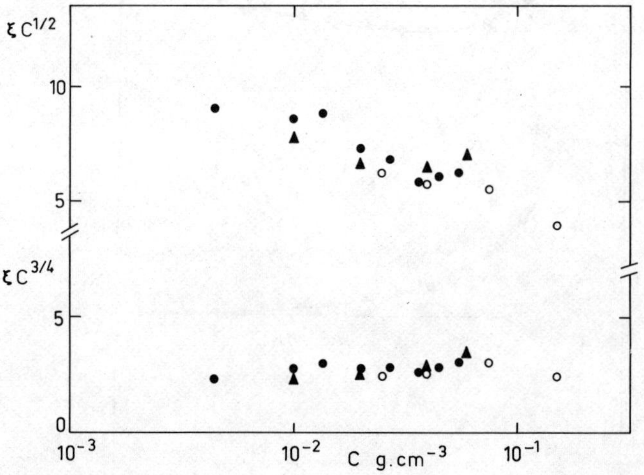

Fig. 5. Correlation length versus concentration in the demi-dilute regime. The same data are plotted on both curves.

The number N_o of quasi monomers is

$$N_o = (X/\xi)^{1/\nu}$$

and the size of a chain in the semi dilute regime is given by

$$R^2 \simeq d\, N_o\, \xi$$

2.3. *Cross-over domain*

If the chains are not very long, or if the solvent is not very good, we are in the cross-over domain. Then, the situation may be very complex. In fact, the behaviour of a segment of chain in a semi-dilute solution may change with the length of the segment. Small segments can be rather rigid ($\nu \simeq 1$) ; when they are larger they may be nearly Brownian ($\nu = 0,5$) ; if they are even larger the excluded volume regime can be reached ($\nu \simeq 0.59$) ; finally if they are larger than the screening length ξ, they become Brownian again.

Thus, comparisons between theories and experiments can be very difficult.

2.4. *Neutron scattering experiments*

The preceding semi qualitative picture of the configurations of a polymer solution has been tested by neutron scattering experiments at small angle ; these experiments have been made at the Laboratoire Léon Brillouin at Saclay during the last ten years[18].

The polymer used is polystyrene made at the Centre de Recherches sur les Macromolécules (Strasbourg) and dissolved in CS^2 which a good solvent and transparent for neutrons.

A few chains may be deuterated and this gives an additional contrast.

In this way, it is possible to measure both the form factor S(q) of all the chains and the form factor $S_1(q)$ of one given chain among the others ones. For an isolated chain, scaling laws show than in the asymptotic domain

$$S_1(q) \propto \frac{1}{q^{1/\nu}}$$

However, when the chains overlap, there is a screening and this simple scaling law is not applicable. This is called the concentration cross-over. Thus, when the solution is semi-dilute, it is convenient to write S(q) as a sum of two parts[19]

$$S_1(q) = S_1^A(q) + S_1^B(q)$$

Part A corresponds to segments with a small number of links ($n < n^*$; excluded volume regime), part B to segments with a large number of links ($n > n^*$; quasi Brownian regime).

Different scaling laws are used for estimating each part. The cross-over occurring for n^* links induces a cross-over in the q space. The cross-over value of q is $q^* = 1/\langle r^2[n^*]\rangle^{1/2}$ where $[n^*]$ is the end to end distance of a segment with n^* links.

Thus, various semi-phenomenological expressions can be written for $S_1(q)$ and S(q).

In particular, in the high concentration range, we expect for S(q) an asymptotic behaviour of the form

$$S(q) \propto \frac{1}{q^2 + \xi^{-2}} \qquad (q \ll q^*)$$

Accordingly, in the semi dilute region, the results of the experiments can be fitted by semi-phenomenological expressions of the form[19]

$$S_1(q) = F_A(q/q^*) + F_B(q)$$

$$S(q) = \frac{F(q/q^*)}{q^2 + \xi^{-2}}$$

where q^* and ξ depend on the concentration of monomers in the solution.

3. DYNAMICS OF POLYMERS IN SOLUTIONS

3.1. *Models and relaxation times*

In a viscous fluid, the velocities of small particles can be assumed to be proportional to the forces which are applied to them. Let $\vec{r}^{\,i}$ be the position vector of a particle i ($r^{i\alpha}$ is a component) and $\vec{F}_j$ the force which is applied to the particle j ($F_{j\beta}$ is a component). We have

$$\dot{r}^{i\alpha} = \sum_{j\beta} D^{i\alpha,j\beta} F_{j\beta} \tag{3.1}$$

In general, the matrix D is not diagonal because hydrodynamic interactions exist ; the motion of a particle in the fluid generates a current which interacts with the other particles.

Thus for a dilute assembly of spherical particles of radius "a", $D^{i\alpha,j\beta}$ is

$$D^{i\alpha,j\beta} = \frac{1}{6\pi\eta}\left[\frac{\delta^{ij}\delta^{\alpha\beta}}{a} + \frac{3}{4r^{ij}}\left(\delta^{\alpha\beta} + \frac{r_{ij}^{\alpha}\, r_{ij}^{\beta}}{r_{ij}}\right)\right] \tag{3.2}$$

where η is the viscosity of the fluid and $r_{ij} = |r_j - r_i|$. The second term is the Oseen's tensor.

In this formalism, the retardation effects are neglected. However, the theory could be modified to include them. The forces can be written as the sum of three terms

$$F_{j\beta} = -\frac{\partial U}{\partial r^{j\beta}} + f_{j\beta}(t) + \mathcal{F}_{j\beta}(t) \tag{3.3}$$

U is the potential vector describing the interaction of the particles, $\vec{f}_j(t)$ is a random force which acts on the particle j and depends on the temperature of the system, $\vec{\mathcal{F}}_j(t)$ is an external force.

It can be assumed that the probability law of the random forces is gaussian. This law is determined by the property

$$\langle f_{i\alpha}(t)\, f_{j\beta}(t')\rangle = [D^{-1}]_{i\alpha,j\beta}\,\delta(t-t')$$

This scheme applies immediately to a polymer if we write U as a sum of two terms

$$U = U_c + U_I$$

where U_c is a chain potential which tells us that each particle is linked to the next one and where U_I defines the excluded volume interaction

$$U_c = \sum_{i=1}^{\mathbb{N}} f(r_{i-1,i}) \qquad U_I = \sum_{ij} U(r_{ij})$$

$$r_{ij} = |\vec{r}_j - \vec{r}_i| \ .$$

Setting (3.2) in (3.1), we obtain a Langevin equation.

A consequence of this equation is that, in the absence of external forces, the probability distribution $P(t,\{r\})$ of the positions of the particles at a given time obeys a Fokker-Planck equation[20]

$$\frac{\partial P}{\partial t} = K P \tag{3.4}$$

$$K = \sum \frac{\partial}{\partial r^{i\alpha}} D^{i\alpha,j\beta} \left(\frac{\partial}{\partial r^{j\beta}} + \frac{\partial U}{\partial r^{j\beta}}\right) \tag{3.5}$$

The solution at equilibrium $\frac{\partial P_o}{\partial t} = 0$ is $P_o \propto e^{-\beta U}$. (Boltzmann's law ; $T = \frac{1}{k_B \beta}$) Equation (3.4) describes the dynamics of the system.

The relaxation times are the eigenvalues of the equation

$$K P_n = \frac{1}{\tau_n} P_n$$

and the function $P_n e^{-t/\tau_n}$ is a solution of Eq. (3.4).

There are short relaxation times which correspond to the relaxation of a few monomers on the chain ; they not depend on the number N of links of the chain. There are long relaxation times which correspond to relaxations of large parts of the chain. Here we shall be concerned only with these collective relaxations which become slower when the size of the chain increases.

The N dependence of the average relaxation time τ of an isolated chain can be found by simple qualitative arguments[21,22].

In the free draining limit, the first term only is kept in Eq. (3.2), and

$$\tau \propto N^{2\nu+1} \qquad \text{when } N \to \infty$$

In the hydrodynamic limit which corresponds to the physical situation, the first term in Eq. (3.2) is neglected, because the second one is dominant and

$$\tau \propto N^{3d} \quad (d=3) \quad \text{when } N \to \infty \tag{3.6}$$

This result can be obtained by a very simple argument. The preceding equations show clearly that τ must be proportional to $\eta\beta$ and dimensional analysis shows that

$$\frac{\tau}{\eta\beta} = L^3 \qquad \text{where L is a length .}$$

In the asymptotic limit, a polymer chain is characterized only by one length R. Therefore, L has to be proportional to R. Hence, we find

$$\tau \propto \eta \beta R^3$$

and since $R \propto N^{\nu}$, we obtain immediately (3.6). Calculations of the dynamics of the Brownian chain are consistent with these results. For instance, the self-correlation function

$$S(q,\omega) = \frac{1}{2\pi} \int_{-\infty}^{+\infty} dt \; \langle \exp\{iq\,[r_j(t) - r_j(0)]\}\rangle$$

has been studied[23,24], and the results appear in the form

$$\frac{S(q,\omega)}{S(q,0)} = \varphi(\omega/\omega_q)$$

For the Rouse model (free draining limit), P.G. de Gennes has found $\omega_q \propto q^4$, in agreement with the scaling formulas

$$\omega \sim \tau^{-1} \sim N^{-2} \quad , \quad q \sim R^{-1} \sim N^{-1/2}$$

For the Zimm-model (hydrodynamic limit), E. Dubois-Violette and P.G. de Gennes have found $\omega_q \propto q^3$, in agreement with the scaling formulas

$$\omega \sim \tau^{-1} \sim N^{-3\nu} \qquad q \sim R^{-1} \sim N^{-\nu} \quad .$$

3.2. The dynamical anomalies

The current experimental results contradict these simple views. The values of ν which are obtained from dynamical experiments are too small. The relaxation times seem shorter than they should be. A great dispersion appears in the results, but the results suggest the existence of a dynamical critical index ζ. According to this view[22,25], for large N, we would get

$$\tau \propto N^{3\nu-\zeta}$$

For instance, anomalies are found in the measurement of the diffusion of polymers in solutions. The diffusion constant D of an isolated polymer should be of the form

$$D \propto \frac{R^2}{\tau} \propto \frac{1}{\eta\beta}\,\frac{1}{R}$$

which gives $D \propto N^{-\nu} \simeq N^{-0.59}$ for large N.

However, for large N (250 < N < 6500), M. Adam and M. Delsanti have found,[26]

$$D = D_o \, N^{-0.55 \pm 0.02} \qquad (3.7)$$

On the other hand, a careful theoretical investigation of this problem has shown that, in the asymptotic limit, the relaxation times cannot be shorter than expected[27] : if it exists, the anomalous index ζ cannot be positive.

An other explanation had to be found and actually it is now clear that these dynamical anomalies are related to cross-over effects[28]. A close investigation shows that the dynamical quantities which are measured depend roughly up in two different lengths namely the radius of gyration R_G and the dynamical radius R_D

$$R_G = \frac{1}{2N^2} \sum_{i,j} \langle r_{ij}^2 \rangle$$

$$\frac{1}{R_D} = \frac{1}{2N^2} \sum_{i,j} \left\langle \frac{1}{r_{ij}} \right\rangle \qquad (3.8)$$

Asymptotically, $R_G \propto N^{\nu}$ and $R_D \propto N^{\nu}$ but R_G reaches the asymptotic limit much more quickly than R_D because comparatively short segments give a larger contribution to R_D than to R_G.

An average relaxation time can be defined by the equation[28]

$$\tau = 4\pi \beta \eta_o \, R_G^2 \, R_D \qquad (3.9)$$

In the cross-over domain, we can also introduce effective indices

$$\nu_G(N) = \frac{N}{R_G} \frac{\partial R_G}{\partial N}$$

$$\nu_D(N) = \frac{N}{R_D} \frac{\partial R_D}{\partial N} \qquad (3.10)$$

When $N \to \infty$, $\nu_G \to \nu$ and $\nu_D \to \nu$ but the variation of ν is very slow and the dynamical experiments measure only the effective indices.

Thus, the approximate dependence of a dynamical constant with respect to N can be found by expressing this constant in terms of lengths and times in the most logical way ; then a length is replaced by R_G and a time by τ where τ is approximately given by (3.9).

For instance, the diffusion constant is obtained by dividing the square a length by a time

$$D \propto \frac{L^2}{t} \propto \frac{R_G^2}{\tau} \propto \frac{1}{\beta\eta_o R_D} \tag{3.11}$$

Thus, in their experiments Adam and Delsanti[26] have been really measuring $\nu_D(N)$ (compare (3.8) and (3.11)).

3.3. *The cross-over problem : a simple approach*

In principle, the cross-over effects could be calculated exactly in the frame of the two parameter theory and certainly, it will be done in the future. However, the main effects can be roughly described by using a very simple model.

Neglecting all fluctuations of r_{ij}, we write

$$R_G = \frac{1}{2N^2} \sum_{ij} r_{ij}^2$$

$$\frac{1}{R_D} = \frac{1}{2N^2} \sum_{ij} \frac{1}{r_{ij}}$$

and using appropriate units, we assume the scaling laws

$$r_{ij} = |i - j|^{1/2} \qquad N < N_c$$

$$r_{ij} = N_c^{1/2} \left|\frac{i-j}{N_c}\right|^{\nu} \qquad N > N_c \qquad (\nu = 0.59)$$

(in the two parameter model N_c correspond to a value z_c).

A calculation of R_G and R_D can be easily made and, applying (3.10), we obtain the indices $\nu_G(N)$ and $\nu_D(N)$ as functions of (N/N_c).

We verify (see Fig.6) that the variation of $\nu_D(N)$ is very slow.

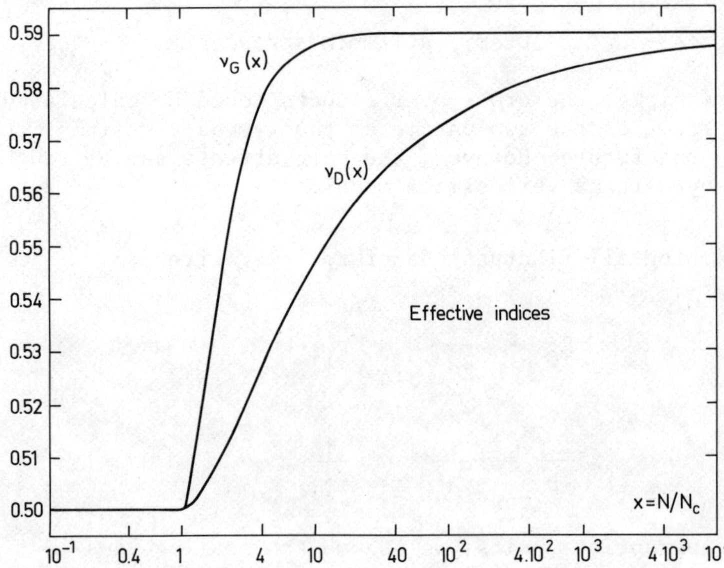

Fig.6. The effective indices $\nu_G(N)$ and $\nu_D(N)$ N_c may correspond to 200 links and therefore for polystyrene to a mass M_c = 20 000.

Thus, these results show that it is practically impossible to measure ν from dynamical measurements. *All measurements of dynamical properties of polymer solutions must be interpreted by taking into account cross-over effects.*

However, it is possible to observe a slow variation of the effective index $\nu_D(N)$ and this effect has been recently observed by Adam and Delsanti[29].

We may note also that very precise measurements of the properties of polystyrene in transdecalin have been made recently by B. Chu and T. Nose[30] ; their interpretation should bring use new interesting information.

4. CONCLUSION

The theory of polymer solutions has made great progress during the recent years and simultaneously more and more precise experiments have been made. The quality and the definition of the polymer samples are also now much better.

The perturbation theory applied to the two parameter model provides *a unified approach* and direct renormalization enables to determine properly the asymptotic properties of the solutions. However, much work remains to be done.

The experimental situations may be very complex but we feel that we have now a good physical picture of the situation.

We have now to make precise comparisons between the theory and the experiments. In particular, it will be necessary in the future, to make simultaneously several experiments on the same polymer and the same solvent in order to determine the unknown relevant parameters (for instance S and z). These quantitative verifications will be made by eliminating all arbitrary parameters and comparing pure numbers.

REFERENCES

1. See also H. Yamakawa, "Modern Theory of Polymer solutions" Harper & Row (1971).
2. P.G. de Gennes, Phys. Lett. 38A; 339 (1072).
3. B. Duplantier, C.R. Acad. Sci. Paris 290 B; 199 (1980).
4. J. des Cloizeaux, J. Physique Lettres 41 ; L-151 (1980).
5. J.C. Le Guillou and J. Zinn-Justin, Phys. Rev. Lett. 39 ; 95 (1977). See also
G.A. Baker, B.G. Nickel, D.I. Meiron, Phys. Rev.B. 17 ; 1285 (1978).
6. Y. Miyaki, Y. Einaga, H. Fujita, Macromolecules 11, 1180 (1978).

7. M. Fukuda, M. Fukutomi, T. Hashimoto, J. Polym. Sci. 12 (1974).
8. A. Yamamoto, M. Fuji, G. Tanaka, H. Yanakawa, Polym. J. 2, 799 (1971)
9. J.P. Cotton, J. Physique Lett. 41, L-231 (1980).
10. J. des Cloizeaux, J. Physique 36, 281 (1975).
11. P.G. de Gennes, Scaling Concept in Polymer Physics - Cornell University Press -1979.
12. J. des Cloizeaux, Phys. Rev. 10, 1665 (1974).
13. M.E. Fisher, J. Chem. Phys. 44, 616 (1966).
14. T.A. Witten, L. Schäfer, J. Phys. A 11, 1843 (1978).
15. C. Domb et F.T. Hioe, J. Chem. Phys. 51, 223 (1969).
16. J. des Cloizeaux, J. Physique 41, 223 (1980).
17. J. des Cloizeaux (First term : ref. 4 ; second term unpublished).
18. These experiments, the results and the interpretations have been described in numerous articles. The basic one is : M. Daoud, J.P. Cotton, B. Farnoux, G. Jannink, G. Sarma, H. Benoit, C. Duplessix, C. Picot and P.G. de Gennes, Macromolecules 8, 804 (1975).
19. B.F. Farnoux, F. Boué, J.P. Cotton, M. Daoud, G. Jannink, N. Nierlich, P.G. de Gennes, J. Physique 39, 77 (1978).
20. J.G. Kirkwood, Macromolecules, edited by P.L. Auer, Gordon and Breach 1967.
R. Zwanzig, Ecole d'Eté de Physique Théorique, Les Houches 1973 - Molecular Fluids - Editor Gordon and Breach - 1976.
21. P.G. de Gennes, Macromolecules 9,587-594 (1976).
22. J. des Cloizeaux, J. Physique, coll. C-2, 137 (1978).
23. P.G. de Gennes, Physics 3, 3 (1967).
24. E. Dubois-Violette and P.G. de Gennes, Physics 3, 181 (1967).
25. D. Jasnow and M.A. Moore, J. Physique Lett. 38, L-463 (1977).
26. M. Adam and M. Delsanti, J. Physique 37, 1045 (1976).
27. J. des Cloizeaux, J. Physique Lett. 39, L-151 (1978); See also G.F. Al Noaimi, G.G. Martinez Mekler and C.A. Wilson, J. de Physique Lett. 39, L-373 (1978).
28. G. Weill and J. des Cloizeaux, J. Physique 40, 99 (1979).
29. M. Adam and M. Delsanti, J. Physique (to be published) (1980).
30. B. Chu, Macromolecules 13, 122 (1980).
B. Chu and T. Nose, Macromolecules 12, 599 (1979).
T. Nose and B. Chu, Macromolecules 12, 590-1122 (1979).

MONTE CARLO RENORMALIZATION GROUP

Robert H. Swendsen

IBM Zurich Research Laboratory
8803 Rüschlikon
Switzerland

I. INTRODUCTION

In 1976, Ma[1] made the suggestion of combining Monte Carlo (MC) computer simulations with a real-space renormalization-group (RG) analysis to calculate critical exponents at second-order phase transitions. Since then, numerous authors[2-14] have presented various ways of implementing Ma's idea to produce a useful theoretical tool. In these lectures, I will discuss a particular Monte Carlo renormalization-group (MCRG) method that I and several coworkers have been using.[7-14] The method is still in the early stages of development, but it has a number of advantages over older methods, and has already produced excellent results for some systems of interest.

A prime advantage of this approach is that we can treat a very general class of lattice models. The "spins", σ_i, at each site i of the lattice can assume discrete values as in the Ising ($\sigma_i = \pm 1$) or q-state Potts ($\sigma_i = 1, 2...q$) models, or continuous values as in the Gaussian model. The σ_i's can also be vectors as in the planar, XY, or Heisenberg models, without causing any special difficulties. The only serious restriction is that they be classical operators.

The Hamiltonian can have the general form

$$H = \sum_{\alpha} K_{\alpha} S_{\alpha} , \tag{1.1}$$

where the S_α's are combinations of the σ_i's such as the sum of all nearest-neighbor products

$$S_{nn} = \sum_{<ij>} \sigma_i \sigma_j \ , \tag{1.2}$$

or the magnetization

$$S_h = \sum_i \sigma_i \quad . \tag{1.3}$$

The K_α's are the associated coupling constants and include the usual factor of $-1/k_B T$. The interactions are only restricted to being fairly short-range due to practical limitations in the MC computer simulation. Second-neighbor, three-spin, and four-spin interactions have already been successfully treated.[8,13,14]

In contrast to the field theoretic methods we have heard discussed, no coarse graining is performed to derive an effective Ginsburg-Landau-Wilson Hamiltonian. The model studied can therefore be somewhat closer to a real physical system. An important consequence is that we do not have to assume that the symmetry of the system at criticality and the dimension are the only important characteristics for determining critical exponents, but we can make a direct test of universality.

The lattice constant, a, is an intrinsic part of the problem (as is the lattice structure) and remains finite throughout the calculation. Since we do not take the limit $a \to 0$, we do not have to introduce a "cutoff" or contend with the problem of divergences.

Similarly, no assumption is made that the parameters of the fixed-point Hamiltonian are small. Specifically, we do not have to be close to a Gaussian fixed point to calculate critical exponents.

Naturally, we have to pay for these nice properties. The primary problem is that the RG transformations introduce an infinite number of effective coupling constants into the renormalized Hamiltonian. I will discuss how this and other difficulties are dealt with in some detail to try to give you an idea of the limitations of the method and how we can judge the errors in such calculations.

As an introduction to the MCRG method, I will first discuss the two older methods that form the basis for our calculations. Unfortunately, I will not be able to present a fair description of what has been achieved with each of the older methods because I have to describe the disadvantages of each in detail to explain the advantages of combining them. To restore a balance, I recommend the reviews by Binder for MC computer simulations[15] and

Niemeyer and Van Leeuwen for real-space renormalization-group calculations.[16]

After presenting the MCRG method, I will give some examples of its application to show how the calculations work out in practice.

After (hopefully) demonstrating the usefulness and reliability of the MCRG method, I will discuss its limitations.

The extension and improvement of MCRG to deal with the limitations of the method is based on unpublished work by Wilson and myself[17] during his stay at the IBM Zurich Research Laboratory. The methods were originally developed to attack the problem of the three-dimensional Ising model (for which work is still in progress), but has borne its first fruits in the treatment of tricritical phenomena.[14]

II. MONTE CARLO COMPUTER SIMULATIONS (STANDARD METHOD)

Computer simulations are necessarily limited to finite systems, and the formal expressions for the desired correlation functions just involve finite sums,

$$\langle S_\alpha \rangle = \operatorname*{Tr}_\sigma S_\alpha(\sigma)\, P_{eq}(\sigma) \quad , \tag{2.1}$$

where σ stands for the configuration $\{\sigma_i\}$,

$$P_{eq}(\sigma) = Z^{-1} \exp[H(\sigma)] \tag{2.2}$$

and the partition function is

$$Z = \operatorname*{Tr}_\sigma \exp[H(\sigma)]. \tag{2.3}$$

A naive direct summation is possible in principle, but quickly becomes impractical, since the number of configurations goes as q^{N^d} (where q is the number of states for each spin, N is the length and d is the dimension of the system). Even for a 10×10 Ising system, this gives $\sim 10^{30}$ states, which would require $\sim 10^{17}$ years of computer time with a fast machine.

A simple random selection of configurations is also impractical. Only configurations with energies within a very narrow range make a significant contribution. Most configurations have an unfavorably high energy so that the factor $\exp[H(\sigma)]$ is practically zero. The configurations with an energy lower than the expectation value are so rare that they would probably not occur within the $\sim 10^6$ or fewer configurations that can be calculated in a reasonable amount of computer time.

The Monte Carlo technique is an efficient method for generating a sequence of configurations with the probability distribution of Eq. (2.1). The basic step is to introduce a form of "dynamics", which may or may not have anything to do with the true dynamic behavior of the model, through a Master Equation,

$$P(a,t+\Delta t) = P(a,t) + \sum_b [W(b\to a)\, P(b,t) - W(a\to b)\, P(a,t)] \quad , \tag{2.4}$$

where a and b are configurations, $P(a,t)$ is the probability of finding configuration a at time t, and $W(a\to b)$ is the conditional probability of finding the system in b at time $t+\Delta t$ if it is in a at time t (jump rate). If the jump rates, $W(a\to b)$, then satisfy the condition of detailed balance

$$W(b\to a)\, P_{eq}(a) = W(a\to b)\, P_{eq}(b) \tag{2.5}$$

and any configuration can be reached from any other configuration in a finite number of steps with non-zero probability, then

$$P(a,t) \underset{t\to\infty}{\to} P_{eq}(a) \tag{2.6}$$

and we say that the system goes to equilibrium. Equation (2.5) can be written as

$$\frac{W(b\to a)}{W(a\to b)} = \exp[H(a) - H(b)] \tag{2.7}$$

so that the partition function (which is unknown) cancels out.

Since we are free to choose any $W(a\to b)$ as long as Eq. (2.7) is fulfilled, we usually make the restriction that $W(a\to b) = 0$ unless a and b differ by at most a single spin. For such single-spin flips, the difference, $H(a) - H(b)$, depends only on the local configuration; for nearest-neighbor interactions, only the nearest neighbors of the spin to be changed enter Eq. (2.7). This allows a very efficient algorithm to be constructed on the computer to generate the desired sequence of configurations.

In practice, in the simplest case, as each spin is visited, the probability of changing that spin is compared with a number between

0 and 1 from a pseudo-random-number generator and depending on which is larger, the spin is either changed or left alone.

The "time" for such simulations is proportional to the number of MC steps, and the natural unit of "time" is one MC step/site. Since the configurations are generated with the probability distribution (2.2), correlation functions are just computed as the mean of the values of the operators in each configuration. In the case of nearest-neighbor interactions, we then have the energy

$$E \sim \langle S_{nn} \rangle \tag{2.8}$$

and the magnetization

$$M \sim \langle S_n \rangle . \tag{2.9}$$

To compute the specific heat, we can either calculate the energy for a sequence of temperatures and differentiate numerically, or carry out the formal derivative in Eq. (2.1), obtaining (apart from some important, but uninteresting factors)

$$C \sim \frac{\partial E}{\partial K_{nn}} \sim \langle S_{nn}^2 \rangle - \langle S_{nn} \rangle^2 , \tag{2.10}$$

and simularly for the susceptibility

$$\chi \sim \frac{\partial M}{\partial K_h} \sim \langle S_h^2 \rangle - \langle S_h \rangle^2 \quad . \tag{2.11}$$

This is a very simple, standard trick, but it is worth repeating because it will prove very valuable later in the RG analysis.

It is clear that an MC simulation is essentially a computer experiment. As such, it has several characteristics in common with normal experiments and requires similar care in evaluating the errors.

Statistical errors are obviously important, but systematic errors can also be a problem. These can arise, for example, from an inadequate random-number generator or from biased estimators[18] as in (2.10) and (2.11).

Since we are interested in the critical point, we must also contend with critical slowing-down. As we consider larger systems, we must make sure that our MC simulations are always long compared with the relaxation times.

These problems are well known and, although they do require a certain amount of care, they can be dealt with.[15,19] A more fundamental difficulty is due to the finite size of the systems we can simulate. All simulations I discuss use periodic boundary conditions so there is no surface, but we still have an upper limit to the wavelength of fluctuations that can be taken into account.

If we could calculate the properties of an infinite system, we would get something like the solid curve in Fig. 2.1 for the susceptibility χ as a function of the temperature T. The function diverges at T_c as $\chi \to A(T-T_c)^{-\gamma}$. However, even if we eliminate all errors for a finite system, we will not get the solid curve, but rather something like the dashed curve. The divergence that we are interested in has disappeared.

If we are far enough away from T_c that the correlation length $\xi << N$, the two curves agree. As we approach T_c, the correlations extend around the system and χ deviates from the infinite-system values. In general, the peak is shifted away from the transition temperature, so we cannot even determine T_c directly.

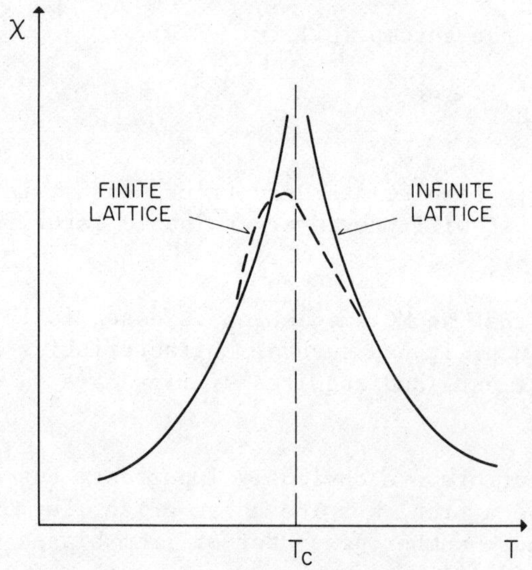

Fig. 2.1. Schematic diagram of typical behavior for the susceptibility χ as a function of temperature T. The solid curve corresponds to an infinite system, and the dashed curve shows the effects of finite size.

The magnitude of the size effect must first be determined by repeating the calculations on different-size lattices. This is *essential* in all MC work, including the MCRG methods. We then consider a range of temperatures away from T_c such that the size effect is negligible and fit the three parameters A, T_c, and γ, hoping that background terms and corrections to scaling can be neglected.

This is the principal limitation preventing the standard MC method from producing highly accurate results. Although one would expect the most information to be found in simulations at or very near T_c, the finite-size effect keeps us from extracting such information efficiently by the standard direct methods.[20]

III. REAL-SPACE RENORMALIZATION-GROUP FORMALISM

Instead of a direct calculation of the thermodynamic properties, the real-space renormalization-group approach seeks to transform the problem in such a way that the critical properties can be expressed in a simple form. Since the critical behavior should not depend in an essential way on the details of the short-wavelength fluctuations, RG transformations integrate out short-range fluctuations, taking their influence into account by changing (renormalizing) the effective coupling constants in the transformed problem.

A particularly simple example of an RG transformation for the two-dimensional square Ising model is to divide the lattice into 3×3 blocks, sum the spins σ_i in each block, and assign a value of ± 1 to a "block spin" σ'_i, depending on whether the sum is positive or negative.

The RG transformation can be written as a transformation of the probabilities

$$P'_{eq}(\sigma') = \operatorname*{Tr}_{\sigma} T(\sigma',\sigma)\, P_{eq}(\sigma) \ , \tag{3.1}$$

where the transformation function for our simple example is a product of δ functions

$$T(\sigma',\sigma) = \prod_{i'} \delta\left(\sigma'_i,\ \mathrm{sgn}\left(\sum_i{}' \sigma_i\right)\right) \tag{3.2}$$

and the sum is over sites in the i' block.

The new probability distribution can be interpreted as arising from an effective Hamiltonian

$$P'_{eq}(\sigma') = \exp[H'(\sigma')]/Z' \quad , \tag{3.3}$$

with new coupling constants $\{K'_\alpha\}$.

The new system of spins $\{\sigma'_i\}$ has a lattice constant a', which is a factor of b (= 3 in our example) larger than the old lattice constant a. Fluctuations over a length scale less than a' have been eliminated.

We now *assume* that the correlation length in units of the *old* lattice constant, a, is unchanged. This can be proven for decimation and linear transformations, but in general it is an assumption. This implies that the new correlation length in units of the *new* lattice constant a' is given by

$$\xi' = \xi/b \quad . \tag{3.4}$$

If the original system were at criticality, $\xi = \infty = \xi'$, and the renormalized system must also be critical. Further iterations of the RG transformation preserve this property, and the transformed systems map onto a sequence of points on a critical hypersurface in the many-dimensional space of coupling constants.

This is illustrated highly schematically in Fig. 3.1, in which the nearest-neighbor coupling is given by the vertical axis, and all other coupling constants are represented by the horizontal axis. I have drawn Fig. 3.1 in accordance with the usual assumption that the RG transformation has a fixed point, that is, the limit of the sequence of renormalized Hamiltonians shown as crosses on the critical hypersurface.

If we make the further assumption that the $K_\alpha^{(n+1)}$'s are analytic functions of the $K_\alpha^{(n)}$'s, a small deviation of $H^{(0)}$ from criticality will result in a small deviation of $H^{(1)}$ from the critical surface. However, this deviation will grow as more RG iterations are performed, in line with Eq. (3.4), as shown by the black dots and triangles.

For temperatures in the original system very close to T_c, the RG trajectories will pass very close to the fixed point before moving off along the dashed line. After a large, but fixed number of RG steps, n, $H^{(n)}$ will lie nearly on the dashed line, at a distance from the fixed point that is proportional to $|T-T_c|$.

From the assumption of analyticity, we can expand the RG transformation about the fixed point $\{K_\alpha^*\}$ and, if we are close to the fixed point, we only need the linear terms

$$(K_\alpha^{(n+1)} - K_\alpha^*) = \sum_\beta T_{\alpha\beta}^* \, (K_\beta^{(n)} - K_\beta^*) \quad , \tag{3.5}$$

where

$$T_{\alpha\beta}^* = \left. \frac{\partial K_\alpha^{(n+1)}}{\partial K_\beta^{(n)}} \right|_{F.P.} \tag{3.6}$$

evaluated at the fixed point.

In the simplest case, the eigenvalue equation

$$\sum_\beta T_{\alpha\beta}^* \, \phi_\beta^r = \lambda \, \phi_\alpha^r \tag{3.7}$$

will yield real eigenvalues, only one of which, λ_T, is "relevant", i.e., $\lambda_T > 1$ (not considering symmetry-breaking fields for the moment). The corresponding eigenvector gives the initial direction of the dotted line in Fig. 3.1.

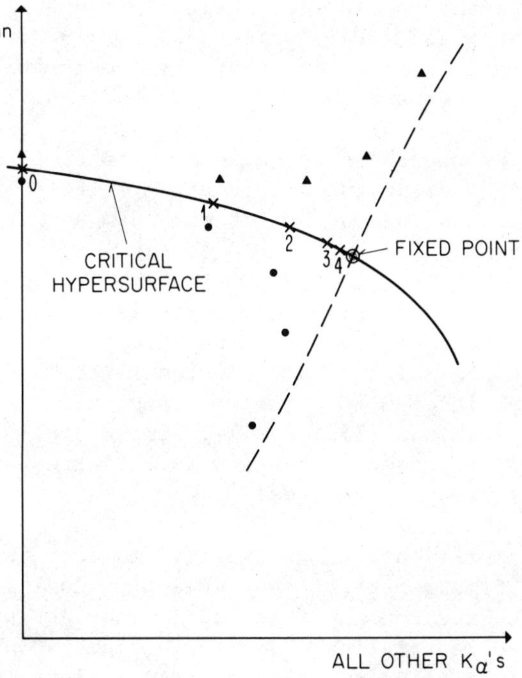

Fig. 3.1. Schematic diagram of RG flows in many-dimensional space of coupling constants.

The remaining eigenvectors have eigenvalues less than one ("irrelevant") and describe the convergence towards the fixed point.

If t is the distance from the fixed point along the relevant direction after some large number of RG iterations, the effect of one more iteration is to increase the distance to

$$t' = \lambda_T t = b^{y_T} t \quad . \tag{3.8}$$

Combining this with Eq. (3.4), we see that

$$\xi^{y_T} t = \xi'^{y_T} t' = \text{constant} \tag{3.9}$$

or

$$\xi \sim t^{-1/y_T} \quad . \tag{3.10}$$

From the definition of the critical exponent ν, we have the simple result

$$\nu = 1/y_T \quad . \tag{3.11}$$

By extending the analysis to include a magnetic field, which breaks the symmetry and is also relevant ($\lambda_h = b^{y_h} > 1$ or $y_h > 0$), expressions for all critical exponents can be derived as described in detail by Niemeijer and Van Leeuven in Ref. 16.

As is well known, the critical exponents so derived *automatically* satisfy all scaling and hyperscaling relations.[16] A violation of hyperscaling could only be detected by demonstrating a breakdown in the RG assumptions.

At this point, we consider the possibility of practical calculations with this formalism, and run into a very serious difficulty. Although the original Hamiltonian might be quite simple (nearest-neighbor interactions, for example) and the RG transformation in Eqs. (3.1) and (3.2) equally simple, the renormalized Hamiltonian will, in general, contain an infinite number of coupling constants and be very complicated.

Since it is difficult to do explicit calculations with an infinite number of parameters, the usual procedure is to "truncate" the renormalized Hamiltonian; that is, to make an approximation that projects the renormalized Hamiltonian into a subspace with a finite number of coupling parameters (usually less than four). Approximate analytic recursion relations are derived, fixed points are found, and critical exponents are calculated. Sometimes the results are extremely good, as determined by comparison with the exactly known critical exponents for special models. But other, equally plausible approximations can give very bad results. Or an approximation can give good results for one model and poor results for another model. It is even possible for an "improvement" in an approximation to lead to worse results.[21]

This uncertainty and the lack of a method of systematically improving the approximations make the results of the usual analytic real-space RG methods somewhat unreliable for the investigation of new models of interest.

IV. MONTE CARLO RENORMALIZATION-GROUP METHOD

The basic idea of combining the Monte Carlo and renormalization-group approaches is illustrated in Fig. 4.1. Starting from a simple Hamiltonian, $H^{(0)}(\sigma^{(0)})$, the direct calculation of the renormalized Hamiltonian, $H^{(1)}(\sigma^{(1)})$, is not practical because of the large number of interactions generated. However, it is straightforward to use the MC method to generate a sequence of configurations, $\{\sigma_i^{(0)}\}_t$, that allows us to calculate any correlation functions of interest for $H^{(0)}$. Since the values of each spin are explicitly stored in the computer, it is then trivial to apply the *exact* RG transformation to the configurations themselves. This produces a sequence of configurations with a distribution given by $H^{(1)}$. Since no approximation is made in the RG transformation, *all* interactions on the finite lattice are correctly taken into account.

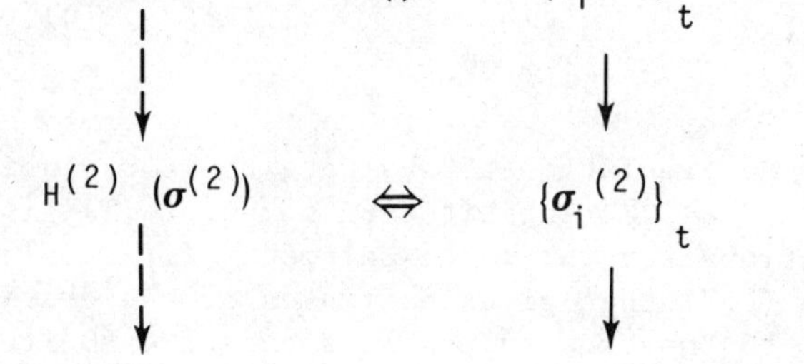

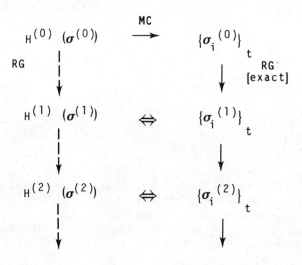

Fig. 4.1. Diagram of MCRG algorithm.

When Ma first suggested combining the MC and RG methods, his specific procedure involved trying to do a direct MC simulation of the fixed point.[1] Although he was able to obtain very encouraging results, his procedure forced him to severely truncate the approximate fixed-point Hamiltonian, and he was only able to include three couplings. This meant that he did not avoid the usual difficulties caused by truncation, and had the added problem of scanning a three-dimensional space for the fixed-point — a very time-consuming and expensive task when each point is obtained from an MC simulation.

I have taken the somewhat different approach of performing the MC simulation on the specific model of interest, which is usually much easier. If the MC simulation is carried out at criticality, successive RG iterations will automatically converge towards the fixed point as shown by the crosses in Fig. 3.1.

To calculate the matrix

$$T^*_{\alpha\beta} = \left. \frac{\partial K_\alpha^{(n)}}{\partial K_\beta^{(n-1)}} \right|_{F.P.} \tag{3.6}$$

we first write the chain-rule equation for the derivative of an arbitrary correlation function

$$\frac{\partial}{\partial K_\beta^{(n-1)}} \langle S_\gamma^{(n)} \rangle = \sum_\alpha \frac{\partial K_\alpha^{(n)}}{\partial K_\beta^{(n-1)}} \frac{\partial}{\partial K_\alpha^{(n)}} \langle S_\gamma^{(n)} \rangle \tag{4.1}$$

and note that we can use the same trick as in Eqs. (2.10) and (2.11) to evaluate the derivatives

$$\frac{\partial}{\partial K_\beta^{(n-1)}} \langle S_\gamma^{(n)} \rangle = \langle S_\gamma^{(n)} S_\beta^{(n-1)} \rangle - \langle S_\gamma^{(n)} \rangle \langle S_\beta^{(n-1)} \rangle \tag{4.2}$$

and

$$\frac{\partial}{\partial K_\alpha^{(n)}} \langle S_\gamma^{(n)} \rangle = \langle S_\gamma^{(n)} S_\alpha^{(n)} \rangle - \langle S_\gamma^{(n)} \rangle \langle S_\alpha^{(n)} \rangle \quad . \tag{4.3}$$

Using the MC simulation to evaluate the correlation functions in Eqs. (4.2) and (4.3), substitution in Eq. (4.1) gives us a set of linear equations for the matrix elements $T_{\alpha\beta}$. Solution of the eigenvalue Eq. (3.7) then gives estimates of the critical exponents, which converge to the true values as successive RG iterations bring us closer to the fixed point.

Although the renormalized Hamiltonians are all on the critical hypersurface, these equations extract information from the fluctuations and allow us to investigate the RG flows in any direction, within or out of the critical hypersurface.

A common source of confusion in Eqs. (4.1)-(4.3) is the question of which of the renormalized Hamiltonians is intended for calculating the expectation values. The answer is simply that it does not matter, since Eq. (3.1) defines the RG transformation in such a way that all renormalized Hamiltonians give the same result. In fact, each quantity in Eqs. (4.2) and (4.3) can be viewed as a (peculiar) correlation function defined on the original spins, $\{\sigma_i^{(0)}\}$.

Note that two consecutive levels of renormalization are used for each estimate of $T_{\alpha\beta}$. Also, we are not treating $\{\sigma_i^{(n-1)}\}_t$ and $\{\sigma_i^{(n)}\}_t$ as independent sequences, but we are explicitly using the correlations between them.

Since these calculations are carried out on finite computers, we have not fully eliminated the truncation of the Hamiltonian. However, it appears in a form very different from that found in the usual real-space RG calculations and the approximations are systematically improvable.

The first of the two places where truncation occurs is in the finite size of the lattice. Although the RG transformation is exact for the finite lattice, a kind of truncation approximation is made in identifying the finite-lattice Hamiltonian with the infinite-lattice limit. This approximation will be good if the range of the renormalized Hamiltonian is small in comparison with the dimensions of the lattice. This can be checked directly by varying the size of the lattice over a large range (for example, from 4×4 to 64×64). Note that the number of interactions included in the effective Hamiltonian increases exponentially with the number of sites in the lattice.

The second place where truncation occurs is in the evaluation of $T_{\alpha\beta}$. In principle, this has an infinite number of components, corresponding to all possible pairs of interactions. Here, we must make a much more severe assumption that only a small part of this matrix is required to extract the necessary information. However, we can test the effect of adding successive interactions to the analysis, and examine both the eigenvectors and eigenvalues to see whether they are affected.

It is clear that the class of interactions we can investigate is very large. For example, in addition to the term

$$K_h S_h = K_h \sum_i \sigma_i , \tag{4.4}$$

we can look at

$$K_{2h} S_h^2 = K_{2h} (\sum_i \sigma_i)^2 \tag{4.5}$$

which gives us the exponent y_{2h} and

$$\gamma = (d - y_{2h})/y_T . \tag{4.6}$$

If scaling is satisfied, $y_{2h} = 2y_h$. Since y_h and y_{2h} are obtained from different correlation functions, this provides a consistency test.

It is also possible to use finite-size scaling[20] to obtain an estimate for η (and consequently, y_h). Again, different correlation functions are used in the analysis and consistent results are obtained.

A few technical points:

The RG transformations contain an ambiguity in the position of blocks; in general, there are b^d ways of defining them. Naturally, all ways are correct and equivalent, but this ambiguity can be exploited by defining the blocks differently each time data is taken for the correlation functions. This improves the independence of the renormalized configurations, and greatly increases the accuracy of the higher RG iterations, without any additional computer time. Unfortunately, this trick was not used in the earlier work and was first noted in Ref. 12.

For correlation functions, $\langle S_\alpha^{(n)} \rangle$ that vanish by symmetry, accuracy can be improved by setting their values exactly equal to zero in Eqs. (4.2) and (4.3) instead of using the generally non-zero values produced by the MC simulation.

Finally, a warning that the results of an MCRG calculation are much more sensitive to deviations from equilibrium than those of standard MC studies. Equilibration runs should generally be longer, and great care should be taken to make sure that the system has reached equilibrium.

V. EXAMPLES OF MCRG CALCULATIONS

With any new calculational tool, it is necessary to test it on problems with known solutions. The first model to try is obviously the $d = 2$ Ising model. Results of MCRG calculations[7,11] for $b = 3$ are shown in Tables 5.1 - 5.3. As with all data I will show, the number of digits does not indicate the accuracy of the results, but are chosen to give some impression of typical fluctuations.

The first point to note is that there is good agreement between the calculations even though they were performed on different-size lattices. In fact, the size effect is only clearly visible in the third iteration for y_T in Table 5.3; this is a rather extreme case in which the lattice has been reduced in size to 3×3 and the effect is still only about 3%. The absence of a size effect in the final calculation is essential, and several size lattices *must* be used as a check.

The magnitude of the size effect depends on both the model and the RG transformation and must be investigated for each case separately.

It is apparent from Tables 5.2 - 5.3 that the convergence is extremely good, and even the second RG iteration is well within the linear region.

Table 5.1. MCRG for $d = 2$ Ising model, $b = 3$ majority rule RG, 45×45 lattice, 9×10^4 MCS/s. Lattice after two RG transformations is 5×5.

RG Iteration	Number of Interactions	y_T	y_h
1	1	0.943	1.859
	2	0.965	
	3	0.965	
2	1	0.987	1.874
	2	1.004	
	3	1.002	
EXACT		1	1.875

The assumption that very few components of the matrix $T_{\alpha\beta}$ are required is also confirmed. In this case, first- and second-neighbor couplings are sufficient and a four-spin interaction has essentially no influence. In the actual calculations, more interactions were included as a check.

Table 5.3 also shows the effect of varying the temperature at which the MC simulation is performed. A 2% change in temperature is more than sufficient to prevent the convergence to the critical fixed point, and the RG iterations move towards the weak- or strong-coupling fixed points.

The column for $K = 0.45$ is especially interesting, because we can see the convergence towards the strong-coupling eigenvalues, $y_T = d - 1 = 1$ and $y_h = d = 2$. The latter eigenvalue is characteristic of a discontinuity fixed point and describes the first-order transition as a function of magnetic field for $T < T_c$.

Table 5.2. Same as Table 5.1, but for a 108×108 lattice, 3×10^4 MCS/s (data taken at intervals of 5 MCS/s.) 3×10^3 MCS/s for equilibration. Lattice after three RG transformations is 4×4.

RG Iteration	Number of Interactions	y_T	y_h
1	1	0.940	1.859
	2	0.955	
	3	0.954	
2	1	1.000	1.873
	2	1.008	
	3	1.006	
3	1	0.977	1.873
	2	1.007	
	3	1.002	
EXACT		1	1.875

Table 5.3. Comparison of different initial temperatures. MCRG same as Table 5.1, but an 81 × 81 lattice with 3×10^4 MCS/s (discarding 3×10^3 MCS/s). Data at intervals of 5 MCS/s lattice after three RG transformations is 3 × 3.

RG Iteration	Number of Interactions	K = 0.43	$K = K_c =$ 0.440687	K = 0.45
		y_T	y_T	y_T
1	1	0.898	0.943	0.955
	2	0.934	0.966	1.004
	3	0.035	0.964	1.002
2	1	0.895	0.986	1.049
	2	0.894	1.005	1.066
	3	0.894	1.005	1.048
3	1	0.559	0.988	1.087
	2	0.627	1.029	1.027
	3	0.634	1.031	1.014
EXACT			1	1 (T=0)
		y_n	y_n	y_n
1		1.839	1.859	1.874
2		1.809	1.873	1.912
3		1.659	1.868	1.959
EXACT			1.875	2 (T=0)

Naturally other RG transformations can be used and we must demand that they *all* give consistent results, since the assumptions and approximations are the same in every case.

Table 5.4 shows the results for a b = 2 transformation. The majority rule must be supplemented to take care of cases in which the sum of spins in a block is zero. In earlier work, I used a particular site in the block as a tie-breaker. However, this disturbs the symmetry of the lattice slightly, and it is preferable

Table 5.4. MCRG for $d = 2$ Ising model, $b = 2$ majority rule RG with random tie-breaker, block-moving, 40×40 lattice, 10^5 MCS/s.

RG Iteration	Number of Interactions	y_T	y_h
1	1	0.923	1.8811
	2	0.972	1.8806
	3	0.976	1.8808
	4	0.976	1.8808
	5	0.976	
	6	0.977	
	7	0.972	
2	1	0.964	1.8730
	2	1.009	1.8734
	3	1.009	1.8735
	4	1.002	1.8734
	5	0.999	
	6	0.999	
	7	1.000	
3	1	0.959	1.8709
	2	0.999	1.8721
	3	1.001	1.8720
	4	0.995	1.8720
	5	0.992	
	6	0.989	
	7	0.988	
EXACT		1	1.875

to assign values of ±1 with equal probability in such cases using a random-number generator, as was done here. The block-moving trick was also used here. Again, the convergence is quite fast, although the first RG iteration is slightly different than for the $b = 3$ transformation. The slight decrease in y_T as the number of interactions increases in the third iteration is a real effect; the longer-range interactions begin to reflect the finite size of the lattice (5×5) and deviate from the infinite-lattice results. This can be checked directly by comparison with simulations on different-size lattices.

Table 5.5 shows the results of an MCRG calculation using decimation. This transformation simply integrates out all spins except those on a particular sublattice. The convergence is substantially worse than for the previous transformations, although it is still consistent with the exactly known exponents. Both the eigenvalues and eigenvectors change each time a new interaction is included, reflecting the importance of long-ranged interactions. These are typical warning signals that the assumption of short-range interactions in the renormalized Hamiltonian has broken down. The MCRG method has provided an internal self-consistency check that tells us that decimation is not suitable for good numerical calculations.

Table 5.5. MCRG for $d = 2$ Ising model, $b = 2$ decimation RG, block-moving, 40×40 lattice 10^4 MCS/s after discarding 10^5 MCS/s (runs of Table 5.4.).

RG Iteration	Number of Interactions	y_T	y_h
1	1	0.974	1.999
	2	1.137	1.941
	3	1.126	1.938
	4	1.159	1.936
	5	1.179	
	6	1.168	
	7	1.175	
2	1	0.902	1.987
	2	1.044	1.927
	3	1.025	1.922
	4	1.101	1.921
	5	1.151	
	6	1.146	
	7	1.164	
3	1	0.677	1.924
	2	0.847	1.885
	3	0.826	1.882
	4	0.979	1.874
	5	1.067	
	6	1.056	
	7	1.060	
EXACT		1	1.875

The scale factor does not have to be an integer. A particularly useful transformation with $b = \sqrt{5}$ was suggested by Van Leeuwen, in which the blocks are formed from a central spin and its four *second*-nearest neighbors. Krinsky and I have used this transformation[8] to analyze the Baxter model[22]

$$H_{Bax} = K_2 \sum_{nnn} \sigma_i \sigma_j + K_4 \sum \sigma_i \sigma_j \sigma_k \sigma_l \quad , \tag{5.1}$$

where the first sum is over *second*-neighbor pairs. The critical exponents depend on the ratio of the coupling constants, so that the model is described by a fixed line instead of a fixed point. The MCRG analysis correctly finds the variation of the critical exponents, although this is quite difficult with the usual real-space methods. Table 5.6 shows a typical case,[8] with $K_4 = 0.2$. The exponent y_M is associated with three-spin operators and produces confluent singularities in the magnetization and susceptibility. y_S is associated with nearest-neighbor interactions and is obtained quite accurately with practically no additional computer time.

Table 5.6. MCRG for the Baxter model with $K_2 = 0.314091$, $K_4 = 0.2$, $b = \sqrt{5}$ RG, no block-moving, 100×100 lattice averaging over 3×10^4 MCS/s after discarding 4×10^3 MCS/s.

RG Iteration	Number of Couplings	y_T	y_h	y_m	y_s
1	1	1.22	1.860	-	1.800
	2	1.18	1.862	0.645	1.800
	3	1.20	1.862	0.682	1.800
	4	1.20	1.862	0.738	1.800
2	1	1.26	1.866	-	1.801
	2	1.25	1.867	0.808	1.801
	3	1.26	1.867	0.830	1.801
	4	1.26	1.867	0.849	1.801
3	1	1.26	1.868	-	1.805
	2	1.25	1.871	0.837	1.805
	3	1.27	1.871	0.877	1.805
	4	1.27	1.871	0.892	1.805
EXACT		1.248	1.875	0.875	1.812

Table 5.7. MCRG for q = 4 Potts model, b = 2 majority rule with tie-breaker, block-moving, 96 × 128 lattice, 4×10^4 MCS/s at intervals of 10 MCS/s after discarding 6×10^3 MCS/s.

RG Iteration	Number of Interactions	y_T	y_h
1	1	1.18	1.863
	2	1.21	1.863
	3	1.21	1.862
	4	1.21	1.862
2	1	1.26	1.862
	2	1.27	1.861
	3	1.27	1.861
	4	1.27	1.861
3	1	1.28	1.861
	2	1.30	1.860
	3	1.30	1.859
	4	1.30	1.859
4	1	1.32	1.863
	2	1.34	1.854
	3	1.33	1.853
	4	1.34	1.854
EXACT (?)		1.5	1.875

Convergence is not always rapid, even when the assumptions of the MCRG are valid. This is illustrated by results[12] for the q = 4 Potts model[23] shown in Table 5.7,

$$H_{POTTS} = K \sum_{nn} \delta_{\sigma_i,\sigma_j} \qquad \sigma_i = 1, 2, \ldots q = 4 \,. \tag{5.2}$$

This model is expected to be in the same universality class as the Baxter-Wu model[24]

$$H_{BW} = \sum \sigma_i \sigma_j \sigma_k \qquad \sigma = \pm 1 \quad , \tag{5.3}$$

(where the sum is over elementary triangles on the triangular lattice). Exact results are available for the Baxter-Wu model

($y_T = 1.5$), but the convergence for the $q = 4$ Potts model is much too slow to either confirm or deny the same exponents. This was puzzling at first, because the Baxter-Wu model shows excellent convergence with the MCRG method.[13] The difference has been explained by the existence of a marginal ($y = 0$) eigenvalue for the $q = 4$ Potts model, which slows the convergence.[12,25-27] This also has the effect of introducing logarithmic corrections,[28,29] which are absent in the Baxter-Wu model, so that the slow MCRG convergence has pointed to a physical difference in the critical behavior of the two models.

The examples I have given have all dealt with discrete "spin" variables. This has just been a matter of convenience, since faster algorithms can be constructed for MC simulations of discrete systems. In fact, it is particularly easy to construct reasonable RG transformations for continuously varying vectors by doing a vector sum of spins in a block, and rescaling the length of the result to unity.

VI. EXTENSION OF MCRG

Although the results of MCRG calculations performed so far have been very encouraging, some problems still remain with the method as I have presented it so far. As might be expected, the problems all concern the finite size of the lattices that can be used, and the finite computer time available for calculations.

Since the MC simulation must be performed at a critical point for the RG transformations to move towards the fixed point, it is very important to know the location of the critical point accurately. The convergence of the MCRG eigenvalues can, and has, been used as a criterion for locating T_c,[8] but it would be much better if this could be reserved as a check of consistency, and T_c calculated independently. Using the convergence to find T_c is also very time consuming, since several MCRG calculations must be carried out at different temperatures. For locating a multicritical point, a two-dimensional space of coupling constants would have to be scanned and the method ceases to be practical.

The finite size of the lattice implies a finite number of RG iterations — usually three or four. This puts a limit on how close we can come to the fixed point. We have seen that this limitation is quite serious for the $q = 4$, $d = 2$ Potts model. In principle, we can always make the lattice larger, but the increased correlation times due to critical slowing-down make statistical errors a serious problem if the lattice is too large.

Finally, we can only look at a finite part of the $T_{\alpha\beta}$ matrix. We can check the effect of adding one more interaction, but the effect of including a thousand interactions has not been discussed.

This was pretty much the state of affairs at the end of 1979, when I began to collaborate with Prof. K. G. Wilson. The ideas presented in the rest of this talk are a combination of Wilson's independent work on MCRG for lattice-gauge theories and the results of our collaboration.

It turns out to be quite easy to check the effect of truncating $T_{\alpha\beta}$. By treating two (or more) RG iterations as a single RG transformation, $(T^2)_{\alpha\beta}$ can be calculated, with eigenvalues $\lambda^2 = b^{2y}$. If the truncation of $T_{\alpha\beta}$ is significant, these eigenvalues will differ from the squares of those obtained from a single transformation, revealing the effect of including many interactions.

To attack the other problems, we have to return somewhat to Ma's original idea of simulating the fixed point.[1] Naturally we can only make an approximation to the fixed point, since we can only treat a finite number of interactions, but this can bring us systematically closer to the fixed point and improve convergence without going to larger systems. The difficulty is in calculating the renormalized coupling constants (which we did not need to obtain $T_{\alpha\beta}$).

We begin with Wilson's method[30] of simulating *two* systems of different sizes. If the linear dimensions of the larger system differ from those of the smaller by the scale factor b, one RG transformation of the larger will make both the same size. The differences in the correlation functions will then reflect the changes in the coupling constants under renormalization. Although using small lattices strongly affects the correlation functions, this effect is the same for both systems; if all correlation functions are the same for both systems, $H^{(1)} = H^{(0)}$ and we have a fixed point.

Instead of trying to adjust the coupling constants by trial and error, the changes in the K_α's due to renormalization can be calculated direct by solving

$$\begin{aligned} \delta\langle S_\alpha^{(1)}\rangle &\equiv \langle S_\alpha^{(1)}\rangle_L - \langle S_\alpha^{(0)}\rangle_S \\ &= \sum_\beta \frac{\partial\langle S_\alpha^{(\ell)}\rangle}{\partial K_\beta^{(\ell)}}\,\delta K_\beta \end{aligned} \tag{6.1}$$

for the δK_β's. The matrix $\partial \langle S_\alpha \rangle / \partial K_\beta$ is evaluated using Eq. (4.3) and can be obtained from either the larger or smaller lattice (taking the average of the two includes quadratic terms).

Application of this idea to successive RG iterations allows us to follow the renormalized Hamiltonians to the fixed point and confirm that the differences

$$\partial K_\alpha^{(n)} = K_\alpha^{(n)} - K_\alpha^{(n-1)} \tag{6.2}$$

go to zero with increasing n, if we have started at criticality.

Since we have also calculated the matrix $T_{\alpha\beta}$ with its eigenvalues and left- and right-eigenvectors, it is straightforward to obtain a systematically improvable estimate for the location of the fixed point from Eq. (6.1).

We can then perform the next MC simulation for a Hamiltonian that is "closer" to the fixed point. The criterion for "closeness" that seems most appropriate is to choose the m coupling constants used in the MC simulation to make the deviations from the fixed point in the direction of the right-eigenvectors corresponding to the m largest eigenvalues vanish. The required values are easily obtained by solving a set of m coupled linear equations.

Wilson has suggested a particularly simple and efficient method for calculating fixed points using MC simulations of large and small systems.[31] The change in a correlation function after n RG transformations of the larger lattice due to changes $\delta K_\beta^{(0)}$ of the original coupling constants is given by

$$\delta \langle S_\alpha^{(n)} \rangle_L = \sum_\beta \frac{\partial \langle S_\alpha^{(n)} \rangle_L}{\partial K_\beta^{(0)}} \delta K_\beta^{(0)} \tag{6.3}$$

with a similar expression for $\delta \langle S_\alpha^{(n-1)} \rangle_S$ on the smaller lattice,

$$\frac{\partial \langle S_\alpha^{(n)} \rangle}{\partial K_\beta^{(0)}} = \langle S_\alpha^{(n)} S_\beta^{(0)} \rangle - \langle S_\alpha^{(n)} \rangle \langle S_\beta^{(0)} \rangle . \tag{6.4}$$

By solving the set of linear equations

$$\langle S_\alpha^{(n)} \rangle_L - \langle S_\alpha^{(n-1)} \rangle_S = \sum_\beta \left[\frac{\partial \langle S_\alpha^{(n)} \rangle_L}{\partial K_\beta^{(0)}} - \frac{\partial \langle S_\alpha^{(n-1)} \rangle_S}{\partial K_\beta^{(0)}} \right] \delta K_\beta^{(0)} \tag{6.5}$$

for the δK_β's, we can find what changes in the original couplings would be necessary to make the correlation functions, and hence the renormalized Hamiltonians, equal. In principle, this allows us to calculate the location of the fixed point in a single step if we are in the linear region.

As the number of RG transformations increases, the deviations of $H^{(n)}$ from the fixed point in the *irrelevant* directions decrease. The differences $H^{(n)} - H^{(n-1)}$ in the irrelevant directions become small, as do the corresponding differences in the correlation functions. After two or three RG transformations, these small differences become very difficult to compute with any accuracy. The best determination of the deviation from the fixed point in the irrelevant directions is found from a single RG transformation.

On the other hand, if we are not exactly at criticality, the deviations in the *relevant* directions will be amplified by the RG transformations, and the corresponding differences in correlation functions will be relatively easy to measure. This property makes Eq. (6.5) very well suited to calculate critical temperatures. Since this feature is characteristic of all relevant operators, the method can be used without additional effort to locate multicritical points.

This determination of T_c is independent of the convergence of the usual MCRG analysis so that convergence can be used as a consistency check.

Landau and I have applied this method successfully to find tricritical points, and determine tricritical exponents in two dimensions.[14] The convergence is quite good and we were able to use relatively small systems, which enabled us to make longer runs with small statistical errors.

For a Blume-Capel model with first- and second-neighbor exchange

$$H_{BC} = K_1 \sum_{nn} \sigma_i \sigma_j + K_2 \sum_{nnn} \sigma_i \sigma_j + K_3 \sum_i \sigma_i^2 \qquad \sigma_i = 0, \pm 1, \qquad (6.6)$$

a 16 × 16 lattice is adequate for accurate results. We also did the calculation for a 32 × 32 lattice and those results are shown in Table 6.1.[14]

The RG transformation used 2 × 2 blocks (b=2) and a "plurality" rule, treating all states equally. Ties were decided with a random-number generator rather than choosing a particular site as a tie-breaker.

Table 6.1. Extended MCRG analysis of the tricritical point in a Blume-Capel model on a 32 × 32 lattice using 2.24×10^5 MCS/s after discarding 1.6×10^4 MCS/s. ($K_1 = 1.25$, $K_2 = 0.28$, $K_3 = -2.99$.) Ten even and six odd interactions were used for each of the values given. The exact (?) results are from the conjectures in Refs. 25, 32 - 34.

RG Iteration	y_T	y_H
1	1.785	1.934
2	1.799	1.934
3	1.805	1.929
EXACT (?)	1.8	1.925

Similar results were obtained for a second-neighbor Ising antiferromagnet in a magnetic field, confirming both universality and the conjectured values for y_T and y_h within the accuracy of the calculation.[14]

VII. CONCLUSION

The MCRG method has already been applied to more systems than I have had time to discuss, and its potential usefulness has been greatly enhanced by the extensions described in the previous section. As I mentioned at the beginning, this method is still in the early stages of development and it is difficult to predict just how far it can take us.

The major limitation to the accuracy remains the statistical errors from the Monte Carlo simulation, as Ma pointed out in his original paper. However, in the light of the results obtained so far and the continuing improvements in the speed and cost of computers, I think we can be fairly optimistic about the future.

ACKNOWLEDGEMENTS

I would like to thank H. W. J. Blöte, V. J. Emery, H. R. Jauslin, S. Krinsky, D. P. Landau, M. Novotny, C. Rebbi and especially K. G. Wilson for many interesting discussions and important contributions.

REFERENCES

1. S.-K. Ma, Renormalization Group by Monte Carlo Methods, *Phys. Rev. Lett.*, 37:461 (1976).
2. S.-K. Ma, Alternative Approach to the Dynamic Renormalization Group, *Phys. Rev. B*, 19:4824 (1979).
3. Z. Friedman and J. Felsteiner, Kadanoff Block Transformation by the Monte Carlo Technique, *Phys. Rev. B*, 15:5317 (1977).
4. A. L. Lewis, Lattice Renormalization Group and the Thermodynamic Limit, *Phys. Rev. B*, 16:1249 (1977).
5. H. Müller-Krumbhaar, Real Space Renormalization for Arbitrary Single-Site Potential, *Z. Phys. B*, 35:339 (1979).
6. P. J. Reynolds, H. E. Stanley, and W. Klein, A Large Cell Monte Carlo Renormalization Group for Percolation, *Phys. Rev. B*, 21:1223 (1980).
7. R. H. Swendsen, Monte Carlo Renormalization Group, *Phys. Rev. Lett.*, 42:859 (1979).
8. R. H. Swendsen and S. Krinsky, Monte Carlo Renormalization Group and Ising Models with $n \geq 2$, *Phys. Rev. Lett.*, 43:177 (1979).
9. H. W. J. Blöte and R. H. Swendsen, First-Order Phase Transitions and the Three-State Potts Model, *Phys. Rev. Lett.*, 43:799 (1979).
10. H. W. J. Blöte and R. H. Swendsen, Critical Behavior of the Three-Dimensional Ising Model, *Phys. Rev. B*, 20:2077 (1979).
11. R. H. Swendsen, Monte Carlo Renormalization Group Studies of the $d = 2$ Ising Model, *Phys. Rev. B*, 20:2080 (1979).
12. C. Rebbi and R. H. Swendsen, Monte Carlo Studies of the q-state Potts Model in Two Dimensions, *Phys. Rev. B*, 21:4094 (1980).
13. M. Novotny, D. P. Landau, and R. H. Swendsen, Monte Carlo Renormalization Group Study of the Baxter-Wu Model, preprint.
14. D. P. Landau and R. H. Swendsen, Tricritical Universality in Two Dimensions, preprint.
15. K. Binder, Monte Carlo Investigations, *in* "Phase Transitions and Critical Phenomena," C. Domb and M. S. Green, eds., Academic, New York (1976).
16. Th. Niemeijer and J. M. J. Van Leeuwen, Renormalization: Ising-Like Spin Systems, *in* "Phase Transitions and Critical Phenomena," C. Domb and M. S. Green, eds., Academic, New York (1976).

17. K. G. Wilson and R. H. Swendsen, unpublished.
18. M. Fisz, "Probability Theory and Mathematical Statistics", Wiley, New York (1963).
19. H. Müller-Krumbhaar and K. Binder, Dynamic Properties of the Monte Carlo Method in Statistical Mechanics, *J. of Stat. Phys.*, 8:1 (1973).
20. It is possible to use finite-size scaling to extract information from the height of the peak ($\chi_{max} \sim N^{\gamma\lambda}$), but this introduces a fourth parameter, and we still have not accounted for corrections to scaling. See, M. E. Fisher, *in* Proceedings of the Enrico Fermi International School of Physics, M. S. Green, ed., Academic Press, New York (1971); M. E. Fisher and A. E. Ferdinand, Interfacial, Boundary, and Size Effects at Critical Points, *Phys. Rev. Lett.*, 19:169 (1967); A. E. Ferdinand and M. E. Fisher, Bounded and Inhomogeneous Ising Models. I. Specific-Heat Anomaly of a Finite Lattice, *Phys. Rev.*, 185:832 (1969).
21. R. H. Swendsen and R. K. P. Zia, The Surprising Effectiveness of the Migdal-Kadanoff Renormalization Scheme, *Phys. Lett.*, 69A:382 (1979).
22. R. J. Baxter, Partition Function of the Eight-Vertex Lattice Model, *Ann. Phys. (N.Y.)*, 70:193 (1972).
23. R. B. Potts, *in* Proceedings of the Cambridge Philos. Soc., 48:106 (1952).
24. R. J. Baxter and F. Y. Wu, Exact Solution of an Ising Model with Three-Spin Interactions, *Phys. Rev. Lett.*, 31:1294 (1973).
25. B. Nienhuis, A. N. Berker, E. K. Riedel, and M. Schick, First- and Second-Order Phase Transitions in Potts Models: Renormalization-Group Solution, *Phys. Rev. Lett.*, 43:737 (1979).
26. R. J. Baxter, Potts Model at the Critical Temperature, *J. Phys. C.*, 6:L445 (1973).
27. M. Nauenberg and D. J. Scalapino, Singularities and Scaling Functions at the Potts-Model Multicritical Point, *Phys. Rev. Lett.*, 44:837 (1980).
28. F. J. Wegner, Corrections to Scaling Laws, *Phys. Rev. B*, 5:4529 (1972).
29. F. J. Wegner and E. K. Riedel, Logarithmic Corrections to the Molecular-Field Behavior of Critical and Tricritical Systems, *Phys. Rev. B*, 7:248 (1973).
30. K. G. Wilson, Monte Carlo Calculations for the Lattice Gauge Theory, Lectures at Les Houches, France (1980).
31. K. G. Wilson, private communication.
32. M. P. M. den Nijs, A Relation Between the Temperature Exponents of the 8-Vertex and q-State Potts Model, *J. Phys. A.*, 12:1857 (1979).
33. R. B. Pearson, to be published.
34. B. Niehnuis, E. K. Riedel, and M. Schick, to be published.

CRITICAL BEHAVIOUR IN INTERFACES

D. J. Wallace

Physics Department
The University of Edinburgh
Edinburgh EH9 3JZ
Scotland

1. INTRODUCTION

In these lectures I am going to talk about problems involving an interface between two discrete thermodynamic phases. In the lectures of Dr. Beysens and Dr. Moldover we saw examples of such interfaces and they are very familiar objects in at least two contexts: the almost planar interface between two phases in equilibrium and the almost spherical interface associated with the decay of a metastable state by the formation of critical droplets of the favoured phase. In theoretical terms we will be dealing with one-component order parameters so the ideas should be applicable to the liquid-vapour system, to binary fluids and, with some possible reservations, to other systems such as domain walls in uniaxial magnets. There are two main characteristic effects of thermal fluctuations on which I wish to focus.

(i) Local fuzziness : As the title of the lectures indicates we are going to be interested in properties of the interface in the critical region. It has been appreciated since the work of Van der Waals that as the temperature T is increased towards the critical temperature T_c the interface becomes increasingly diffuse; it has a local width, over which the mean density changes from e.g. that of the liquid to that of the vapour.

In a scaling theory, the bulk correlation length ξ is a measure of this local fuzziness - the decay of density correlation from one ordered phase to the other takes place with the same length scale as the decay of any bulk correlation. As T is increased to T_c, the interface disappears in the sense that this width diverges as $(T_c-T)^{-\nu}$ and the density difference tends to zero as $(T_c-T)^{\beta}$, as indicated in Fig. 1. (Strictly speaking I should use the exponent ν' of the correlation length below T_c; in the subsequent field theory calculations all scaling and hyperscaling laws will be satisfied so we drop the distinction). One important point to bear in mind is that, given this scaling picture, if we are probing structure within the interface we are studying phenomena at length scales less than ξ. Therefore, modulo corrections to scaling (due to the effective short distance cut-off Λ), within the interface the fluctuations are critical; the title of the lectures is also a statement of fact.

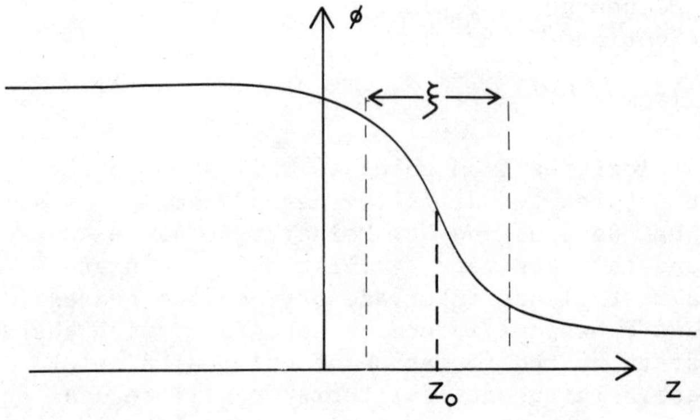

Fig. 1. Characteristic form of the density profile through an interface located at height z_o.

(ii) Wandering : The second characteristic in which we will be interested is connected with the wandering of the locally fuzzy interface i.e. the collective displacements in density away from e.g. planar. The significance of such fluctuations is that, in the absence of an interaction such as gravity which explicitly breaks the symmetry under the Euclidean group of rotations and translations, the position of the planar interface breaks spontaneously the Euclidean group E(d) of the d bulk dimensions down to the group E(d-1) of the (d-1) dimensions of the interface. The deviations from planar are the Goldstone modes of this spontaneously broken E(d) symmetry; they have a mass gap only if there are explicit symmetry breaking effects such as gravity or finite volume. If we could switch off gravity then the existence of these gapless Goldstone modes implies that the interface is a critical problem for all $T<T_c$ - a third reason for the choice of title. In a dynamical context, these modes are just the familiar surface waves. In this article I do not discuss dynamics; the Goldstone modes are fluctuations away from planar in the ensemble sense of equilibrium statistical mechanics. It is useful to think in analogy with spin waves in an isotropic ferromagnetic; they are the fluctuations transverse to the direction of spontaneous magnetisation in spin space and are gapless because of the spontaneous breaking of the continuous symmetry (e.g. O(n)) in spin space below T_c. Here we are going to meet similar effects in n = 1 systems associated with the breaking of Euclidean invariance by an interface.

Let me review briefly the extent to which these characteristics have been observed experimentally. The most detailed and refined experiments appear to have been carried out by Huang and Webb (1969) and by Wu and Webb (1973), who study light reflection from the liquid-vapour interface in sulphurhexafluoride, SF_6. The basic formula for reflectivity R for this kind of experiment is the Born approximation

$$R = \left|(n_\ell + n_v)^{-1} \int_{-\infty}^{\infty} \frac{dn(z)}{dz} e^{-2ikz} dz\right|^2 \qquad (1)$$

where n is the refractive index, subscripts ℓ and v refer to liquid and vapour and k is the wave-vector of the reflected light. The temperature range in this kind of experiment is limited because R goes to zero very rapidly as T increases to T_c but accurate results are obtained in the range $10^{-4} \leqslant t \equiv (T_c - T)/T_c \leqslant 10^{-2}$. Assuming the validity of the Lorentz-Lorenz relation in the neighbourhood of the critical point, the refractive index profile dn/dz is related to the mean density profile $d\phi/dz$ permitting comparison of theory and experiment.

Some results of the Wu and Webb experiments can be summarised as follows. (i) The mean-field scaling form of the density profile, which we shall discuss in the next section, can not fit the data (even allowing for arbitrary critical indices). (ii) Other scaling forms can fit the data e.g. an error function density profile (Gaussian $d\phi/dz$) fits data well with $\nu = 0.62 \pm 0.01$, $\beta = 0.333 \pm 0.008$. With the increased knowledge of the importance of corrections to scaling discussed in Dr. Sengers' lectures on fluids, we should presumably be careful in interpreting the errors quoted for these exponents; nevertheless they provide clear evidence that the interface is a useful vehicle for studying critical behaviour. (iii) The damped surface waves corresponding to the Goldstone modes are observed in quasi-elastic light scattering.

One part of these lectures is therefore to discuss some of the theoretical background to this kind of experiment. The second part is concerned with similar ideas in the context of almost spherical droplets. Here we are going to be probing the nature of the singularity at a first order phase transition and the role of the critical droplet in the metastable state. In order to compare theoretical predictions with experiments on nucleation and limits of metastability important dynamical effects must be taken into account; I shall not attempt to review this subject (see for example Binder and Stauffer (1976), Langer and Turski (1973) and Langer (1980)). Instead I shall concentrate on theoretical aspects of the nature of the singularity in free energy at a first order phase transition. I shall review field theory arguments for the existence of an essential singularity with universal features at first order phase transitions and show that the results of series expansions for the Ising model on a square lattice are in good agreement with these predictions. The contribution of non-spherical droplets is an important factor in the field theory calculations.

Finally in this introduction I remark that I will say little about (d-1) dimensional phase transitions which occur in surfaces and interfaces, or are associated with them; for references on this topic see for example Wortis (1980) and Lawrie and Lowe (1980).

2. THE ALMOST PLANAR INTERFACE

The Field-theory Model

Our starting point is the Ginzburg-Landau-Wilson (GLW) model for a one component field $\phi(x)$ representing e.g. magnetisation density, or fluid density etc. The general form for the reduced Hamiltonian is taken as

$$\mathcal{H} = \int d^dx\left(\tfrac{1}{2}(\nabla\phi(x))^2 + V(\phi(x))\right) \qquad (2)$$

and the partition function and correlation functions are to be calculated by functional integration over all configuration $\phi(x)$:

$$Z = \int D\phi\; e^{-\mathcal{H}} \qquad (3)$$

$$\langle\phi\rangle = \int D\phi\;\phi\; e^{-\mathcal{H}}/Z \qquad (4)$$

etc.

Since we are interested in the almost planar interface between coexisting phases, $V(\phi)$ has the form of a double potential well Many of the arguments of this section will go through for a general $V(\phi)$ of the form in Fig. 2. For specific calculations we take the prototype double well potential :

$$\mathcal{H} = \int d^dx\;\left(\tfrac{1}{2}(\nabla\phi)^2 - \tfrac{1}{2}\,t\,\phi^2 + \tfrac{1}{4}\,g\,\phi^4\right). \qquad (5)$$

Here t is the reduced temperature $(T_c - T)/T_c$, taking for convenience of notation t as positive for $T<T_c$. For fluids the field ϕ in (5) must be interpreted as the density difference $\rho-\rho_c$; for further comments see the lectures of Dr. Sengers.

A cut-off Λ roughly the inverse of the mean molecular spacing should also be understood in (2) and (5); its presence increases the complexity of specific calculations if renormalized perturbation theory is not used. Further derivative coupling terms $(V(\phi, \partial\mu\phi, \ldots))$ in (2) similarly complicate matters but should not invalidate general arguments provided the couplings are local i.e. short range.

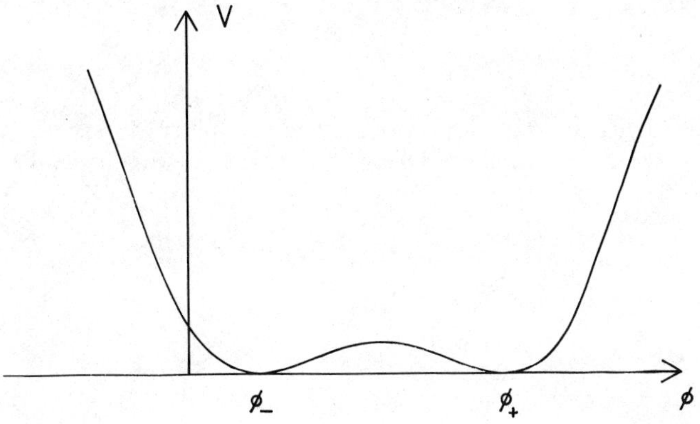

Fig. 2. Form of potential $V(\phi)$ permitting two phases of classical densities ϕ_+ and ϕ_- .

Van der Waals Theory

The mean-field theory of the interface was described by Van der Waals (1894); for more recent references see Cahn and Hilliard (1958), Widom (1972), Evans (1979). As described by Dr. Brézin in his lectures it is obtained in the context of the LGW model (2-4) by expanding the field ϕ about an extremum of (2) i.e. a solution of the field equations

$$\frac{\delta \mathcal{H}}{\delta \phi} = 0 \quad : \quad \nabla^2 \phi = \frac{\partial V}{\partial \phi} \quad . \tag{6}$$

The mean-field approximation is obtained by neglecting fluctuations i.e. replacing the functional integral in (3) and (4) by the maximum value of the integrand.

The existence of two coexisting homogeneous phases is guaranteed by the two minima of V at $\phi_\pm$ as described above. To describe the planar interface we must start from a solution of (6) which depends on a single coordinate z, say, with the boundary conditions

$$\phi(z) \to \phi_\pm \quad \text{as} \quad z \to \pm\infty \quad . \tag{7}$$

For a $V(\phi)$ as in Fig. 2, our experience of motion in one dimension tells us that a solution satisfying the boundary conditions (7) always exists, because we can write (6) in the form

$$\frac{d^2\phi}{dz^2} = - \frac{\partial \tilde{V}}{\partial \phi} \quad ; \tag{8}$$

this is identical to the equation of motion of a particle of coordinate ϕ moving at time z in a potential $\tilde{V}$ which is the inverted form of the V of Fig. 2:

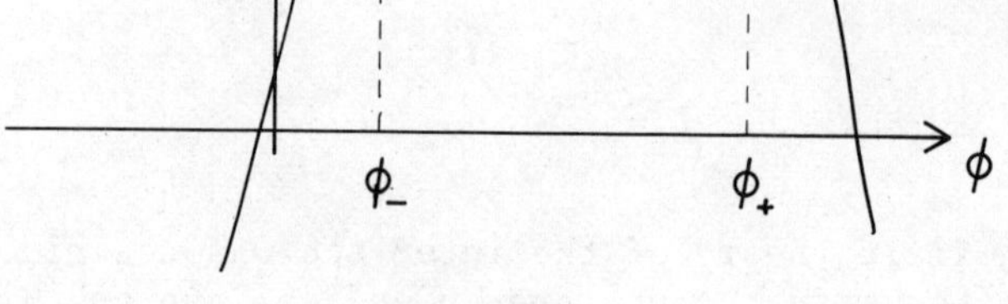

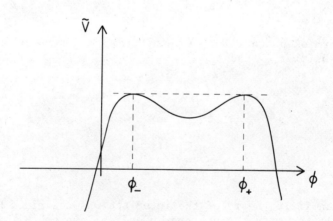

Fig. 3. The potential $\tilde{V}$ for the equation of motion (8).

The solution we want is the familiar kink solution which starts at the top of one hill at time $z = -\infty$, swoops across the gap and just reaches the other top at $z = +\infty$.

For the particular case of the ϕ^4 model it is straightforward to integrate (8) to obtain the kink solution

$$\phi_c(z) = \sqrt{t/g}\ \tanh\left(\sqrt{t/2}\ (z-z_0)\right) . \tag{9}$$

Expanding the configuration $\phi(x)$ in (3) and (4) about the solution $\phi_c(z)$

$$\phi = \phi_c(z) + \hat{\phi} \tag{10}$$

and neglecting the fluctuations $\hat{\phi}$ gives mean field results. The free energy excess over the homogeneous phase, $\mathcal{H}(\phi_+)$, is extensive in the area i.e. defines a surface tension:

$$\begin{aligned} \mathcal{H}(\phi_c) - \mathcal{H}(\phi_+) &= \int d^{d-1}x\left[\int dz\left(\tfrac{1}{2}(d\phi_c/dz)^2 + V(\phi_c) - V(\phi_+)\right)\right] \\ &= \int dz\,(d\phi_c/dz)^2 \int d^{d-1}x \end{aligned} \tag{11}$$

where the second equality comes from the first integral ("conservation of energy") of (8). Similarly the mean density profile is

$$\langle\phi\rangle = \phi_c(z) \quad . \tag{12}$$

Noting that in the mean field approximation for (5) the bulk correlation length ξ is given by $\xi^{-2} = 2t$ we can write (9) in the scaling form

$$\langle\phi\rangle = \tfrac{1}{2}(\phi_+ - \phi_-)\ \tanh\left(\tfrac{1}{2}(z-z_0)/\xi\right) \tag{13}$$

where the order parameter difference in the two phases is proportional to t^β. Equation (13) describes a planar interface located at height z_0 and with a width determined by ξ.

Note that the exponential approach to the ordered phases as $|z| \to \infty$ illustrated in the tanh function in (13) is a general feature of interfaces in $n = 1$ systems. It should be contrasted with the uniform rotation of order parameter which characterises models with continuous internal symmetry when non-trivial boundary conditions such as (7) are applied; for further discussion of the latter see e.g. Fisher et al (1973). The localisation of the interface implicit in (13) and exploited below implies that models whose order parameter has a continuous symmetry are specifically excluded from the following discussions.

Fluctuations

Let us begin now to look at the effect of fluctuations. The substitution of (10) into (3) is equivalent to translating the integration variable ϕ by the fixed field ϕ_c and one obtains formally

$$Z = e^{-\mathcal{H}(\phi_c)} \int D\hat{\phi} \exp - \int d^d x(\tfrac{1}{2} \hat{\phi}(x)\mathcal{M}\hat{\phi}(x) + O(\hat{\phi}^3)). \tag{14}$$

The term linear in $\hat{\phi}$ in the Taylor expansion of $\mathcal{H}(\phi)$ vanishes because ϕ_c is an extremum (equation (6)). The differential operator $\mathcal{M}$ in the quadratic term is given by

$$\begin{aligned} \mathcal{M} &= - \nabla^2 + \frac{\partial^2 V}{\partial \phi^2} (\phi_c) \\ &= - \nabla^2 - t + 3t \tanh^2((t/2)^{\frac{1}{2}}(z-z_0)) \end{aligned} \tag{15}$$

for the ϕ^4 potential (5). As outlined in Edouard Brézin's lectures one can now formally set up perturbation theory with a free propagator G which is the Green function of $\mathcal{M}$: $\mathcal{M}G(x,y) = \delta^d(x-y)$. Perturbation theory will be useful if there are no critical modes. From the form of (15) it is clear that as $T \to T_c$, $(t \to 0)$ we have the usual critical problem of a vanishingly small gap; successive orders in perturbation theory make unboundedly larger contributions to the mean profile $\langle\phi\rangle$ as $t \to 0$. We can resolve this problem in e.g. $4-\varepsilon$ dimensions using the renormalization group equation described by Brézin applied to the one-point function $\langle\phi\rangle$. This type of calculation is included in the papers by Ohta and Kawasaki (1977), Rudnick and Jasnow (1978) and Jasnow and Rudnick (1978), and results in the typical scaling generalisation of (13):

$$\langle\phi\rangle = t^{\beta} \, \Phi \, ((z-z_0)/\xi) \tag{16}$$

where

$$\Phi(u) = \tanh(u/2) + O(\varepsilon) \quad .$$

However I want to review the fact that, as these authors recognize, there are more subtleties in this problem which require close attention. These originate in the fact that ϕ_c in (9) breaks e.g. the translation invariance of the original Hamiltonian (5). In an infinite volume system there is a solution ϕ_c for all values of the interface position z_0; differentiating (6) with respect to z_0 gives immediately

$$\mathcal{M} \, \partial_{z_0} \phi_c = 0 \tag{17}$$

i.e. $\partial_{z_0}\phi_c$ $(\equiv - \partial_z\phi_c)$ is an eigenfunction of $\mathcal{M}$ with zero eigenvalue for all fixed t ($T < T_c$ of course). The eigenfunction $\partial_z\phi_c$ represents the Goldstone mode of the spontaneously broken translation invariance. According to (10) the field configuration corresponding to a fluctuation of the mode $\partial_{z_0}\phi_c$ with small amplitude a is

$$\phi(x) = \phi_c(z) + a\,\partial_{z_0}\phi_c$$
$$\simeq \phi_c(z-a)$$

i.e. represents an interface translated by an amount a. Equation (17) is just the infinitesimal statement that in the model (2) it doesn't cost us any energy to do this.

Further if we denote by x the (d-1) coordinates orthogonal to z, with q a (d-1) component wave vector, it follows immediately from the general form of $\mathcal{M}$(equation 15) and (17) that

$$\mathcal{M} e^{iq.x}\,\partial_{z_0}\phi_c(z) = q^2\,e^{iq.x}\,\partial_{z_0}\phi_c(z). \tag{18}$$

Thus we have a continuous spectrum of eigenfunctions of $\mathcal{M}$ with no gap in the limit $q \to 0$. The field configuration corresponding to a superposition of these modes is

$$\phi(x) = \phi_c(z) + \sum_q a_q\,e^{iq.x}\,\partial_{z_0}\phi_c(z)$$
$$= \phi_c(z-f(x)) \tag{19}$$

where we have defined the field f(x), depending on the (d-1) coordinates x, by

$$f(x) = \sum_q a_q\,e^{iq.x} + O(a^2)\;. \tag{20}$$

Since the configuration (19) represents an interface translated locally by an amount f(x), we see that the Goldstone modes (18) of the spontaneously broken Euclidean invariance are the collective displacements of the interface away from planar.

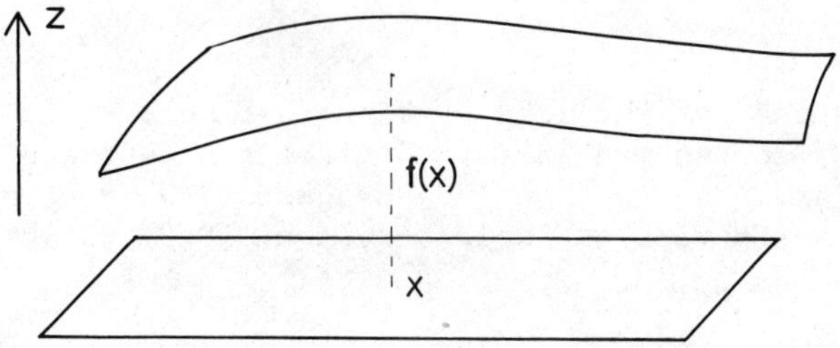

Fig. 4. The field f(x) describing the configuration (19) of the collective displacement from planar.

It should now be clear that it is the modes (18) which are responsible for the second important characteristic of the interface, the wandering from planar. Before we turn to further discussion of their importance, we should remark that for a generic potential V all other modes will have a mass gap of order t i.e. ξ^{-2}. For example for the ϕ^4 model (5) the next eigenvalues are $q^2+3t/2$. Although important in the critical region these modes are harmless compared to (18) for fixed t.

Interface Wandering

We can now begin to get some idea of the effect of interface wandering due to the modes (18), by substituting $\hat{\phi} = f(x)\partial_{z_0}\phi_c(z)$ in equation (14). This results in an effective energy for the f field, using equation (18),

$$\mathcal{H}_{eff}(f) = \frac{\sigma}{kT} [\int d^{d-1}x[1 + \tfrac{1}{2}(\nabla f)^2 + ?]. \qquad (21)$$

where $\sigma/kT \equiv \int dz(d\phi_c/dz)^2$. The ? reminds us that we have neglected the effect of the other modes of $\mathfrak{M}$ with a fixed mass gap of order $t \sim \xi^{-2}$ and also higher orders in f, to which we must return.

The additional contribution to the width due to the deviation from planar is embodied in the correlation function $\langle f^2\rangle$. The lowest order contribution to this object in perturbation theory is the graph

$\langle f^2\rangle =$ 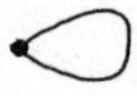 $= \frac{kT}{\sigma}\int^{\xi^{-1}} \frac{d^{d-1}q}{(2\pi)^{d-1}} \frac{1}{q^2} + ?\ . \qquad (22)$

Here we use the free propagator from (21), put a cut-off ξ^{-1} on the allowed wave-vectors because of our neglect of the other fluctuations and a question mark because of the effect of higher powers of f in (21) which give higher orders in perturbation theory.

Changing to radial and angular coordinates we see that the integral in (22) is divergent at q = 0 for $d \leqslant 3$, suggesting

$$\langle f^2\rangle = \begin{cases} \text{finite} & d > 3 \\ \infty & d \leqslant 3 \end{cases} \qquad (23)$$

as a consequence of the long wavelength fluctuations of the Goldstone field f. Of course many physical effects make the planar interface more familiar in $d \leqslant 3$ than (23) suggests. We mention here gravity (for a general review see Moldover et al 1979) which adds energy

$$\tfrac{1}{2}\, g\, \Delta\, \rho \int d^{d-1}x\, f^2 \qquad (24)$$

to expression (21), representing the excess gravitational potential energy of a column of fluid of excess density $\Delta\rho$ in a gravitational acceleration g. Expression (24) gives a mass gap in the propagator in (22): $q^{-2} \to (q^2+m^2)^{-1}$ where $m^2 = kTg\Delta p/\sigma$. The integral (22) is then finite behaving as m^{d-3} for m small (d<3). Similarly in a finite volume L^d with the interface pinned by the boundary condition f=0, the allowed eigenfunctions in (21) are $\Pi_{i=1}^{d-1}\sin(q_i x_i)$ with $q_i = \pi n_i/L$ where the integers n_i are non-zero. The integral in (22) is then replaced by a finite sum which behaves as L^{3-d} for large L(d<3).

A problem still remains however. If the higher order terms in the effective energy (21) contain for example f^4, then the very weak logarithmic effects in three dimensions predicted from the lowest order graph are overwhelmed by the catastrophic infra-red behaviour of integrals arising at higher order in perturbation theory. For example the graph in Fig. 5 would then

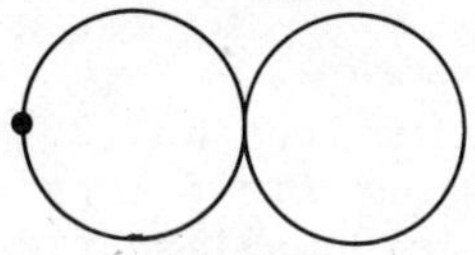

Fig. 5. Possible higher order contribution to $\langle f^2 \rangle$ from an interaction of order f^4.

have a quadratic dependence on m or L in three dimensions because there is a loop with two propagators. We would be faced with the breakdown of the usefulness of perturbation theory typical of a massless field theory in the two (transverse) dimensions, as reviewed in Edouard Brézin's lectures.

In order to control what is going on it is extremely important therefore to identify properly the effective energy for the long wavelength fluctuations of the field f. Now the long wavelength condition just means we are interested in smooth slowly varying (in space) deviations from planar and we know on physical grounds what the appropriate energy is - surface tension. To answer the question posed in (21) we need only write down the expression for the (hyper) surface area of the surface of Fig. 4. Recognising that the local normal to the surface of Fig. 4 is the unit vector $(-\nabla f,1)(1+(\nabla f)^2)^{-\frac{1}{2}}$ and has direction cosine $(1+(\nabla f)^2)^{-\frac{1}{2}}$ with the z axis, we see that the result is

$$\mathcal{H}_{\text{eff}}(f) = \frac{\sigma}{kT} \int d^{d-1}x(1+(\nabla f)^2)^{\frac{1}{2}} \quad . \tag{25}$$

The first two terms in the expansion of the square root agree with those in (21) obtained by explicit calculation from the GLW model (2). There are fourth and higher order terms in f but they have the characteristic that <u>each</u> power of f appears with a derivative and hence each line at a vertex in a Feynman diagram has a factor q.

A proper assessment of Fig. 5 shows therefore that its singular behaviour is no worse than that of the lowest order diagram in (22). In fact for any d>1, the effect of higher order diagrams from (25) is to change only the amplitude of the singularity in (22). (The systematics of this statement will be exposed more fully in the next subsection). The situation is parallel to a perturbative calculation of mean field critical behaviour in ordinary ϕ^4 for d>4, as Brézin reviewed.

Some remarks are in order.

(i) The relevance of the Hamiltonian (25) for the statistical mechanics of surface waves has long been recognised (Mandelstam 1913, Buff et al 1965). The only possibly new feature in the above is in looking beyond the free approximation (21) to see the generality of some of its predictions.

(ii) The use of the Hamiltonian (25) to control the singularities induced by the Goldstone field f parallels the similar application of spin wave theory, or non-linear σ models, to obtain the singularities due to Goldstone modes on the coexistence curve in models with a continuous symmetry. See Holstein and Primakoff (1940), Vaks et al (1968), Wallace and Zia (1975) and references therein.

(iii) The effective Hamiltonian (25) can be derived from the GLW model (2) in the limit of (a) long wavelengths and (b) a deep well potential, so that fluctuations with a gap O(t) are entirely neglected. The spirit is equivalent to the derivation of the non-linear σ models from the corresponding GLW model with continuous symmetry. The basic idea is that we want to get the effective Hamiltonian for fluctuations involving f(x) by averaging over all the others - since these have a mass gap there should be no infra-red problems in this partial trace. Of course we cannot do this partial trace exactly- approximation (b) above means we aim to get the answer "classically", leaving undone a functional integral (Gaussian plus higher order) which would be negligible if the wells are deep and narrow, c.f. expression (14). There is still a problem however because a density configuration ϕ corresponding to an arbitrary displacement f(x) is not an extremum of $\mathcal{H}$ and therefore for general f we have no maximum of the integrand about which to expand. However we can find a configuration which is close enough to a solution, given approximation (a). Consider the refinement of equation (10):

$$\phi = \phi_c((z-f(x))(1+(\nabla f)^2)^{-\frac{1}{2}}) + \hat{\phi} \qquad (26)$$

where the modes (18) are to be excluded from $\hat{\phi}$ because they are contained in f. Substituting (26) into (3) gives (up to Jacobian factors which do not affect the arguments)

$$e^{-\mathcal{H}_{eff}(f)} = e^{-\mathcal{H}(\phi_c((z-f)(1+(\nabla f)^2)^{-\frac{1}{2}}))} \times \int D\hat{\phi}\, e^{-\frac{1}{2}\int d^d x \hat{\phi}(x) \mathcal{M} \hat{\phi}(x) - \int d^d x \hat{\phi}(x) J(f) + O(\hat{\phi}^3)} . \quad (27)$$

The major difference between (27) and (14) is the appearance of the linear term because we are no longer expanding about an extremum of $\mathcal{H}$. However if f is up to linear in x, the source term J must vanish because we are then expanding about an extremum of $\mathcal{H}$ - a tilted but still planar interface. Hence J contains second or higher derivatives of f and is effectively zero for long wavelength fluctuations - it gives rise to curvature and higher derivative effects. Finally the substitution $z' = (z-f(x))(1+(\nabla f)^2)^{-\frac{1}{2}}$, $x' = x$ in $\mathcal{H}(\phi_c((z-f)(1+(\nabla f)^2)^{-\frac{1}{2}}))$ yields (25), again up to negligible higher derivative effects. The additional $(1+(\nabla f)^2)^{-\frac{1}{2}}$ factor in (26) is the direction cosine mentioned previously and just represents the local apparent dilation of the interface profile if we pass along the z direction through the locally tilted interface - it is just the Euclidean analogue of the Lorentz contraction of a moving extended object. Formal semiclassical techniques have been used by Diehl et al (1980) and M.J. Lowe (unpublished) to obtain the same result.

(iv) What is the symmetry which fixes the form (25) and for example the coefficient of $((\nabla f)^2)^2$ to be - 1/8? It is clearly not a symmetry of linear transformations on f because this would not relate $O(f^2)$ and $O(f^4)$. The answer is that, like all good Goldstone fields, f(x) carries a non-linear realization of the full original symmetry group E(d) under which $\mathcal{H}_{eff}(f)$ is invariant. Intuitively this is obvious: the area of any given surface is invariant under any rotations or translations. Mathematically, it is easy to check that $\mathcal{H}_{eff}(f)$ is invariant under the transformations

$$\text{(a)} \quad f \to f+a, \quad \text{(28a)}$$

translation by a in z direction

$$\text{and (b)} \quad f \to f-\theta(f\partial_i f + x_i) + O(\theta^2), \quad \text{(28b)}$$

rotation by infinitesimal angle θ in the (x_i-z) plane. The invariance under (b) holds only up to total derivatives which reflect the importance of boundary conditions. This aspect of (25) is recognised for example in Jhon et al (1978) and stressed in Günther et al (1980). (In contrast to the non-linear σ models, where there is a Goldstone field for each "broken" generator of the symmetry group, here only one field f(x) does the entire job; this arises because the original action of E(d) on the d coordinates (x,z) is already a representation on the coset space of the Euclidean group factored by rotations).

(v) As we have discussed, $\langle f^2 \rangle$ is expected to be only logarithmically divergent in three dimensions. It is natural to anticipate that factors more subtle than finite size or gravity can control this very weak divergence, and this is the case. Dobrushin (1972) and van Beijeren (1975) showed that the finite step energy in the three dimensional lattice Ising model was sufficient to keep a finite interface width at all temperatures up to at least the transition temperature of the two dimensional Ising model. There is now evidence that in the 3-d model there is a roughening temperature T_R ($0<T_R<T_c$) at which the interface width changes from finite to infinite. A selection of references is Burton et al(1951),Weeks et al (1973), Leamy et al (1973), Chui and Weeks (1976), Kosterlitz (1977), Knops (1977), Swendsen (1977), Ohta and Kawasaki (1978), van Biejeren (1977), Avron et al (1980). As T is increased through T_R the free energy per unit length of a step in the interface is expected to vanish, and the surface fluctuations to behave as in the continuum model (25), with the subsequent divergence in $\langle f^2 \rangle$ restoring translation invariance despite the imposition of non-translation-invariant boundary conditions attempting (and failing) to pin the interface. The divergence of the interface width and the restoration of the Euclidean symmetry at all finite temperatures is rigorously established in the two-dimensional Ising model (Aizenman 1979 and references therein). (The width diverges as $L^{\frac{1}{2}}$ as in the field theory.) Thus $T_R = 0$ in two dimensions for the Ising model with homogeneous couplings (but see also Abraham 1980).

Of course the roughening transition and restoration of Euclidean symmetry is eliminated in the presence of e.g. a finite gravitational field. In the next section we discuss how one can further exploit the almost Goldstone modes which still exist in a small gravitational field for d < 3.

Statistical Mechanics of $\mathcal{H}(f)$

So far we have established the importance of $\mathcal{H}_{eff}(f)$ for controlling potentially catastrophic infra-red behaviour due to the field f. Implicit in this use is a cut-off ξ^{-1} on the allowed wavevectors in f as we explained above. In this section we wish to explore the consequences of attempting to remove this "short-distance" cut-off and thereby setting up a full statistical mechanics for the model (25). The work of this section is contained in Wallace and Zia (1979); it strongly parallels the exploitation of non-linear σ models in $2+\varepsilon$ dimensions (Polyakov 1975, Migdal 1975, Brézin and Zinn-Justin 1976 a,b, Brézin et al 1976 a, b, Nelson and Pelcovits 1976, 1977, Hikami and Brézin 1978, McKane and Stone 1980). The technicalities of the calculations such as dimensional regularisation and minimal subtraction were in part reviewed in Brézin's lectures; see the above and Amit (1978) for further references and background.

We consider the partition function

$$Z = \int D f \, e^{-\mathcal{H}(f)} , \tag{29}$$

$$\mathcal{H}(f) = \frac{1}{T_0} \int d^{d-1}x \left((1+(\nabla f)^2)^{\frac{1}{2}} + \tfrac{1}{2} m^2 f^2 \right) \tag{30}$$

where for convenience of notation we have introduced a bare temperature $T_0 = kT/\sigma$ and m^2 is the mass term from gravity (expression 24). A conventional diagrammatic perturbation theory is obtained from the expansion

$$(1+(\nabla f)^2)^{\frac{1}{2}} = 1 + \tfrac{1}{2}(\nabla f)^2 - \frac{1}{8}((\nabla f)^2)^2 + O(f^6). \tag{31}$$

The free propagator is $T_0(q^2+m^2)^{-1}$ and expansion in powers of T_0 is an expansion in number of loops. Dimensional analysis shows that $[T_0] = [L]^{d-1} = [q]^{-(d-1)}$. Thus each extra loop integral accompanying each extra power of T_0 has a complementary dimension $[q]^{d-1}$ i.e. for $d > 1$ more powers of q in the numerator than the denominator. This is the basis for our previous statement of trivial infra-red ($q \to 0$) behaviour for $d > 1$; we now face the inevitable strong coupling ultra-violet problem of (30).

Our aim is to control this u-v problem by conventional renormalisation theory. In regulating the theory at intermediate stages of the calculation we should not put in a cut-off Λ because we are then in danger of losing the Euclidean symmetry properties of (30). It is particularly convenient in this problem to use dimensional regularisation, in which the integrals are defined by analytic continuation from the dimensions in which they are convergent; this gives poles in d (at d = 1) instead of the $\ln\Lambda$ terms of a conventional cut-off. It is convenient to remove these poles by a minimal subtraction in which bare parameters are expressed as power series in a dimensionless renormalised coupling T e.g.

$$T_0 = \kappa^{-(d-1)} T \sum_{n=0} a_n(d) T^n \tag{32}$$

where a_n are pure (multi) poles at d = 1 whose residues are fixed to cancel off the poles in the explicit calculation of any correlation function. In this scheme it is the freedom of choice of the momentum scale κ in (32) which gives the renormalisation group equations for the renormalised correlation functions.

A further technical remark is in order. Implicit in what I have said is the assumption that the naive functional measure Df behind the perturbation expansion in T is also invariant under E(d) transformations. Under a transformation $f \to f'$, we have formally

$$Df \to Df' = Df \det \frac{\delta f'}{\delta f} = Df \exp \operatorname{tr} \ln(\delta f'/\delta f) \tag{33}$$

The reader may check that, manipulating formally with δ functions, the Jacobian factor in (33) for the transformations (28) is 1 i.e. the measure is indeed invariant (again up to a total derivative for (28b)). This is possibly the least rigorous remark made at this Institute, involving $\delta(0)$ terms as it does; however such terms would vanish anyway in dimensional regularisation so we should be safe enough.

The results of explicit calculations of the vertex functions with two and four external legs can be summarised as follows.

(i) No field renormalisation is required; the wave function renormalisation Z is 1. We (Wallace and Zia 1979) first obtained this result by explicit calculation of one and two loop graphs and subsequently proved it to all orders in perturbation theory using a Ward Identity which expresses the transformation property of the generating functional for all vertex functions under the rotation transformation (28b). This is still probably not the most efficient way of establishing the result; since, according to (28b) the field f transforms into a term involving $\partial_i f^2$, we must introduce an infinite hierarchy of sources for f^2, f^3, f^4 etc. in order to get closure under the rotation transformation. However by simple power counting each of these extra operators requires no renormalisation near d = 1 and the Ward Identity reduces to the statement that the divergent part of the generating functional is an invariant proportional to $(1+(\nabla f)^2)^{\frac{1}{2}}$. See Brézin et al (1976 a) for corresponding remarks in non-linear σ models. The intuitive content of this argument is trivial; unlike say, the original density ϕ, the field f really is a length which can "mix" with the coordinate x under a rotation. It cannot therefore develop an anomalous dimension η and remains unrenormalised.

(ii) A corollary of part of this argument is that the field f^2 also requires no renormalisation. The only renormalisation of m^2 is that induced by the renormalisation of T_0 in (32). Thus we can introduce a renormalised mass through

$$m_R^2 = \kappa^{-(d-1)} T\, m^2/T_0 \quad . \tag{34}$$

(iii) The only nontrivial renormalisation is therefore that of the bare temperature. The vertex functions $\Gamma^{(n)}$ expressed in terms of T, m_R and κ obey the renormalisation group equation

$$\left(\kappa \frac{\partial}{\partial\kappa} + \beta(T) \frac{\partial}{\partial T} + \gamma_1(T) m_R^2 \frac{\partial}{\partial m_R^2}\right)\Gamma^{(n)}(q;T,m_R,\kappa) = 0 \tag{35a}$$

where

$$\beta(T) = \kappa \frac{\partial T}{\partial \kappa}\bigg|_{T_0,m^2} ; \quad m_R^2\, \gamma_1(T) = \kappa \frac{\partial m_R^2}{\partial\kappa}\bigg|_{T_0,m^2} \quad . \tag{35b}$$

The general result (34) implies

$$\gamma_1(T) = -(d-1) + \beta(T)/T \tag{36}$$

and a one loop calculation of $\Gamma^{(2)}$ gives the coefficient a_1 in (32) from which one finds

$$\beta(T) = (d-1)T - T^2 + O(T^3). \tag{37}$$

We have for convenience defined the normalisation of T to absorb the factor $I = -\int d^{d-1}q(2\pi)^{-d+1}\, q^2(q^2+1)^{-1}$ which can be extracted from each loop integral. (Note $I = +1+O(d-1)$ - such is the way of dimensional regularisation.)

The analysis of (37) follows exactly the same lines as for $\beta(g) = (d-4)g + g^2 + O(g^3)$ of the usual ϕ^4 models (5), but the interpretation will be very different because g is an irrelevant coupling near $g^* = (4-d) + O(4-d)^2$ whereas T is a relevant coupling. According to (37) the fixed points of $\dot{T} = \beta(T)$ are

$T = 0$; infra-red stable for all $d > 1$ - the renormalisation group statement that long distance behaviour of (30) is, as we have seen, trivially calculable in perturbation theory.

$T_c = (d-1) + O(d-1)^2$; ultra-violet stable - the high q behaviour of (30) is not uncontrollable for $d - 1$ small because the effective coupling at short distances has the finite value T_c.

The interpretation of T_c as the phase transition temperature in Ising-like systems in $1+\varepsilon$ dimensions has the following support and implications:

(i) T_c has the correct stability properties for a critical temperature. The usual analysis thinks in terms of infra-red stability of perturbations of the irrelevant coupling g from g^*, and corresponding instability of the relevant coupling T from T_c. Running the flow equations backwards to short distances therefore implies $T \to T_c$ as above. Of course "short" distances here are still, in this continuum model, long compared to the ultimate cut-off of mean molecular spacing.

(ii) We have already one length scale m^{-1} in the problem associated with the gravitational field. We can introduce a new physical length scale ξ, independent of m. Dimensional analysis demands $\xi = \kappa^{-1}\,\zeta(T)$ and if ξ is to be a physical quantity invariant under a renormalisation group transformation it must obey $(\kappa\, \partial/\partial\kappa + \beta(T)\partial/\partial T)\xi = 0$. The solution of this equation, with suitable normalisation, is

$$\xi = \kappa^{-1}\ T^{1/(d-1)} \exp \int_0^T (1/\beta(T') - T'^{-1}(d-1)^{-1})dT' \tag{38}$$

Clearly for $d>1$, $\xi \to 0$ as $T \to 0$. For $T \to T_c$ the integral in (38) diverges because of the zero of β, so that

$$\xi \propto (T_c - T)^{-\nu} \; ; \quad \nu = -1/\beta'(T_c) \quad .$$

The exponent ν thus identified has according to (37) the expansion $\nu = (d-1)^{-1}(1+O(d-1))$. This agrees at leading order with the result from lattice renormalisation group methods (See Migdal 1975) but the field theory provides a systematic, albeit perturbative, scheme; we find $\nu = (d-1)^{-1}(1-\frac{1}{2}(d-1)+O(d-1)^2)$. It is therefore tempting and natural to identify ξ as a measure of the bulk correlation length in the absence of gravity. The origin of ξ in the Hamiltonian (30) is of course because T_0 also has a dimension; in fact solving (35b) for T_0 as a function of T yields $1/T_0 = \xi^{-(d-1)}$. i.e. the surface tension in (30) behaves as expected in hyperscaling (Widom 1972).

(iii) An attractive intuitive picture is that although each configuration of the model (30) has a sharp interface the strong coupling at short distances creates statistically an interface locally fuzzy over a distance ξ. At distances shorter than ξ the effective coupling runs up to T_c : within the interface as we remarked at the beginning we have critical fluctuations whose effective temperature is T_c.

(iv) It remains to explore the implications for the mean density profile. A suitable starting point is the definition of a probability distribution P(z) for the position z of the centre of the interface, by identifying the moments of P with expectation values $\langle f^{2n}(0)\rangle$

$$\langle f^{2n}\rangle = \int_{-\infty}^{\infty} P(z) z^{2n} dz \quad . \tag{39}$$

The $\langle f^{2n}\rangle$ obey the renormalisation group equation (35a). If they are rewritten as functions of ξ rather than T, the absence of re-normalisations other than that of T implies a scaling form for $\langle f^{2n}\rangle$ identical to what one would obtain from naive dimensional analysis with the identification $T_0 = \xi^{d-1}$ discussed above:

$$\langle f^{2n}\rangle = \xi^{2n}\ p_{2n}(y) \tag{40}$$

where y is the scaling variable

$$y = \frac{m_R^2}{\kappa^{-(d-1)}T}\ \xi^{d+1} \equiv m^2\ \xi^2 . \tag{41}$$

The scaling function $p_{2n}(y)$ is calculable in perturbation theory through the graphs of Fig. 6.

Fig. 6. Low order graphs contributing to f^{2n}.

Inspection shows that this gives $p_{2n}(y)$ as a power series with successive corrections of order $y^{(d-1)/2}$ i.e. although we have a scaling form, as in $2+\varepsilon$ calculations (Brézin and Zinn-Justin 1976 b) it is not useful for all values of the scaling variable y. As emphasised in Brézin's lectures, ε expansions for correlation functions make sense only as a double power series in the coupling and ε. Nevertheless if we allow ourselves the freedom of keeping d fixed in the loop integrals, the $\langle f^{2n}\rangle$ do then correctly contain the wandering divergences as $y \to 0$.

Given $\langle f^{2n}\rangle$, it is straightforward to use (39) to obtain the probability distribution P(z), in a scaling form. Identifying a renormalised "smeared" interface profile by

$$\frac{d\langle\phi\rangle}{dz} \propto P(z) \tag{42}$$

gives one a theoretical prediction to compare with experiment. To first order a Gaussian form is obtained for P(z). The two terms at next order give a P(z) which can be written in the form

$$P(z) \propto \xi^{-1} \exp\{-\tfrac{1}{2}\, y^{(3-d)/2}(z/\xi)^2[1-(A-1)y^{(d-1)/2} + \frac{1}{6} Ay(z/\xi)^2 + \ldots]\} \quad . \tag{43}$$

The number A is governed by the ratio of the second and third graphs in Fig. 6.

We are now in a position to return to the question of the comparison of data with the various theoretical predictions for the profile. As we remarked in the introduction, the mean field tanh profile does not fit the data. Wu and Webb (1973) did fit their data successfully with two forms, a profile obtained by Fisk and Widom (1969) from a modified scaling form of the equation of state, and an error function profile corresponding to the Gaussian approximation to (43). The motivation for the latter choice lay in the work of Buff et al (1965) who did effectively the calculation above using the free form (21) of $\mathcal{H}(f)$ along with a cut-off ξ^{-1} on allowed q's, as in (22). The result of the renormalisation group analysis of this model is to show that (a) the Buff et al calculation can be viewed as the first term of a systematic expansion in $1+\varepsilon$ dimensions with corrections of the form (43) and (b) if one attempts to remove the smearing effect of the ξ^{-1} cut-off, the strong coupling short distance effects of a sharp interface appear to try statistically to resurrect the fuzziness. As regards the $4-\varepsilon$ expansions, as recognised by Ohta and Kawasaki (1977), Jasnow and Rudnick (1978), one has also to compromise a strict ε expansion with the need to work in three dimensions to obtain correctly the wandering effects; numerically the result is close enough to the error function to be indistinguishable with present data (Jasnow and Rudnick, 1978).

Much still remains to be clarified. On the theoretical side, it would be more satisfying if one could calculate the exponent η in $1+\varepsilon$ and had some clean control over the contradictory demands of an ε expansion and the need to work in fixed $d = 3$. Generalisations of the surface tension model are also of interest (Lowe and Wallace 1980 a). On the experimental side, as remarked by Jasnow and Rudnick (1978), experiments with light of longer wavelengths could provide useful discrimination of the various models, and refinement of binary fluid experiments should also be very revealing.

3. ALMOST SPHERICAL DROPLETS

In the above we studied problems of two coexisting phases separated by an almost planar interface. In this final section we consider behaviour in the neighbourhood of the coexistence curve where we shall be concerned with similar ideas involving large almost spherical droplets. The basic model now is the GLW Hamiltonian (2) with for example an additional term $-H\int d^dx\phi(x)$ which tilts the double well potential of Fig. 2. to favour $\phi_+(\phi_-)$ when $H>0$ $(H<0)$. As H goes through zero the equilibrium system will undergo the first order phase transition across the coexistence curve. We are going to be interested in the nature of the free energy (and its derivatives) in the limit $H \to 0$ i.e. as we approach the coexistence curve at fixed temperature. The discussion here is primarily concerned not with behaviour near the critical point, although we expect the results to be valid there also.

Our aim is to indicate the case for the existence of an essential singularity at $H=0$ for all $T<T_c$. Rigorous results on this problem are as yet rather limited (Lanford and Ruelle 1969) although an analogous singularity in the percolation problem is rigorously established (Kunz and Souillard 1978 a,b). (We should note however the special case of no singularity in the limit of very weak long range forces (Kac et al 1963, Lebowitz et al 1966, Lieb 1966)). The first theoretical evidence for an essential singularity came from droplet models (Andreev, 1964, Fisher 1967 and references therein).

A real space renormalisation group approach (Klein et al 1976) seems limited by the difficulty of setting up systematic recursion formulae near H = 0 = T (R.K.P. Zia, unpublished). The transfer matrix approach is also a possible vehicle for exhibiting the essential singularity (Newman and Schulman 1977, McCraw and Schulman 1978) but is as yet of limited quantitative power. The approach which we exploit here is the field theory method introduced in an important paper by Langer (1967). The technicalities are similar to those used in high order estimates in perturbation theory (Zinn-Justin 1979, Parisi 1979, Wallace 1978). Our contribution (Günther et al 1980, Lowe and Wallace 1980b) is to draw attention to universal features of this essential singularity which appear to be in good agreement with series expansion results from the two dimensional Ising model. We start by considering a droplet model which is a good introduction for the field theory calculations.

Primitive Droplet Model

In conventional field theory calculations (e.g. those in $4-\varepsilon$ dimensions) we take into account small fluctuations about one (an absolute) minimum of the Hamiltonian (e.g., ϕ_+ for $H > 0$). The droplet models attempt to assess the effect of field configurations which are not included in these calculations, namely configurations corresponding to large regions, or droplets, of the unfavoured phase. In a simple droplet model, a single such spherical droplet of radius R gives a term of the form $\exp-(hR^d+R^{d-1})$ to the partition function Z. We have dropped here all inessential factors; R^{d-1} represents the surface tension and hR^d is the excess bulk energy of the unfavoured phase inside the droplet, with $h \propto H$ at fixed T. The presence of a gas of such droplets, assumed sufficiently dilute that they do not interact, just exponentiates this result. Summing over droplets of all radii gives a free energy per unit volume with the structure

$$F(h) = \int_0^\infty dR \, \exp-(hR^d+R^{d-1}) \tag{44}$$

For the present let us ignore the many potential defects of this model and look at its implications.

It is clear that F(h) has a singularity at h = 0 because the expansion of (44) as a power series in h has zero radius of convergence:

$$F \sim \sum_{L=0}^{\infty} F_L(-h)^L \tag{45}$$

where

$$F_L = \frac{1}{\Gamma(L+1)} \int_0^\infty dR\, R^{Ld} \exp-R^{d-1}$$

$$= \frac{\Gamma((Ld+1)/(d-1))}{(d-1)\Gamma(L+1)} \qquad (46)$$

where the Euler Γ function is obtained by substituting $t = R^{d-1}$. All derivatives of F exist at h = 0 but the series (45) diverges no matter how small h is.

To elucidate the nature of this singularity, we look at the analytic continuation of F(h) through complex values of h to h real negative. We must ensure that the integral (44) remains convergent at infinity i.e. $\mathrm{Re}(hR^d)>0$. Hence to reach $\arg h = \pm\pi$, we should rotate the contour of integration through a counter angle, to $\arg R = \mp \pi/d$. Since for h negative the contours are now in the complex plane, we must expect that F contains an imaginary part. The leading behaviour of the imaginary part for $|h|$ small is obtained by distorting the contour of integration (the integrand is an entire function) to follow the contour of steepest descent. This passes along the real axis to the saddle point of (44) at $R_c = (d-1)/(d|h|)$ where it descends into the complex plane as shown in Fig. 7.

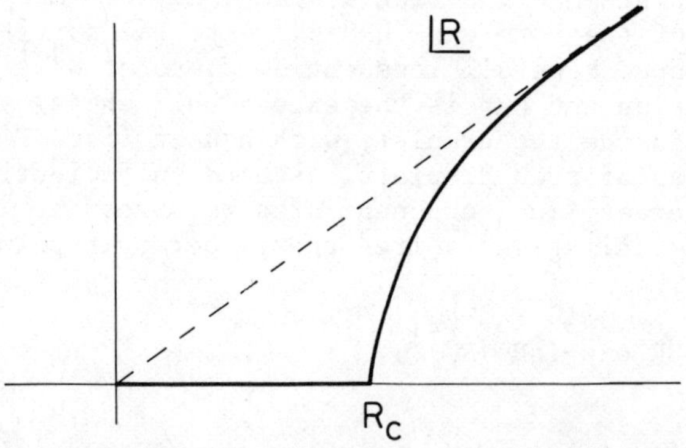

Fig. 7. Continued integration and steepest descent integral for $\arg h = -\pi$ in (44).

The imaginary part of F receives its leading contribution for h small from the saddle point R_c. A standard calculation yields

$$\mathrm{Im}F(\arg h = \pm\pi) = \mp B|h|^{(d-3)/2}[\exp-A|h|^{-(d-1)}](1+O(|h|^{+d-1})) \qquad (47a)$$

where

$$A = (d-1)^{d-1} d^{-d}; \quad B = (\pi\, d^{d-3} (d-1)^{-(d-2)} 2^{-1})^{\frac{1}{2}}. \tag{47b}$$

The dominant exponential term is the value of the integrand at the saddle point; the prefactor comes from the Gaussian integral away from R_c.

Equation (47) shows clearly that F has a branch point at h = 0 and gives the discontinuity across the tip of the cut along the negative h axis. This information is sufficient in fact to obtain the asymptotic behaviour of the coefficients F_L in (46), for large L. A Cauchy integral formula with the contour C in Fig. 8 yields

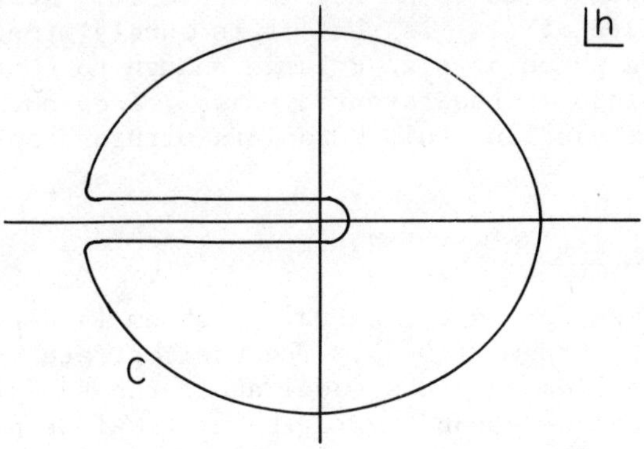

Fig. 8. Cauchy integral contour for equation (48).

$$F(h) = \frac{1}{2\pi i} \int_C \frac{F(h')}{h'-h}\, dh'$$

$$= \frac{1}{\pi} \int_{-\infty}^{0} \frac{\mathrm{Im}F(h'; \arg h'=\pi)}{h'-h}\, dh' \tag{48}$$

discarding the contour at infinity. Expanding formally in h and substituting expression (47) yields

$$F_L = (2\pi)^{-\frac{1}{2}}\, d^{(d-3)/2} (d-1)^{-d/2} [d^d (d-1)^{-(d-1)}]^{(L+\frac{3}{2}-\frac{1}{2}d)/(d-1)}$$
$$\cdot\, \Gamma((L+\frac{3}{2}-\frac{d}{2})/(d-1))[1+O(L^{-1})]. \tag{49}$$

The corrections of order L^{-1} come from $O(|h|^{d-1})$ in (47). Stirling's formula establishes the equivalence of (49) and the exact calculation (46) for L large.

The physical interpretation of R_c is important: it is the radius of the unstable critical droplet of the metastable phase - if $R<R_c$ surface tension wins and the energy drops, and for $R>R_c$

the increasingly negative bulk energy wins. The descent into the complex plane in Fig. 7 is precisely the reflection of this instability.

Of course the defects in the original model (44) are many. The assumption of spherical droplets requires refinement. The prefactor power law in (47) is clearly not reliable because for example we do not know if we should sum over droplets of all radii ($\int dR$) or volumes ($\int dR\, R^{d-1}$). I believe that the assumption of non-interacting droplets is probably more reliable than one might at first guess: these droplets should not be confused with clusters in e.g. the Ising model since the phase in each droplet has its own almost equilibrium distribution of plus and minus spins. (For the coarse-graining aspects see Binder (1976), also Glimm et al (1976)). These problems seem to me to be not insuperable. However one more problematic feature of (44) is that it is surely wrong for droplets of the unfavoured phase of size R large enough to contain critical (or larger) droplets of the favoured phase. I do not yet see a convincing way of handling this "droplets within droplets" problem.

Field Theory Calculation

A prescription for bypassing this problem is described in an important paper by Langer (1967). The basic strategy involves a perturbation expansion in a GLW model about the spherically symmetric classical solution corresponding to the critical droplet in the metastable state. Thus we have to repeat the steepest descent calculation required to obtain expression (47), but now for a functional integral. The advantage of Langer's approach is that by working in the metastable phase one replaces a many-length-scale problem (sum over droplets of all scale sizes) by a one length problem (perturbation about a spherically symmetric solution of a unique scale size for a given external field). The analytic continuation to the "wrong sign" external field cannot yet be done rigorously. However the necessity for rigorous continuation can be sidestepped by a very plausible prescription for how to interpret the fluctuations about this classical solution: the unstable mode corresponding to dilation of the critical droplet is integrated into the complex plane, as in Fig. 7.

The major result from this approach is that some entropy effects of the contributions of non-spherical droplets can be systematically calculated. We shall be discussing these surface wobbles in a spirit similar to that used for deviations from planar in section 2.

The starting point is a GLW model (2) with an additional term $-H\int d^dx\phi(x)$ in the potential V so that $\phi_+(\phi_-)$ is the stable phase for $H>0(H<0)$ as in Fig. 9. We imagine taking the free energy, say, of

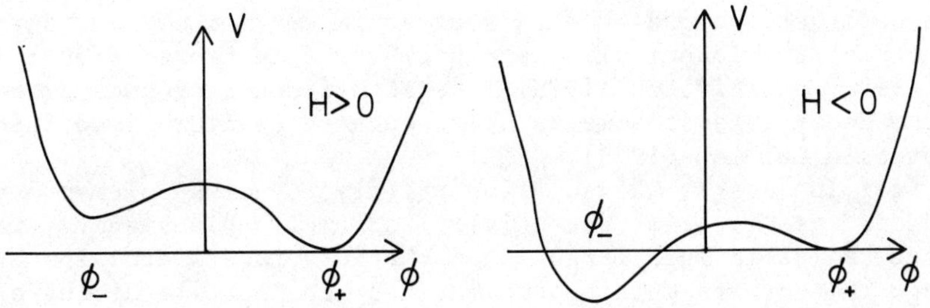

Fig. 9. Potential for discussing a first order phase transition.

the stable phase with positive $\langle\phi\rangle$ for H>0 and continuing that function to H<0. The word continuation implies a smooth change in the state of the system; the expectation value of the field remains positive.

Arguments similar to those in eqs. (6) to (9) can now convince us of the existence of a radially symmetric solution of the field equations. We need the decomposition

$$\nabla^2 = \frac{d^2}{dr^2} + \frac{d-1}{r}\frac{d}{dr} + \frac{1}{2}\frac{L^2}{r^2} \quad ; \tag{50a}$$

here $L^2 = L_{ij}L_{ij}$ (50b)

is the total angular momentum operator and $L_{ij} \equiv x_i\partial/\partial x_j - x_j\partial/\partial x_i$. For a function ϕ which depends only on the radial variable r, the classical field equation takes the form

$$\frac{d^2\phi}{dr^2} + \frac{d-1}{r}\frac{d\phi}{dr} = -\frac{\partial\tilde{V}}{\partial\phi}, \tag{51}$$

where $\tilde{V} = -V$. We can interpret (51) as the equation of a particle of "coordinate" ϕ in "time" r, moving in a potential $\tilde{V}$ with a damping force $(d-1)r^{-1}d\phi/dr$ which decreases with increasing time r. For H<0, $\tilde{V}$ is as shown in Fig. 10. The critical droplet is a solution which starts at r = 0 sufficiently close to the stable phase ϕ_- that in its subsequent motion it "dissipates enough energy" to reach ϕ_+ at $r = \infty$ - the boundary condition imposed by the continuation to the metastable region. As $(V(\phi_+) - V(\phi_-))$ decreases (i.e. $H \to 0^-$) the field must spend an increasing time near ϕ_- to reduce appropriately the energy loss due to the damping force.

The above gives a qualitative picture of the existence of a

classical solution ϕ_c corresponding to the critical droplet. We wish to set up the perturbation expansion about $\phi_c(r)$ to take into account fluctuation effects. We remark below on some additional technicalities, beyond those discussed in section 2. Further discussions of the technicalities can be found in Langer (1967), Günther et al(1980) and Affleck (1979); the significance of such calculations for the decay rate of a metastable state in quantum field theory is reviewed in Coleman (1979).

(i) In general the solution of (51) is not an elementary function, even for a ϕ^4 potential. For small $|H|$ however, (when the droplet radius is much larger than the interface width) the profile through the interface is approximated by the profile of the planar interface. Thus Eq. (13) for the ϕ^4 potential suggests

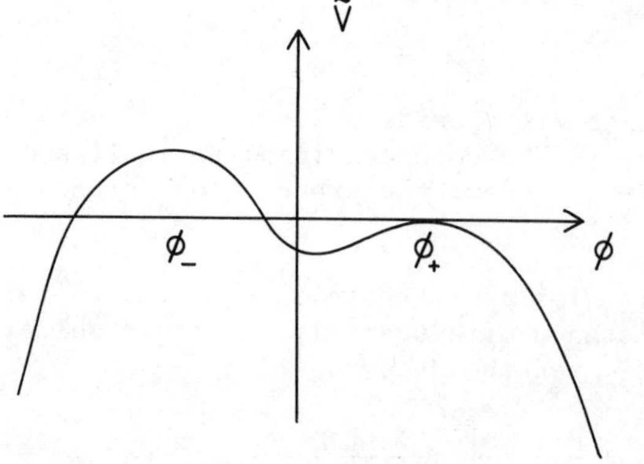

Fig. 10. Potential $\tilde{V}$ for the equation of motion (51).

$$\phi_c \simeq \tfrac{1}{2}(\phi_+ + \phi_-) + \tfrac{1}{2}(\phi_+ - \phi_-)\tanh[(r-R_c)/(2\xi)] \tag{52}$$

where $R_c \propto |H|^{-1}$ is determined, for $|H| \to 0$, by the competition between surface and volume terms as in (44) et seq. Systematic corrections to the approximation (52) for ϕ_c and the classical energy $\mathcal{H}(\phi_c)$ are obtainable as an asymptotic power series in H (neglecting exponentially small corrections $\propto \exp-(\text{const}/|H|)$). In fact, if in the absence of the interaction $-H\int d^dx\phi(x)$ the potential is symmetric under $\phi \to -\phi$, then the correction terms contain no odd powers in H i.e.

$$\mathcal{H}(\phi_c) = A|H|^{-(d-1)}(1+O(H^2)); \tag{53}$$

the value of A depends upon the details of the potential V.

(ii) To study the effects of fluctuations we expand the field ϕ about the classical solution, as in (9). The classical contribution $\exp-\mathcal{H}(\phi_c)$ will give, according to (53), an exponential factor as in (47a). To go further one must look at the small oscillations as in (14). Using the decomposition(50) the differential operator $\mathcal{M}$ has the form

$$\mathcal{M} = -\frac{d^2}{dr^2} - \frac{(d-1)}{r}\frac{d}{dr} - \frac{1}{2}\frac{L^2}{r^2} + U(r) \tag{54}$$

where $U(r) = \frac{\partial^2 V}{\partial \phi^2}(\phi_c)$. Since the droplet centre can be chosen arbitrarily,

$$\mathcal{M}\partial_\mu \phi_c = 0 \tag{55}$$

i.e. $(x_\mu/r)\phi_c'(r)$ is an eigenfunction of $\mathcal{M}$ with zero eigenvalue. This d-fold degenerate $\ell = 1$ eigenfunction of $\mathcal{M}$ is the Goldstone mode of the translation invariance broken spontaneously by the centre of the droplet. Since $\phi_c'(r)$ is localised near $r = R_c$ (c.f. (52)), by standard quantum mechanical perturbation theory, (54) shows that $Y_\ell^\alpha(\eta)\phi_c'(r)$ is an approximate eigenfunction of $\mathcal{M}$ with eigenvalue dominated by the angular momentum barrier:

$$E_\ell = \frac{(\ell-1)(\ell+d-1)}{R_c^2}\left(1 + O\left(\frac{\ell^2\xi^2}{R_c^2}\right)\right). \tag{56}$$

Here $Y_\ell^\alpha(\eta)$ is the spherical harmonic function of d dimensions, $\ell=1$ being the exact zero mode (55). In deriving (56) one uses the eigenvalues of L^2:

$$L^2\, Y_\ell^\alpha(\eta) = -2\ell(\ell+d-2)Y_\ell^\alpha(\eta). \tag{57}$$

The degeneracy for a given angular momentum ℓ is

$$\nu_\ell(d) = \frac{(2\ell+d-2)}{\Gamma(d-1)}\frac{\Gamma(\ell+d-2)}{\Gamma(\ell+1)} . \tag{58}$$

(iii) The significance of this particular subset of eigenfunctions is that in the limit of interest ($R_c \to \infty$) they become a band of soft modes describing deviations from the critical bubble ϕ_c:

$$\phi = \phi_c(r) + \sum_{\ell,\alpha} a_{\ell\alpha}\, Y_\ell^\alpha(\eta)\, \phi_c'(r)$$

$$\simeq \phi_c(r + f(\eta)) \tag{59}$$

where

$$f(\eta) = \sum_{\ell\alpha} a_{\ell\alpha}\, Y_\ell^\alpha(\eta) . \tag{60}$$

(iv) The effect of these soft wobbles is summarised by the result for the imaginary part of the free energy (Günther et al, 1980):

$$\mathrm{Im}F(\arg H=\pm\pi) = \mp B\, |H|^b \exp-[A|H|^{-(d-1)}(1+O(H^2))] \tag{61a}$$

where

$$b = \begin{cases} (3-d)d/2 & 1<d<5 \; ; \; d \neq 3 \\ -7/3 & d = 3 \end{cases} \tag{61b}$$

This is the primary result of the field theory calculation. The prefactor to the exponential comes from the first correction to the classical result. It is obtained by neglecting $O(\phi^3)$ in expression (14) and is given, formally, by $(\mathrm{Det}\,\mathcal{M})^{-\frac{1}{2}}$. The $\ell = 0$ mode in (56) has negative energy and will give a factor $\sqrt{-1}$ - hence we are calculating an imaginary part. This mode represents the instability of the droplet under dilation. It should be understood in the sense of analytic continuation and steepest descent integration as in the primitive droplet model (44); the sign in (61) is taken from (47). The integration over the $\ell = 1$ zero mode must be replaced by an exact integration over all possible droplet positions. This gives a factor of the volume of the system along with a Jacobian

$$J = [\mathcal{H}(\phi_c)]^{d/2} \; . \tag{62}$$

This yields using (53) a contribution - $(d-1)d/2$ to the exponent b in (61). The other contribution to b comes from the $\ell \neq 1$ modes (56) in the determinant of $\mathcal{M}$; a special contribution arises in $d = 3$, as in the planar interface (c.f. 22). The result of these calculations of the contribution to the partition function due to a single droplet is expression (61), with a volume factor. The one remaining ingredient is the recognition that the contribution of a dilute gas of almost critical droplets exponentiates to give the stated result (61) for the free energy per unit volume.

Test of the Essential Singularity

In principle the result (61) may be tested experimentally in metastable systems since it is one of the factors contributing to the lifetime of the metastable state. Several effects of a dynamical nature (see e.g. Binder and Stauffer 1976) also enter into the calculation of the lifetime. As yet sufficient uncertainty remains in the calculation of these dynamical factors that it seems very unlikely that the exponent b in the prefactor of (61a) can be reliably tested by experiment. The present status of the theory of the lifetime is reviewed in Langer (1980).

It is possible however to test the structure (61) by comparison with series expansions in lattice models. As we saw in the simple droplet model (Eqs.47 - 49) the imaginary part of the free energy for H small governs the asymptotic nature of the perturbation expansion in H. The use of the Cauchy integral (48) for the field theory result can not be rigorously justified; we must at this stage make the assumption that (61) is indeed the dominant singularity at $H = 0$. With this proviso, it is straightforward to substitute (61) into (48) to obtain

$$F_L = \frac{BA^{b/(d-1)}}{\pi(d-1)} A^{-L/(d-1)} \Gamma\left(\frac{L-b}{d-1}\right) [1 + O(L^{(d-3)/(d-1)})] \ . \tag{63}$$

The term of order $L^{(d-3)/(d-1)}$ comes from the correction of order H^2 in (61a).

This prediction (63) is for the growth of the perturbation expansion in H away from the coexistence curve at fixed T. It can be compared directly with numerical estimates of the coefficients F_L (or other thermodynamic quantities) obtained from series expansions in lattice models. The principal idea here is that, although A and B depend upon the nature of the potential V and hence are not directly comparable with lattice models, the value of b depends only on the dimension d and appears to be universal. For example it should be the same for all lattices of the same d.

A comprehensive analysis of series expansions to test this prediction has not yet been carried out, but early results are rather encouraging. Consider the case $d = 2$ and the coefficients M_L of the expansion of the magnetisation M as a power series in H in the form (we follow the notation of Baker and Kim (1980))

$$M \equiv \frac{\partial F}{\partial H} \sim \sum_{L=0} M_L \ (-2H)^L \tag{64}$$

Then eq. (62) predicts

$$\frac{M_L}{M_{L-1}} = \frac{L}{2A} \left(1 + \frac{1-b}{L} + O(L^{-2})\right) \tag{65}$$

The linear growth with L of the ratio of coefficients is characteristic of two dimensions. It is convincingly displayed in the plots of Baker and Kim (1980) who obtain estimates for M_L from the low-temperature series expansions of the square Ising model (Baxter and Enting 1979, Sykes et al 1973,1975). For their values of M_L, we show in Table 1 estimates for

$$b_L = M_L/(LM_{L-1}) \tag{66}$$

(at a temperature $\exp\text{-}4\beta J = 0.1 \exp\text{-}4\beta_c J$). The approach to a constant value for large L is rather convincing. This is a direct test of the structure of the exponential factor in (61).

Table 1: Coefficients b_L (eq. 66) for the square Ising model from the estimates in Baker and Kim (1980); errors are at most 1 in the last decimal place unless otherwise indicated.

L	b_L	L	b_L
3	0.150500	14	0.0970292(8)
4	0.109435	15	0.0969156(12)
5	0.099248	16	0.096827(4)
6	0.099087	17	0.096754(6)
7	0.099753	18	0.096697(5)
8	0.0993156(2)	19	0.09664(2)
9	0.0985395(3)	20	0.09661(3)
10	0.0979570(3)	21	0.9658(5)
11	0.0975968(3)	22	0.09657(6)
12	0.0973559(3)	23	0.0966(2)
13	0.0971733(2)	24	0.0966(4)

In order to test the structure of the prefactor in (61) (i.e. the exponent b) we define

$$c_L = L^2(b_L - b_{L+1})/b_{L+1} \quad . \tag{67}$$

According to (65) and (66), the field theory calculation predicts

$$c_L = (1-b) + O(L^{-1}) \quad . \tag{68}$$

Since, according to (61), b = 1 in two dimensions, this equation predicts that if c_L is plotted against 1/L, then for 1/L small we expect a straight line with intercept at the origin. This plot using b_L from Table 1 is shown in Fig. 11. A straight line indeed appears to be setting in for $1/L \leqslant 1/11$ and extrapolation to the axis indicates

$$b = 1 \pm 0.1 \quad . \tag{69}$$

This seems to provide an excellent test of the entropy effect of surface wobbles; if these were neglected and we incorporated only the Jacobian factor (62), in two dimensions we would have b = -1.

A number of concluding remarks are in order.

(i) Similar specific predictions for higher dimensions are given in Lowe and Wallace (1980). The leading behaviour for large L is

$$M_L/M_{L-1} \sim \text{const } L^{1/(d-1)} \quad . \tag{70}$$

One can envisage that further analysis of lattice models will refine the estimate (69) for b in two dimensions and at least test the leading behaviour (70) for higher d.

(ii) In the critical region, renormalisation group or scaling arguments enable one to cast (61a) into a scaling form, with each factor of H scaled by $\xi^{\beta\delta/\nu} \equiv t^{\beta\delta}$ and an overall factor ξ^{-d} appropriate to the scaling of F. (Langer 1980, Houghton and Lubensky 1980).

(iii) Although we have stressed the universal characteristics of the result (61), there are certainly other universality classes of first order phase transitions. As we remarked, if the potential $V(\phi)$ does not exhibit a $\phi \to -\phi$ symmetry in the limit $H \to 0$, then the correction terms in (61a) will be $O(H)$ instead of $O(H^2)$; this will change the structure of the correction terms to the leading behaviour (70). Further modifications must be expected if in the first order phase transition a continuous symmetry is broken (San Miguel and Gunton 1980).

(iv) One must be aware of the possibility that the comparison between the field theory calculation and lattice models in three

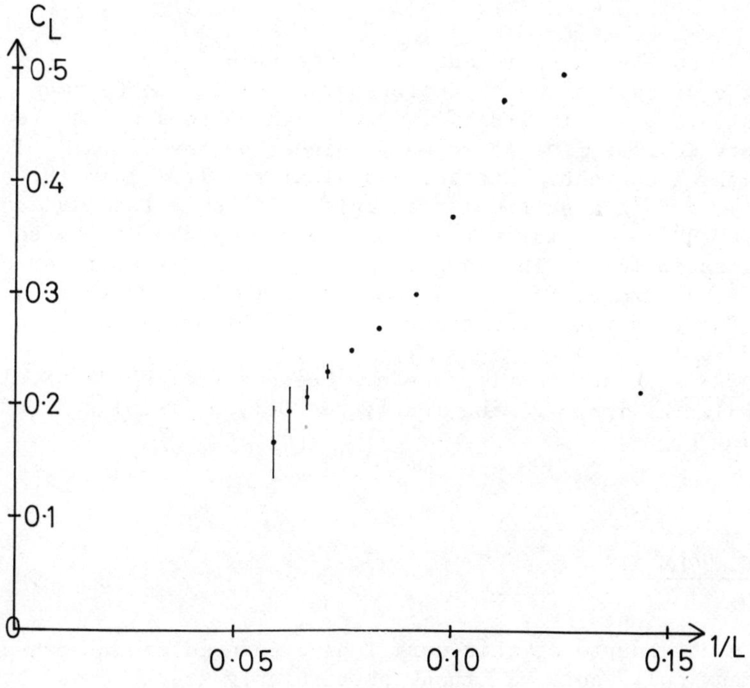

Fig. 11. Plot of c_L (from Table 1 and (66)) against 1/L.

dimensions may be influenced by the roughening transition in the latter. Since diagonal interfaces are known to be as rough as in the continuum models (The transverse width diverges as the logarithm of the surface area (Pearson 1980).) an intuitive guess indicates that roughening will not influence ImF in (61). This would suggest that ImF is not then a back door through which the roughening transition can appear as a singularity in the <u>bulk</u> properties of lattice models. Clearly further work is required here.

(v) Finally, we should stress that (61) is the result of a calculation of classical and one-loop effects. On what basis can we neglect effects of higher orders in perturbation about the critical droplet ϕ_c? This is analogous to the question we posed after (23) and has a similar answer: we must at the least examine the nature of the effective interaction of the Goldstone field (as we did there following (25)). Here the Goldstone field is $f(\eta)$,describing the deviations from the critical droplet as in (59) and (60). The effective energy (Günther et al 1980) is just the generalisation of the primitive droplet model (44) to allow for non-spherical droplets. It contains a bulk term and a surface tension term (c.f. the notation of (30))

$$\mathcal{H} = \frac{1}{T}\{H \int d\Omega(R_c+f)^d \, d^{-1} + \int d\Omega(R_c+f)^{d-1}[1+(L_{ij}f)^2/(R_c+f)^2]^{\frac{1}{2}}\} , \tag{71}$$

where Ω represents solid angle. Since $R_c \propto H^{-1}$, rescaling f by R_c shows that the dimensionless expansion parameter in the Feynman graph expansion of (71) is $T|H|^{d-1}$. Thus higher order corrections due to the soft wobbles $f(\eta)$ are down by higher powers of $|H|^{d-1}$. The special logarithmic correction to the classical result $\mathcal{H}(\phi_c) \propto |H|^{-d+1}$ evident in (61) is just a particular example. The expected two loop correction of order H^{d-1} affects in two dimensions only the $O(L^{-2})$ correction in the ratio (65). This argument is the basic step in establishing that higher order effects do not change (61). Explicit calculations from (71) to verify this claim should be illuminating.

<u>Note Added</u>: Exact results are now being established for cluster probabilities in the Ising model: see Delyon (1980) and references therein.

ACKNOWLEDGEMENTS

In the course of this work I have enjoyed useful conversations with many colleagues. I thank particularly K.J. Barnes, P.H. Dondi, M.E. Fisher, P.W. Higgs, D. Jasnow, J.S. Langer, M.J. Lowe, D.A. Nicole, L.S. Schulman, M. Stone and R.K.P. Zia.

REFERENCES

Abraham, D.B., 1980, Phys. Rev. Lett. 44:1165.
Affleck, I.K., 1979, Ph.D. thesis, Harvard University.
Aizenman, M., 1979, Phys. Rev. Lett. 43: 407.
Amit, D.J., 1978,"Field Theory, the Renormalisation Group and Critical Phenomena", McGraw-Hill, New York.
Andreev, A.F., 1964, Sov. Phys. JETP 18: 1415.
Avron, J.E., Balfour, L.S., Kuper, C.G., Landau, J., Lipson,S.G. and Schulman, L.S., 1980, Phys. Rev. Lett. 45: 814.
Baker, G.A. Jr. and Kim, D., 1980, J. Phys. A: Math. Gen. 13:L103.
Baxter, R.J. and Enting, I.G., 1979, J. Statist. Phys. 21: 103.
Binder, K., 1976, Ann. Phys. 98: 390.
Binder, K. and Stauffer, D., 1976, Adv. Phys. 25: 343.
Brézin, E. and Zinn-Justin, J. 1976a, Phys. Rev. Lett. 36: 691.
Brézin, E. and Zinn-Justin, J. 1976b, Phys. Rev. B14: 3110.
Brézin, E., Zinn-Justin, J. and le Guillou, J.C., 1976a, Phys.Rev. D14: 2615.
Brézin, E., Zinn-Justin, J. and le Guillou, J.C., 1976b, Phys. Rev. B14: 4976.
Buff, F.P., Lovett, R.A. and Stillinger, F.H., 1965, Phys. Rev. Lett. 15: 621.
Burton, W.K., Cabrera, N. and Frank, F.C., 1951, Phil. Trans. R. Soc. 243A: 299.
Cahn, J.W. and Hilliard, J.E., 1958, J. Chem. Phys. 28: 258.
Coleman, S., 1979 in : "The Whys of Subnuclear Physics", Zichichi, ed., Plenum, New York.
Chui, S.T. and Weeks, J.D., 1976, Phys. Rev. B14: 4978.
Delyon, F., 1980, preprint, Ecole Polytechnique.
Diehl, H.W., Kroll, D.M. and Wagner, H., 1980, Ziet. Phys. B36: 329.
Dobrushin, R.L., 1972, Theory Prob. Appl. (USSR) 17:582.
Evans, R., 1979, Adv. Phys. 28: 143.
Fisher, M.E., 1967, Physics 3: 255.
Fisher, M.E., Barber, M.N. and Jasnow, D. 1973, Phys. Rev. A8: 1111.
Fisk, S. and Widom, B., 1969, J. Chem. Phys. 50: 3219.
Glimm, J., Jaffe, A. and Spencer, T., 1976, Ann. Phys. 101:610 and 631.
Günther, N.J., Nicole, D.A. and Wallace, D.J., 1980, J. Phys.A.:Math. Gen. 13: 1755.
Hikami, S. and Brézin E., 1978, J. Phys. A: Math. Gen. 11:1141.
Holstein, T. and Primakoff, H., 1940, Phys. Rev. 58: 1098.
Houghton, A. and Lubensky, T.C., 1980, preprint, Brown University.
Huang, J.S. and Webb, W.W., 1969, J. Chem. Phys. 50: 3677.
Jasnow, D. and Rudnick, J., 1978, Phys. Rev. Lett. 41: 698.
Jhon, M.S., Desai, R.C. and Dahler, J.S., 1978, J. Chem. Phys. 68: 5615.
Kac, M., Uhlenbeck, G.E. and Hemmer, P.C., 1963, J. Math. Phys. 4: 216.
Klein, W., Wallace, D.J. and Zia, R.K.P., 1976, Phys. Rev. Lett. 37: 639.

Knops, H.J.F., 1977, Phys. Rev. Lett. 39:766.
Kosterlitz, J.M., 1977, J. Phys. C.: Solid St. Phys. 10: 3753.
Kunz, H. and Souillard, B., 1978a, Phys. Rev. Lett. 40: 133.
Kunz, H. and Souillard, B., 1978b, J. Statist. Phys. 19: 77.
Lanford, O.E. and Ruelle, D., 1969, Commun. Math. Phys. 13: 194.
Langer, J.S., 1967, Ann. Phys. 41: 108.
Langer, J.S., 1980, in "Proceedings of the Sitges International School of Statistical Mechanics", to be published.
Langer, J.S. and Turski, L.A., 1973, Phys. Rev. A8: 3230.
Lawrie, I.D. and Lowe, M.J., 1980, J. Phys. A., to be published.
Leamy, J.H., Gilmer, G.H., Jackson, K.A. and Bennema, P., 1973, Phys. Rev. Lett. 30: 601.
Lebowitz, J.L., Penrose, O. and Hemmer P.C., 1966, J. Math. Phys. 7: 98.
Lieb, E., 1966, J. Math. Phys. 7: 1016.
Lowe, M.J. and Wallace, D.J., 1980a, Phys. Lett. B93: 433.
Lowe, M.J. and Wallace, D.J., 1980b, J. Phys. A : Math. Gen. 13:L381.
Mandelstam, L.I., 1913, Ann. Phys. (Leipzig) 41: 609.
McGraw, R.J. and Schulman, L.S., 1978, J. Statist. Phys. 18: 293.
McKane, A.J. and Stone, M., 1980, Nucl.Phys. B163:169.
Migdal, A.A., 1975, Sov. Phys. JETP 42: 743.
Moldover, M.R., Sengers, J.V., Gammon, R.W. and Hocken, R.J., 1979, Rev. Mod. Phys. 51: 79.
Nelson, D.R. and Pelcovits, R.A., 1976, Phys. Lett. A57: 23.
Nelson, D.R. and Pelcovits, R.A., 1977, Phys. Rev. B16: 2191.
Newman, C.M. and Schulman, L.S., 1977, J. Math. Phys. 18: 23.
Ohta, T. and Kawasaki, K., 1977, Prog. Theor. Phys. 58: 467.
Ohta, T. and Kawasaki, K., 1978, Prog. Theor. Phys. 60: 365.
Parisi, G., 1979, in: "Hadron Structure and Lepton-Hadron Interactions" Levy et al eds., Plenum, New York.
Pearson, R.B., 1980, preprint, Institute for Theoretical Physics, Santa Barbara.
Polyakov, A.M., 1975, Phys. Lett. B59: 79.
Rudnick, J. and Jasnow, D., 1978, Phys. Rev. B17: 1351.
San Miguel, M. and Gunton, J.D., 1980, preprint, Temple University.
Swendsen, R.H., 1977, Phys. Rev. B15: 689.
Sykes, M.F., Gaunt, D.S., Essam, J.W., Mattingley, S.R. and Elliot, C.J., 1973, J. Phys. A: Math. Gen. 6: 1507.
Sykes, M.F., Watts, M.G. and Gaunt, D.S., 1975, J. Phys. A: Math.Gen. 8: 1448.
Vaks, V.G., Larkin, A.I. and Pikin, S.A., 1968, Sov. Phys. JETP 26: 647.
van Beijeren, H., 1975, Comm. Math. Phys. 40: 1.
van Beijeren, H., 1977, Phys. Rev. Lett. 38: 993.
Van der Waals, J.D., 1894, Z. Phys. Chem. 13: 657; English translation in : Rowlinson, J.S. 1979, J. Statist. Phys. 20: 197.
Wallace, D.J. and Zia, R.K.P., 1975, Phys. Rev. D12: 5340.
Wallace, D.J., 1978, in : "Solitons and Condensed Matter Physics", Bishop and Schneider, eds., Springer Verlag, Berlin.
Wallace, D.J. and Zia, R.K.P., 1979, Phys. Rev. Lett. 43: 808.

Weeks, J.D., Gilmer, G.H. and Leamy, H.K., 1973, Phys. Rev. Lett. 31: 549.
Widom, B., 1972, in "Phase Transitions and Critical Phenomena", Vol. 2, Domb and Green, eds., Academic Press, New York.
Wortis, M., 1980, in: Proceedings of the International Conference on Ordering in Two Dimensions (Lake Geneva, Wisconsin) to appear.
Wu, E.S. and Webb, W.W., 1973, Phys. Rev. A8: 2065 and 2077.
Zinn-Justin, J., 1979, in "Hadron Structure and Lepton-Hadron Interactions", Levy et al,eds., Plenum, New York.

INDEX